Applied Probability
Control
Economics
Information and Communication
Modeling and Identification
Numerical Techniques
Optimization

Applications of Mathematics

10

W. Murray Wonham

Linear Multivariable Control: a Geometric Approach

Second Edition

With 27 Figures

Springer-Verlag
New York Heidelberg Berlin

W. Murray Wonham
Department of Electrical Engineering
University of Toronto
Toronto, Ontario M5S 1A4
Canada

AMS Subject Classifications: 49Exx, 93B05, 93B25

Library of Congress Cataloging in Publication Data
Wonham, W. Murray, 1934-
Linear multivariable control.
(Applications of mathematics ; 10)
Bibliography: p.
Includes indexes.
1. Control theory. 2. Algebras, Linear.
I. Title.
QA402.3.W59 1979 629.8'312 79-11423

The first edition was published by Springer-Verlag as Vol. 101 of the Lecture Notes in Economics series.

Printed in the United States of America

9 8 7 6 5 4 3 2 1

ISBN 0-387-90354-2 Springer-Verlag New York
ISBN 3-540-90354-2 Springer-Verlag Berlin Heidelberg

For
Anne, Marjorie and Cynthia

Preface

In writing this monograph my aim has been to present a "geometric" approach to the structural synthesis of multivariable control systems that are linear, time-invariant and of finite dynamic order. The book is addressed to graduate students specializing in control, to engineering scientists engaged in control systems research and development, and to mathematicians with some previous acquaintance with control problems. The present edition of this book is a revision of the preliminary version, published in 1974 as a Springer-Verlag "Lecture Notes" volume; and some of the remarks to follow are repeated from the original preface.

The label "geometric" in the title is applied for several reasons. First and obviously, the setting is linear state space and the mathematics chiefly linear algebra in abstract (geometric) style. The basic ideas are the familiar system concepts of controllability and observability, thought of as geometric properties of distinguished state subspaces. Indeed, the geometry was first brought in out of revulsion against the orgy of matrix manipulation which linear control theory mainly consisted of, not so long ago. But secondly and of greater interest, the geometric setting rather quickly suggested new methods of attacking synthesis which have proved to be intuitive and economical; they are also easily reduced to matrix arithmetic as soon as you want to compute. The essence of the "geometric" approach is just this: instead of looking directly for a feedback law (say $u = Fx$) which would solve your synthesis problem if a solution exists, first characterize solvability as a verifiable property of some constructible state subspace, say $\mathscr{S}$. Then, if all is well, you may calculate F from $\mathscr{S}$ quite easily. When it works, the method converts what is usually an intractable nonlinear problem in F, to a straightforward quasilinear one in $\mathscr{S}$. The underlying mathematical idea is to exploit the semilattice structure of suitable families of subspaces of the state space.

By this means the first reasonably effective structure theory has been given for two control problems of longstanding interest: regulation, and noninteraction. It should, nevertheless, be emphasized that our major concern is with "synthesis" as distinguished from "design." In our usage of these terms, "synthesis" determines the structure of the feedback control, while "design" refers to the numerical massaging (ideally, optimization) of free parameters within the structural framework established by synthesis. In this sense, design as such is not explored in detail; it is, in fact, an active area of current research.

The book is organized as follows. Chapter 0 is a quick review of linear algebra and selected rudiments of linear systems. It is assumed that the reader already has some working knowledge in these areas. Chapters 1–3 cover mainly standard material on controllability and observability, although sometimes in a more "geometric" style than has been customary, and at times with greater completeness than in the literature to date. The essentially new concepts are (A, B)-invariant subspaces and (A, B)-controllability subspaces: these are introduced in Chapters 4 and 5, along with a few primitive applications by way of motivation and illustration. The first major application—to tracking and regulation—is developed in leisurely style through Chapters 6–8. In Chapters 6 and 7 purely algebraic conditions are investigated, for output regulation alone and then for regulation along with internal stability. Chapter 8 attacks the problem of structural stability, or qualitative insensitivity of the regulation property to small variations of parameters. The result is a simplified, "generic" version of the general algebraic setup, leading finally to a structurally stable synthesis, as required in any practical implementation. In part, a similar plan is followed in treating the second main topic, noninteracting control: first the algebraic development, in Chapters 9 and 10, then generic solvability in Chapter 11. No description is attempted of structurally stable synthesis of noninteracting controllers, as this is seen to require adaptive control, at a level of complexity beyond the domain of fixed linear structures; but its feasibility in principle should be plausible. The two closing Chapters 12 and 13 deal with quadratic optimization. While not strongly dependent on the preceding geometric ideas the presentation, via dynamic programming, serves to render the book more self-contained as the basis for a course on linear multivariable control.

The framework throughout is state space, only casual use being made of frequency domain descriptions and procedures. Our viewpoint is that time and frequency domains each enjoy their proper role in multivariable control theory, and we do not insist, let alone demonstrate, that problems and results in the one domain necessarily dualize to the other. On the other hand, frequency interpretations of our results, especially by means of signal flow graphs, have been provided when they are readily available and seem helpful. Further research along this line might well be fruitful.

A word on computation. The main text is devoted to the geometric structure theory itself. To minimize clutter, nearly all routine numerical examples have been placed among the exercises at the end of each chapter. In this way each of the major synthesis problems treated theoretically is accompanied by a skeleton procedure for, and numerical illustration of, the required computations. With these guidelines, the reader should easily learn to translate the relatively abstract language of the theory, with its stress on the qualitative and geometric, into the computational language of everyday matrix arithmetic.

It should be remarked, however, that our computational procedures are "naive," and make no claim to numerical stability if applied to high-dimensional or ill-conditioned examples. Indeed, one of the strengths of the "geometric approach" is that it exhibits the structure theory in basis-independent fashion, free of commitment to any particular technique of numerical computation. The development of "sophisticated" computing procedures, based on state-of-the-art numerical analysis, is a challenging topic of current research, to which the reader is referred in the appropriate sections of the book.

On this understanding, it can be said that our "naive" procedures are, in fact, suitable for small, hand computations, and have been programmed successfully in APL by students for use with the book. The exercise of translating between the three levels of language represented by geometric structure theory, matrix-style computing procedures, and APL programs, respectively, has been found to possess considerable pedagogical value.

The present edition differs from the first mainly in Chapter 8, which has been rewritten to better exhibit the role of transversality as the geometric property underlying structurally stable linear regulation and the "Internal Model Principle." For the rest, some minor errors in the first edition have been corrected and some improvements made in exposition: for this it is a pleasure to acknowledge the suggestions and criticisms of Bruce Francis, Huibert Kwakernaak, Alan Laub, Bruce Moore and Jan Willems.

I decided against attempting to include in the book everything that is currently known within the geometric framework, two notable omissions being the results on decentralized control and on "generalized dynamic covers," due respectively to Morse and to Silverman and their coworkers. However, the reader who has completed Chapter 5 of the book should be well prepared to explore the journals.

Finally, thanks are due once more to Professor A. V. Balakrishnan and Springer-Verlag for their encouragement and assistance; and to Mrs. Rita de Clercq Zubli for her expert typing of the manuscript.

Toronto
July, 1978

W. M. WONHAM

List of Figures

Contents

Mathematical Preliminaries 0

For the reader's convenience we shall quickly review linear algebra and the rudiments of linear dynamic systems. In keeping with the spirit of this book we emphasize the geometric content of the mathematical foundations, laying stress on the presentation of results in terms of vector spaces and their subspaces. As the material is standard, few proofs are offered; however, detailed developments can be found in the textbooks cited at the end of the chapter. For many of the simpler identities involving maps and subspaces, the reader is invited to supply his own proofs; an illustration and further hints are provided in the exercises. It is also recommended that the reader gain practice in translating geometric statements into matrix formalism, and vice versa; for this, guidance will also be found in the exercises.

0.1 Notation

If k is a positive integer, $\mathbf{k}$ denotes the set of integers $\{1, 2, \ldots, k\}$. If Λ is a finite set or list, $|\Lambda|$ denotes the number of its elements. The real and imaginary parts of a complex number etc. are written $\Re e$, $\Im m$, respectively. The symbol $:=$ means equality by definition.

0.2 Linear Spaces

We recall that a linear (vector) space consists of an additive group, of elements called *vectors*, together with an underlying field of *scalars*. We consider only spaces over the field of real numbers $\mathbb{R}$ or complex numbers $\mathbb{C}$.

The symbol $\mathbb{F}$ will be used for either field. Linear spaces are denoted by script capitals $\mathscr{X}, \mathscr{Y}, \ldots$; their elements (vectors) by lower case Roman letters $x, y, \ldots$; and field elements (scalars) by lower case Roman or Greek letters. The symbol 0 will stand for anything that is zero (a number, vector, map, or subspace), according to context.

The reader will be familiar with the properties of vector addition, and multiplication of vectors by scalars; for instance, if $x_1, x_2 \in \mathscr{X}$ and $c_1, c_2 \in \mathbb{F}$, then

$$c_1 x_1 \in \mathscr{X}, \qquad c_1(x_1 + x_2) = c_1 x_1 + c_1 x_2,$$
$$(c_1 + c_2)x_1 = c_1 x_1 + c_2 x_1, \qquad (c_1 c_2)x_1 = c_1(c_2 x_1).$$

Let $x_1, \ldots, x_k \in \mathscr{X}$, where $\mathscr{X}$ is defined over $\mathbb{F}$. Their *span*, written

$$\mathrm{Span}_{\mathbb{F}}\{x_1, \ldots, x_k\} \quad \text{or} \quad \mathrm{Span}_{\mathbb{F}}\{x_i, i \in \mathbf{k}\}$$

is the set of all linear combinations of the x_i, with coefficients in $\mathbb{F}$. The subscript $\mathbb{F}$ will be dropped if the field is clear from context. $\mathscr{X}$ is *finite-dimensional* if there exist a (finite) k and a set $\{x_i, i \in \mathbf{k}; x_i \in \mathscr{X}\}$ whose span is $\mathscr{X}$. If $\mathscr{X} \neq 0$, the least k for which this happens is the *dimension* of $\mathscr{X}$, written $d(\mathscr{X})$; when $\mathscr{X} = 0$, $d(\mathscr{X}) := 0$. If $k = d(\mathscr{X}) \neq 0$, a spanning set $\{x_i, i \in \mathbf{k}\}$ is a *basis* for $\mathscr{X}$.

Unless otherwise stated, all linear spaces are finite dimensional; the rare exceptions will be some common function spaces, to be introduced only when needed.

A set of vectors $\{x_i \in \mathscr{X}, i \in \mathbf{m}\}$ is (*linearly*) *independent* (*over* $\mathbb{F}$) if for all sets of scalars $\{c_i \in \mathbb{F}, i \in \mathbf{m}\}$, the relation

$$\sum_{i=1}^{m} c_i x_i = 0 \tag{2.1}$$

implies $c_i = 0$ for all $i \in \mathbf{m}$. If the x_i $(i \in \mathbf{m})$ are independent, and if $x \in \mathrm{Span}\{x_i, i \in \mathbf{m}\}$, then the representation

$$x = c_1 x_1 + \cdots + c_m x_m$$

is unique. The vectors of a basis are necessarily independent. If $m > d(\mathscr{X})$, the set $\{x_i, i \in \mathbf{m}\}$ must be *dependent*, i.e. there exist $c_i \in \mathbb{F}$ $(i \in \mathbf{m})$ not all zero, such that (2.1) is true.

Let $d(\mathscr{X}) = n$ and fix a basis $\{x_i, i \in \mathbf{n}\}$. If $x \in \mathscr{X}$ then $x = c_1 x_1 + \cdots + c_n x_n$ for unique $c_i \in \mathbb{F}$. For computational purposes x will be represented, as usual, by the $n \times 1$ column vector $\mathrm{col}[c_1, \ldots, c_n]$. As usual, addition of vectors, and scalar multiplication by elements in $\mathbb{F}$, are done componentwise on the representative column vectors.

In most of our applications, vector spaces $\mathscr{X}$ etc. will be defined initially over $\mathbb{R}$. It is then sometimes convenient to introduce the *complexification* of $\mathscr{X}$, written $\mathscr{X}_{\mathbb{C}}$, and defined as the set of formal sums

$$\mathscr{X}_{\mathbb{C}} = \{x_1 + ix_2 \colon x_1, x_2 \in \mathscr{X}\},$$

i being the imaginary unit. Addition and scalar multiplication in $\mathscr{X}_{\mathbb{C}}$ are done in the obvious way. In this notation if $x = x_1 + ix_2 \in \mathscr{X}_{\mathbb{C}}$ then $\mathfrak{Re}\ x := x_1$ and $\mathfrak{Im}\ x := x_2$. Note that $d(\mathscr{X}_{\mathbb{C}}) = d(\mathscr{X})$, because if $\{x_i, i \in \mathbf{n}\}$ is a basis for $\mathscr{X}$, so that

$$\mathscr{X} = \operatorname{Span}_{\mathbb{R}}\{x_i, i \in \mathbf{n}\},$$

then

$$\mathscr{X}_{\mathbb{C}} = \operatorname{Span}_{\mathbb{C}}\{x_i, i \in \mathbf{n}\},$$

and clearly $x_1, \ldots, x_n$ are independent over $\mathbb{C}$.

0.3 Subspaces

A (*linear*) *subspace* $\mathscr{S}$ of the linear space $\mathscr{X}$ is a subset of $\mathscr{X}$ which is a linear space under the operations of vector addition and scalar multiplication inherited from $\mathscr{X}$: namely $\mathscr{S} \subset \mathscr{X}$ (as a set) and for all $x_1, x_2 \in \mathscr{S}$ and $c_1, c_2 \in \mathbb{F}$ we have $c_1 x_1 + c_2 x_2 \in \mathscr{S}$. The notation $\mathscr{S} \subset \mathscr{X}$ (with $\mathscr{S}$ a script capital) will henceforth mean that $\mathscr{S}$ is a subspace of $\mathscr{X}$. If $x_i \in \mathscr{X}$ $(i \in \mathbf{k})$, then $\operatorname{Span}\{x_i, i \in \mathbf{k}\}$ is a subspace of $\mathscr{X}$. Geometrically, a subspace may be pictured as a hyperplane passing through the origin of $\mathscr{X}$; thus the vector $0 \in \mathscr{S}$ for every subspace $\mathscr{S} \subset \mathscr{X}$. We have $0 \le d(\mathscr{S}) \le d(\mathscr{X})$, with $d(\mathscr{S}) = 0$ (resp. $d(\mathscr{X})$) if and only if $\mathscr{S} = 0$ (resp. $\mathscr{X}$).

If $\mathscr{R}, \mathscr{S} \subset \mathscr{X}$, we define subspaces $\mathscr{R} + \mathscr{S} \subset \mathscr{X}$ and $\mathscr{R} \cap \mathscr{S} \subset \mathscr{X}$ according to

$$\mathscr{R} + \mathscr{S} := \{r + s\colon r \in \mathscr{R}, s \in \mathscr{S}\},$$

$$\mathscr{R} \cap \mathscr{S} := \{x\colon x \in \mathscr{R}\ \&\ x \in \mathscr{S}\}.$$

These definitions are extended in the obvious way to finite collections of subspaces. It is well to note that $\mathscr{R} + \mathscr{S}$ is the span of $\mathscr{R}$ and $\mathscr{S}$ and may be much larger than the set-theoretic union; the latter is generally not a subspace. Also, as the zero subspace $0 \subset \mathscr{R}$ and $0 \subset \mathscr{S}$, it is always true that $0 \subset \mathscr{R} \cap \mathscr{S} \neq \varnothing$; that is, two subspaces of $\mathscr{X}$ are never "disjoint" in the set-theoretic sense.

The numerical addition and intersection of subspaces is summarized in Exercise 0.6.

The family of all subspaces of $\mathscr{X}$ is partially ordered by subspace inclusion ($\subset$), and under the operations $+$ and $\cap$ is easily seen to form a *lattice*: namely $\mathscr{R} + \mathscr{S}$ is the smallest subspace containing both $\mathscr{R}$ and $\mathscr{S}$, while $\mathscr{R} \cap \mathscr{S}$ is the largest subspace contained in both $\mathscr{R}$ and $\mathscr{S}$.

Inclusion relations among subspaces may be pictured by a *lattice diagram*, in which the nodes represent subspaces, and a rising branch from $\mathscr{R}$

to $\mathscr{S}$ means $\mathscr{R} \subset \mathscr{S}$. Thus, for arbitrary $\mathscr{R}$ and $\mathscr{S} \subset \mathscr{X}$, we have the diagram shown below.

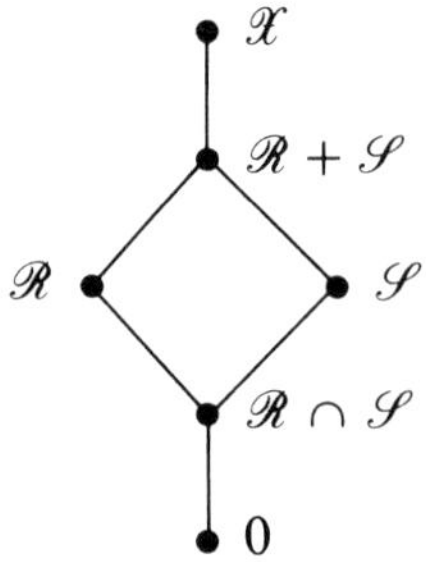

Let $\mathscr{R}, \mathscr{S}, \mathscr{T} \subset \mathscr{X}$ and suppose $\mathscr{R} \supset \mathscr{S}$. Then

$$\mathscr{R} \cap (\mathscr{S} + \mathscr{T}) = \mathscr{R} \cap \mathscr{S} + \mathscr{R} \cap \mathscr{T} \tag{3.1a}$$

$$= \mathscr{S} + \mathscr{R} \cap \mathscr{T}. \tag{3.1b}$$

Equation (3.1) is the *modular distributive rule*; a lattice in which it holds is called *modular*. It is important to realize that the distributive relation (3.1a) need not hold for arbitrary choices of $\mathscr{R}$, $\mathscr{S}$ and $\mathscr{T}$: for a counterexample take three distinct one-dimensional subspaces of the two-dimensional plane $\mathscr{X}$; then, $\mathscr{R} \cap (\mathscr{S} + \mathscr{T}) = \mathscr{R} \cap \mathscr{X} = \mathscr{R}$, but $\mathscr{R} \cap \mathscr{S} = \mathscr{R} \cap \mathscr{T} = 0$. On the other hand, if for some $\mathscr{R}$, $\mathscr{S}$ and $\mathscr{T}$, with no inclusion relation postulated, it happens to be true that

$$\mathscr{R} \cap (\mathscr{S} + \mathscr{T}) = \mathscr{R} \cap \mathscr{S} + \mathscr{R} \cap \mathscr{T}, \tag{3.2}$$

then it is also true that

$$\mathscr{S} \cap (\mathscr{R} + \mathscr{T}) = \mathscr{R} \cap \mathscr{S} + \mathscr{S} \cap \mathscr{T} \tag{3.3a}$$

and (by symmetry)

$$\mathscr{T} \cap (\mathscr{R} + \mathscr{S}) = \mathscr{R} \cap \mathscr{T} + \mathscr{S} \cap \mathscr{T}. \tag{3.3b}$$

For the standard technique of proof of such identities, see Exercise 0.2.

Two subspaces $\mathscr{R}, \mathscr{S} \subset \mathscr{X}$ are (*linearly*) *independent* if $\mathscr{R} \cap \mathscr{S} = 0$. A family of k subspaces $\mathscr{R}_1, \ldots, \mathscr{R}_k$ is *independent* if

$$\mathscr{R}_i \cap (\mathscr{R}_1 + \cdots + \mathscr{R}_{i-1} + \mathscr{R}_{i+1} + \cdots + \mathscr{R}_k) = 0$$

for all $i \in \mathbf{k}$. Note that an independent set of vectors cannot include the zero vector, but any independent family of subspaces remains independent if we adjoin one or more zero subspaces. The following statements are equivalent:

i. The family $\{\mathscr{R}_i, i \in \mathbf{k}\}$ is independent.

ii. $\displaystyle\sum_{i=1}^{k} \left(\mathscr{R}_i \cap \sum_{j \neq i} \mathscr{R}_j \right) = 0.$

iii. $\displaystyle\sum_{i=2}^{k} \left(\mathscr{R}_i \cap \sum_{j=1}^{i-1} \mathscr{R}_j \right) = 0.$

iv. Every vector $x \in \mathscr{R}_1 + \cdots + \mathscr{R}_k$ has a *unique* representation $x = r_1 + \cdots + r_k$ with $r_i \in \mathscr{R}_i$.

If $\{\mathscr{R}_i, i \in \mathbf{k}\}$ is an independent family of subspaces of $\mathscr{X}$, the sum

$$\mathscr{R} := \mathscr{R}_1 + \cdots + \mathscr{R}_k$$

is called an *internal direct sum*, and may be written

$$\begin{aligned}\mathscr{R} &= \mathscr{R}_1 \oplus \cdots \oplus \mathscr{R}_k \\ &= \bigoplus_{i=1}^{k} \mathscr{R}_i.\end{aligned}$$

In general the symbol $\oplus$ indicates that the subspaces being added are known, or claimed, to be independent.

If $\mathscr{R}, \mathscr{S} \subset \mathscr{X}$ there exist $\hat{\mathscr{R}} \subset \mathscr{R}$ and $\hat{\mathscr{S}} \subset \mathscr{S}$, such that

$$\mathscr{R} + \mathscr{S} = \hat{\mathscr{R}} \oplus (\mathscr{R} \cap \mathscr{S}) \oplus \hat{\mathscr{S}}. \tag{3.4}$$

In general $\hat{\mathscr{R}}$ and $\hat{\mathscr{S}}$ are by no means unique (see Exercise 0.3). The decomposition (3.4) does not have a natural extension to three or more subspaces.

If $\mathscr{R}$ and $\mathscr{S}$ are independent, clearly

$$d(\mathscr{R} \oplus \mathscr{S}) = d(\mathscr{R}) + d(\mathscr{S});$$

and from (3.4) we have for arbitrary $\mathscr{R}$ and $\mathscr{S}$,

$$d(\mathscr{R} + \mathscr{S}) = d(\mathscr{R}) + d(\mathscr{S}) - d(\mathscr{R} \cap \mathscr{S}).$$

Let $\mathscr{X}_1$ and $\mathscr{X}_2$ be arbitrary linear spaces over $\mathbb{F}$. The *external direct sum* of $\mathscr{X}_1$ and $\mathscr{X}_2$, written (temporarily) $\mathscr{X}_1 \tilde{\oplus} \mathscr{X}_2$, is the linear space of all ordered pairs $\{(x_1, x_2): x_1 \in \mathscr{X}_1, x_2 \in \mathscr{X}_2\}$, under componentwise addition and scalar multiplication. Writing $\simeq$ for isomorphism (i.e. dimensional equality of linear spaces), we have

$$\mathscr{X}_1 \simeq \{(x_1, 0): x_1 \in \mathscr{X}_1\} \subset \mathscr{X}_1 \tilde{\oplus} \mathscr{X}_2,$$

and we shall identify $\mathscr{X}_1$ with its isomorphic image. The construction extends to a finite collection of $\mathscr{X}_i$ in the obvious way. Evidently the definition makes $\mathscr{X}_1$ and $\mathscr{X}_2$ independent subspaces of $\mathscr{X}_1 \tilde{\oplus} \mathscr{X}_2$, and in this sense we have

$$\mathscr{X}_1 \oplus \mathscr{X}_2 = \mathscr{X}_1 \tilde{\oplus} \mathscr{X}_2,$$

where $\oplus$ denotes the internal direct sum defined earlier. Conversely, if we start with independent subspaces $\mathscr{X}_1, \mathscr{X}_2$ of a parent space $\mathscr{X}$, then clearly

$$\mathscr{X}_1 \tilde{\oplus} \mathscr{X}_2 \simeq \mathscr{X}_1 \oplus \mathscr{X}_2$$

in a natural way. So, we shall usually not distinguish the two types of direct sum, writing $\oplus$ for either, when context makes it clear which is meant. However, if $\mathscr{X}_1 \oplus \mathscr{X}_2$ is an external direct sum it may be convenient to write $x_1 \oplus x_2$ instead of (x_1, x_2) for its elements. Similarly, if $B: \mathscr{U} \to \mathscr{X}_1 \oplus \mathscr{X}_2$ is a map (see below) that sends u to $B_1 u \oplus B_2 u$, we may write $B = B_1 \oplus B_2$.

0.4 Maps and Matrices

Let $\mathscr{X}$ and $\mathscr{Y}$ be linear spaces over $\mathbb{F}$. A function $\varphi: \mathscr{X} \to \mathscr{Y}$ is a *linear transformation* (or *map*, for short) if

$$\varphi(c_1 x_1 + c_2 x_2) = c_1 \varphi(x_1) + c_2 \varphi(x_2) \tag{4.1}$$

for all $x_1, x_2 \in \mathscr{X}$ and $c_1, c_2 \in \mathbb{F}$. Of course, the sum and scalar multiplications on the left (or right) side of (4.1) refer to the corresponding operations in $\mathscr{X}$ (or $\mathscr{Y}$). Maps will usually be denoted by Roman capitals $A, B, \ldots$. An exception may occur when $d(\mathscr{Y}) = 1$, as we may identify $\mathscr{Y} = \mathbb{F}$ and call φ a *linear functional* f' (see Section 0.12, below).

With $\mathscr{X}$ and $\mathscr{Y}$ fixed, consider the set $\mathbf{L}(\mathscr{X}, \mathscr{Y})$ of all (linear) maps $C: \mathscr{X} \to \mathscr{Y}$. This set is turned into a linear space over $\mathbb{F}$ by the natural definitions of addition, and multiplication by a scalar:

$$(C_1 + C_2)x := C_1 x + C_2 x$$

$$(cC_1)x := c(C_1 x),$$

for all $x \in \mathscr{X}$, $c \in \mathbb{F}$, and $C_1, C_2 \in \mathbf{L}(\mathscr{X}, \mathscr{Y})$. It will be seen below that $d(\mathbf{L}(\mathscr{X}, \mathscr{Y})) = d(\mathscr{X})d(\mathscr{Y})$.

Usually we simply write $C: \mathscr{X} \to \mathscr{Y}$ instead of $C \in \mathbf{L}(\mathscr{X}, \mathscr{Y})$.

Let $\{x_i, i \in \mathbf{n}\}$ be a basis for $\mathscr{X}$ and $\{y_j, j \in \mathbf{p}\}$ a basis for $\mathscr{Y}$. If $C: \mathscr{X} \to \mathscr{Y}$ is a map, we have

$$Cx_i = c_{1i} y_1 + c_{2i} y_2 + \cdots + c_{pi} y_p, \qquad i \in \mathbf{n},$$

for uniquely determined elements $c_{ji} \in \mathbb{F}$. Observe that if $x \in \mathscr{X}$ then Cx is completely determined by the Cx_i: linearity does the rest.

The array

$$\text{Mat } C = \begin{bmatrix} c_{11} & \cdots & c_{1n} \\ \vdots & & \vdots \\ c_{p1} & \cdots & c_{pn} \end{bmatrix}$$

is the *matrix* of C relative to the given basis pair. We assume that the rules of matrix algebra are familiar. Matrices are handy in computing the action of maps, but we shall not often need them in developing the theory. Usually we need not distinguish sharply between C and Mat C, and write simply $C = \text{Mat } C$, where an array is exhibited in place of Mat C on the right, and the bases employed are clear from the context.

More fundamentally, one can think of Mat C as a function $\mathbf{p} \times \mathbf{n} \to \mathbb{F}$. The symbol $\mathbb{F}^{p \times n}$ denotes the class of all $p \times n$ matrices with elements in $\mathbb{F}$. It is turned into a linear space over $\mathbb{F}$, of dimension pn, by the usual operations of matrix addition and multiplication of matrices by scalars.

Let $C: \mathscr{X} \to \mathscr{Y}$ be a map. $\mathscr{X}$ is the *domain* of C and $\mathscr{Y}$ is the *codomain*; the array size of Mat C is thus $d(\mathscr{Y}) \times d(\mathscr{X})$. The *kernel* (or *null space*) of C is the subspace

$$\text{Ker } C := \{x: x \in \mathscr{X} \;\&\; Cx = 0\} \subset \mathscr{X},$$

while the *image* (or *range*) of C is the subspace

$$\begin{aligned}\operatorname{Im} C &:= \{y\colon y \in \mathscr{Y} \;\&\; \exists x \in \mathscr{X},\, y = Cx\} \\ &= \{Cx\colon x \in \mathscr{X}\} \subset \mathscr{Y}.\end{aligned}$$

Note the distinction between image and codomain.

If $\mathscr{R} \subset \mathscr{X}$, we write

$$\begin{aligned}C\mathscr{R} &:= \{y\colon y \in \mathscr{Y} \;\&\; \exists x \in \mathscr{R},\, y = Cx\} \\ &= \{Cx\colon x \in \mathscr{R}\};\end{aligned}$$

and if $\mathscr{S} \subset \mathscr{Y}$,

$$C^{-1}\mathscr{S} := \{x\colon x \in \mathscr{X} \;\&\; Cx \in \mathscr{S}\}.$$

Both $C\mathscr{R} \subset \mathscr{Y}$ and $C^{-1}\mathscr{S} \subset \mathscr{X}$ are subspaces. Observe that C^{-1} is the *inverse image function* of the map C, and as such it will be regarded as a function from the set of all subspaces of $\mathscr{Y}$ to those of $\mathscr{X}$. In this usage C^{-1} does not denote a linear map from $\mathscr{Y}$ to $\mathscr{X}$. However, in the special case where $d(\mathscr{X}) = d(\mathscr{Y})$ and the ordinary inverse of C as a map $\mathscr{Y} \to \mathscr{X}$ happens to exist, this map will also be written, as usual, C^{-1}; since the two usages are then consistent no confusion can arise.

As easy consequences of the definitions, we have

$$\begin{aligned}d(C\mathscr{R}) &= d(\mathscr{R}) - d(\mathscr{R} \cap \operatorname{Ker} C), \\ d(C^{-1}\mathscr{S}) &= d(\operatorname{Ker} C) + d(\mathscr{S} \cap \operatorname{Im} C),\end{aligned}$$

and in particular, as $\operatorname{Im} C = C\mathscr{X}$,

$$d(\mathscr{X}) = d(\operatorname{Ker} C) + d(\operatorname{Im} C).$$

Also, for $\mathscr{S} \subset \mathscr{Y}$ there exists $\mathscr{R} \subset \mathscr{X}$, in general not unique, such that

$$d(\mathscr{R}) = d(\mathscr{S} \cap \operatorname{Im} C)$$

and

$$\mathscr{R} \oplus \operatorname{Ker} C = C^{-1}\mathscr{S}.$$

If $C\colon \mathscr{X} \to \mathscr{Y}$ and $\mathscr{R}_1, \mathscr{R}_2 \subset \mathscr{X}$, we have

$$C(\mathscr{R}_1 + \mathscr{R}_2) = C\mathscr{R}_1 + C\mathscr{R}_2;$$

but in general

$$C(\mathscr{R}_1 \cap \mathscr{R}_2) \subset (C\mathscr{R}_1) \cap (C\mathscr{R}_2), \tag{4.2}$$

with equality if and only if

$$(\mathscr{R}_1 + \mathscr{R}_2) \cap \operatorname{Ker} C = \mathscr{R}_1 \cap \operatorname{Ker} C + \mathscr{R}_2 \cap \operatorname{Ker} C. \tag{4.3}$$

Dually, if $\mathscr{S}_1, \mathscr{S}_2 \subset \mathscr{Y}$, we have

$$C^{-1}(\mathscr{S}_1 \cap \mathscr{S}_2) = C^{-1}\mathscr{S}_1 \cap C^{-1}\mathscr{S}_2;$$

but

$$C^{-1}(\mathscr{S}_1 + \mathscr{S}_2) \supset C^{-1}\mathscr{S}_1 + C^{-1}\mathscr{S}_2,$$

with equality if and only if

$$(\mathscr{S}_1 + \mathscr{S}_2) \cap \operatorname{Im} C = \mathscr{S}_1 \cap \operatorname{Im} C + \mathscr{S}_2 \cap \operatorname{Im} C.$$

If $\mathscr{R}_1 \cap \mathscr{R}_2 = 0$, in general

$$C(\mathscr{R}_1 \oplus \mathscr{R}_2) \neq C\mathscr{R}_1 \oplus C\mathscr{R}_2,$$

because the subspaces on the right need not be independent; they are independent if and only if

$$(\mathscr{R}_1 \oplus \mathscr{R}_2) \cap \operatorname{Ker} C = \mathscr{R}_1 \cap \operatorname{Ker} C \oplus \mathscr{R}_2 \cap \operatorname{Ker} C.$$

Essential to any grasp of algebra is a command of Greek adverbs. A map $C: \mathscr{X} \to \mathscr{Y}$ is an *epimorphism* (or C is *epic*) if $\operatorname{Im} C = \mathscr{Y}$. C is a *monomorphism* (or C is *monic*) if $\operatorname{Ker} C = 0$. If C is epic there is a map $\check{C}_r: \mathscr{Y} \to \mathscr{X}$, a *right inverse* of C, such that

$$C\check{C}_r = 1_{\mathscr{Y}}, \tag{4.4}$$

the identity map on $\mathscr{Y}$. If C is monic there is a map $\check{C}_l: \mathscr{Y} \to \mathscr{X}$, a *left inverse* of C, such that

$$\check{C}_l C = 1_{\mathscr{X}},$$

the identity on $\mathscr{X}$. In general $\check{C}_r$ and $\check{C}_l$ are not unique. If C is both epic and monic, C is an *isomorphism*, and this can happen only if $d(\mathscr{X}) = d(\mathscr{Y})$. Then, we write $\mathscr{X} \simeq \mathscr{Y}$ and $C: \mathscr{X} \sim \mathscr{Y}$; in this case $\check{C}_r = \check{C}_l = C^{-1}$, the ordinary inverse of C. Conversely, if $d(\mathscr{X}) = d(\mathscr{Y})$, and if $\{x_i, i \in \mathbf{n}\}$, $\{y_j, j \in \mathbf{n}\}$ are bases for $\mathscr{X}$ and $\mathscr{Y}$, respectively, we can manufacture an isomorphism $C: \mathscr{X} \simeq \mathscr{Y}$ by defining $Cx_i = y_i$ $(i \in \mathbf{n})$.

An arbitrary map $A: \mathscr{X} \to \mathscr{X}$ is an *endomorphism* of $\mathscr{X}$. A is an *automorphism* of $\mathscr{X}$ if A is an isomorphism.

Let $\mathscr{V} \subset \mathscr{X}$, $d(\mathscr{V}) = k$. Since $\mathscr{V}$ can be regarded as a k-dimensional linear space in its own right, a vector $v \in \mathscr{V}$ can be described purely as an element of $\mathscr{V}$, or it can be exhibited also as an element of the ambient space $\mathscr{X}$. To formalize this viewpoint, model $\mathscr{V}$ as $\mathbb{F}^k$, let $\{e_j, j \in \mathbf{k}\}$ be a basis (say the standard unit basis) for $\mathbb{F}^k$, and let $\{x_i, i \in \mathbf{n}\}$ be a basis for $\mathscr{X}$. Since $e_j \in \mathscr{V} \subset \mathscr{X}$, each e_j can be represented uniquely in the form

$$e_j = \sum_{i=1}^{n} v_{ij} x_i, \qquad j \in \mathbf{k}.$$

The $n \times k$ matrix $[v_{ij}]$ determines a unique map $V: \mathscr{V} \to \mathscr{X}$, the *insertion map of $\mathscr{V}$ in $\mathscr{X}$*. Thus,

$$\operatorname{Mat} V = \begin{bmatrix} v_{11} & \cdots & v_{1k} \\ \vdots & & \vdots \\ v_{n1} & \cdots & v_{nk} \end{bmatrix}$$

A vector $v \in \mathscr{V}$ now has two alternative representations: either as a linear combination of the e_j (i.e. as a $k \times 1$ vector: call this v); or, as the corresponding element of $\mathscr{X}$: the $n \times 1$ vector $x = Vv$. Clearly V is monic. The insertion map is represented by any matrix whose column vectors form a basis for $\mathscr{V}$ relative to the given basis for $\mathscr{X}$. This is a standard device for the numerical representation of a subspace.

Let $C: \mathscr{X} \to \mathscr{Y}$, and let $\mathscr{V} \subset \mathscr{X}$ be a subspace with insertion map $V: \mathscr{V} \to \mathscr{X}$. The *restriction of* C *to* $\mathscr{V}$ is the map $C \mid \mathscr{V}: \mathscr{V} \to \mathscr{Y}$ given by

$$C \mid \mathscr{V} := CV.$$

Thus, $C \mid \mathscr{V}$ has the action of C on $\mathscr{V}$ but is not defined off $\mathscr{V}$. Now suppose $\operatorname{Im} C \subset \mathscr{W} \subset \mathscr{Y}$. Occasionally, it is useful to bring in a new map describing the action of C with reduced codomain $\mathscr{W}$. If $W: \mathscr{W} \to \mathscr{Y}$ is the insertion of $\mathscr{W}$ in $\mathscr{Y}$ then our new map, written $\mathscr{W} \mid C: \mathscr{X} \to \mathscr{W}$, is determined by the relation

$$W(\mathscr{W} \mid C) = C.$$

Since $\operatorname{Im} C \subset \operatorname{Im} W$, and each $y \in \operatorname{Im} W$ has a unique pre-image w with $Ww = y$, it is clear that $\mathscr{W} \mid C$ is well-defined.

Let $\mathscr{X} = \mathscr{R} \oplus \mathscr{S}$. Since the representation $x = r + s$ $(r \in \mathscr{R}, s \in \mathscr{S})$ is unique for each $x \in \mathscr{X}$, there is a function $x \mapsto r$, called the *projection on* $\mathscr{R}$ *along* $\mathscr{S}$. It is easy to see that the projection is a (linear) map $Q: \mathscr{X} \to \mathscr{X}$, such that $\operatorname{Im} Q = \mathscr{R}$ and $\operatorname{Ker} Q = \mathscr{S}$; furthermore

$$\mathscr{X} = Q\mathscr{X} \oplus (1 - Q)\mathscr{X}.$$

Note that $1 - Q$ is the projection on $\mathscr{S}$ along $\mathscr{R}$, so that $Q(1 - Q) = 0$, or $Q^2 = Q$. Conversely, if $Q: \mathscr{X} \to \mathscr{X}$ is a map such that $Q^2 = Q$ (the property of *idempotence*) it is easy to show that

$$\mathscr{X} = \operatorname{Im} Q \oplus \operatorname{Ker} Q,$$

i.e. Q is the projection on $\operatorname{Im} Q$ along $\operatorname{Ker} Q$. For computational purposes it is also useful to employ the *natural projection* $\tilde{Q}: \mathscr{X} \to \mathscr{R}$, again defined as the map $x = r + s \mapsto r$, but with $\mathscr{R}$ rather than $\mathscr{X}$ as codomain. Thus, $\tilde{Q} = \mathscr{R} \mid Q$. These seemingly fussy distinctions are essential both for conceptual clarity and for consistency in performing matrix calculations.

0.5 Factor Spaces

Let $\mathscr{S} \subset \mathscr{X}$. Call vectors $x, y \in \mathscr{X}$ *equivalent mod* $\mathscr{S}$ if $x - y \in \mathscr{S}$. We define the *factor space* (or *quotient space*) $\mathscr{X}/\mathscr{S}$ as the set of all equivalence classes

$$\bar{x} := \{y: y \in \mathscr{X}, y - x \in \mathscr{S}\}, \qquad x \in \mathscr{X}.$$

Geometrically, $\bar{x}$ is just the hyperplane passing through x obtained by parallel translation of $\mathscr{S}$. In $\mathscr{X}/\mathscr{S}$, we define

$$\bar{x}_1 + \bar{x}_2 := \overline{x_1 + x_2}, \qquad x_1, x_2 \in \mathscr{X}$$

and

$$c\bar{x} := \overline{cx}, \qquad x \in \mathscr{X},\ c \in \mathbb{F}.$$

It is a straightforward exercise to show that these definitions of sum and scalar multiplication in $\mathscr{X}/\mathscr{S}$ are unambiguous, and turn $\mathscr{X}/\mathscr{S}$ into a vector space over $\mathbb{F}$. One easily sees that

$$d\left(\frac{\mathscr{X}}{\mathscr{S}}\right) = d(\mathscr{X}) - d(\mathscr{S}).$$

Indeed, if $\mathscr{R} \subset \mathscr{X}$ is any subspace such that $\mathscr{R} \oplus \mathscr{S} = \mathscr{X}$, and if $\{r_1, \ldots, r_\rho\}$ is a basis for $\mathscr{R}$, then $\{\bar{r}_1, \ldots, \bar{r}_\rho\}$ is a basis for $\mathscr{X}/\mathscr{S}$, so that $d(\mathscr{X}/\mathscr{S}) = \rho$.

As an application of these ideas we see that if $C: \mathscr{X} \to \mathscr{Y}$ then

$$\operatorname{Im} C = C\mathscr{X} \simeq \frac{\mathscr{X}}{\operatorname{Ker} C}.$$

In particular, if C is monic, $\mathscr{X} \approx C\mathscr{X}$; and if C is epic,

$$\mathscr{Y} \simeq \frac{\mathscr{X}}{\operatorname{Ker} C}.$$

For $x \in \mathscr{X}$ the element $\bar{x} \in \mathscr{X}/\mathscr{S}$ is the *coset of $\mathscr{X}$ mod $\mathscr{S}$*; $\bar{x}$ is sometimes written $x + \mathscr{S}$. The function $x \mapsto \bar{x}$ is a map $P: \mathscr{X} \to \mathscr{X}/\mathscr{S}$ called the *canonical projection of $\mathscr{X}$ on $\mathscr{X}/\mathscr{S}$*. Clearly P is epic, and $\operatorname{Ker} P = \mathscr{S}$.

This terminology sharply distinguishes P from the projections Q and $\tilde{Q}$ defined earlier: note that $\mathscr{X}/\mathscr{S}$ is certainly not a subspace of $\mathscr{X}$, and if $\mathscr{S} \neq 0$, Q is not epic. Concretely, let $\mathscr{R} \oplus \mathscr{S} = \mathscr{X}$ for some $\mathscr{R}$. Make up a basis for $\mathscr{X}$ by taking the union of a basis $\{x_1, \ldots, x_\rho\}$ for $\mathscr{R}$ and of one for $\mathscr{S}$, in that order, and take $\{\bar{x}_1, \ldots, \bar{x}_\rho\}$ as a basis for $\mathscr{X}/\mathscr{S}$. If Q (resp. $\tilde{Q}$) is the projection (resp. natural projection) on $\mathscr{R}$ along $\mathscr{S}$, we have

$$\operatorname{Mat} Q = \begin{bmatrix} 1^{\rho\times\rho} & 0^{\rho\times\sigma} \\ 0^{\sigma\times\rho} & 0^{\sigma\times\sigma} \end{bmatrix},$$

$$\operatorname{Mat} \tilde{Q} = [1^{\rho\times\rho} \quad 0^{\rho\times\sigma}],$$

and

$$\operatorname{Mat} P = [1^{\rho\times\rho} \quad 0^{\rho\times\sigma}]$$

where superscripts indicate matrix dimensions.

If $\mathscr{S} \subset \mathscr{T} \subset \mathscr{X}$ and $P: \mathscr{X} \to \mathscr{X}/\mathscr{S}$ is canonical, we define

$$\frac{\mathscr{T}}{\mathscr{S}} := P\mathscr{T};$$

thus $\mathscr{T}/\mathscr{S}$ is a subspace of $\mathscr{X}/\mathscr{S}$. If $\mathscr{T} \subset \mathscr{X}$ is arbitrary, we have

$$P\mathscr{T} = \frac{\mathscr{T} + \mathscr{S}}{\mathscr{S}}.$$

If $\bar{\mathscr{T}}$ is a subspace of $\mathscr{X}/\mathscr{S}$, then $\mathscr{T} := P^{-1}\bar{\mathscr{T}}$ is the unique subspace of $\mathscr{X}$ with the properties: (i) $\mathscr{T} \supset \mathscr{S}$ and (ii) $P\mathscr{T} = \bar{\mathscr{T}}$. Thus, P^{-1} determines a bijection between the family of subspaces of $\mathscr{X}/\mathscr{S}$ and the family of subspaces $\mathscr{T} \subset \mathscr{X}$ such that $\mathscr{T} \supset \mathscr{S}$.

If $\mathscr{S} \subset \mathscr{U} \cap \mathscr{V}$, then

$$\frac{\mathscr{U}}{\mathscr{S}} + \frac{\mathscr{V}}{\mathscr{S}} = \frac{\mathscr{U} + \mathscr{V}}{\mathscr{S}}$$

and

$$\frac{\mathscr{U}}{\mathscr{S}} \cap \frac{\mathscr{V}}{\mathscr{S}} = \frac{\mathscr{U} \cap \mathscr{V}}{\mathscr{S}}.$$

Finally, if $\mathscr{S} \subset \mathscr{T} \subset \mathscr{X}$, then

$$\frac{\mathscr{X}/\mathscr{S}}{\mathscr{T}/\mathscr{S}} \simeq \frac{\mathscr{X}}{\mathscr{T}};$$

and if $\mathscr{T}$ is arbitrary,

$$\frac{\mathscr{T} + \mathscr{S}}{\mathscr{S}} \simeq \frac{\mathscr{T}}{\mathscr{T} \cap \mathscr{S}}.$$

Now let $C: \mathscr{X} \to \mathscr{Y}$ be a map and let $\operatorname{Ker} C \supset \mathscr{S}$. If $P: \mathscr{X} \to \mathscr{X}/\mathscr{S}$ is the canonical projection we claim there is a unique map $\bar{C}: \mathscr{X}/\mathscr{S} \to \mathscr{Y}$ such that

$$C = \bar{C}P. \tag{5.1}$$

Thus, C "factors through" $\mathscr{X}/\mathscr{S}$. To see this let $\mathscr{X} = \mathscr{R} \oplus \mathscr{S}$, with $\{r_1, \ldots, r_\rho\}$ $\{\bar{r}_1, \ldots, \bar{r}_\rho\}$ bases for $\mathscr{R}$ and $\mathscr{X}/\mathscr{S}$, respectively, and $\bar{r}_i = Pr_i$. Define

$$\bar{C}\bar{r}_i = Cr_i, \qquad i \in \boldsymbol{\rho}. \tag{5.2}$$

As $C\mathscr{S} = 0$, the definition is unambiguous. If $x = r + s$,

$$Cx = C(r + s) = Cr = \bar{C}\bar{r} = \bar{C}Px$$

which verifies (5.1). On the other hand, (5.1) implies (5.2), showing that $\bar{C}$ is unique.

0.6 Commutative Diagrams

Relations among maps and spaces are often displayed by an "arrow diagram"; thus,

$$\mathscr{X} \xrightarrow{C} \mathscr{Y}$$

displays the map $C: \mathscr{X} \to \mathscr{Y}$. A diagram with several connecting arrows, as in

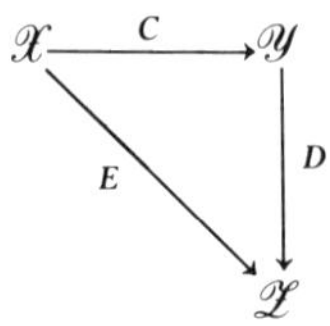

is said to *commute* if the maps composed by following different paths between the same end points are equal, i.e. $DC = E$. A dotted arrow indicates that the corresponding map is asserted to exist and to make the diagram commute. Thus, the result (5.1) on factor spaces can be displayed as

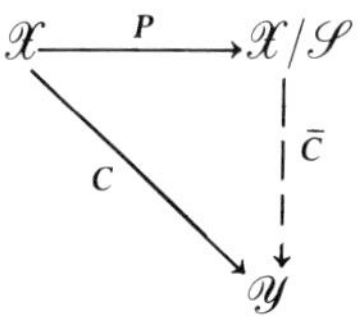

$\operatorname{Ker} C \supset \mathscr{S}$

Commutative diagrams are helpful mnemonic and heuristic devices, and the reader is encouraged to draw one when the occasion arises.

A sequence of maps $B: \mathscr{U} \to \mathscr{X}$, $C: \mathscr{X} \to \mathscr{Y}$ is *exact at* $\mathscr{X}$ if $\operatorname{Im} B = \operatorname{Ker} C$. Thus, B is monic if the sequence (or diagram)

$$0 \longrightarrow \mathscr{U} \xrightarrow{\;B\;} \mathscr{X}$$

is exact at $\mathscr{U}$ (the first arrow represents the map with image $0 \subset \mathscr{U}$), while C is epic if the sequence

$$\mathscr{X} \xrightarrow{\;C\;} \mathscr{Y} \longrightarrow 0$$

is exact at $\mathscr{Y}$ (the second arrow represents the map with image 0, i.e. the zero map).

0.7 Invariant Subspaces. Induced Maps

Let $A: \mathscr{X} \to \mathscr{X}$ and let $\mathscr{S} \subset \mathscr{X}$ have the property $A\mathscr{S} \subset \mathscr{S}$. $\mathscr{S}$ is said to be *A-invariant*. Write $\bar{\mathscr{X}} = \mathscr{X}/\mathscr{S}$ and let $P: \mathscr{X} \to \bar{\mathscr{X}}$ be the canonical projection. We claim that there exists a unique map $\bar{A}: \bar{\mathscr{X}} \to \bar{\mathscr{X}}$ such that $\bar{A}P = PA$, i.e. the diagram commutes.

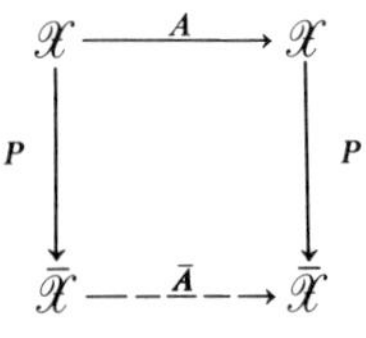

Indeed let $\{\bar{x}_i,\ i \in \boldsymbol{\rho}\}$ be a basis for $\bar{\mathscr{X}}$, where $\bar{x}_i = Px_i$. If the diagram commutes, then

$$\bar{A}\bar{x}_i = \bar{A}Px_i = PAx_i, \qquad i \in \boldsymbol{\rho}. \tag{7.1}$$

On the other hand, since $A\mathscr{S} \subset \mathscr{S}$ and $P\mathscr{S} = 0$, the definition of $\bar{A}\bar{x}_i$ by (7.1) is unambiguous and so actually determines a map $\bar{A}$ with the stated property. $\bar{A}$ is the *map induced in* $\bar{\mathscr{X}}$ *by* A.

Let $\mathscr{R}$ be any subspace such that $\mathscr{S} \oplus \mathscr{R} = \mathscr{X}$, and let $\{r_i,\ i \in \boldsymbol{\rho}\}$ be a basis for $\mathscr{R}$. Choosing a basis $\{s_j,\ j \in \boldsymbol{\sigma}\}$ for $\mathscr{S}$, we see that in the basis $\{s_1, \ldots, r_\rho\}$ for $\mathscr{X}$,

$$\text{Mat } A = \begin{bmatrix} A_1^{\sigma \times \sigma} & A_3^{\sigma \times \rho} \\ 0^{\rho \times \sigma} & A_2^{\rho \times \rho} \end{bmatrix}. \tag{7.2}$$

The matrix of $\bar{A}$ in the basis $\{\bar{r}_i,\ i \in \boldsymbol{\rho}\}$ is simply $A_2^{\rho \times \rho}$.

If $A\mathscr{S} \subset \mathscr{S}$ we denote the *restriction of* A *to* $\mathscr{S}$ with codomain $\mathscr{S}$ simply by $A \mid \mathscr{S}$ (rather than the full notation $\mathscr{S} \mid A \mid \mathscr{S}$). In the basis $\{s_j,\ j \in \boldsymbol{\sigma}\}$, $A \mid \mathscr{S}$ has the matrix $A_1^{\sigma \times \sigma}$. Let $S\colon \mathscr{S} \to \mathscr{X}$ be the insertion of $\mathscr{S}$ in $\mathscr{X}$. The restriction $A \mid \mathscr{S}$ and the induced map $\bar{A}$ on $\mathscr{X}/\mathscr{S}$ are displayed together in the commutative diagram below.

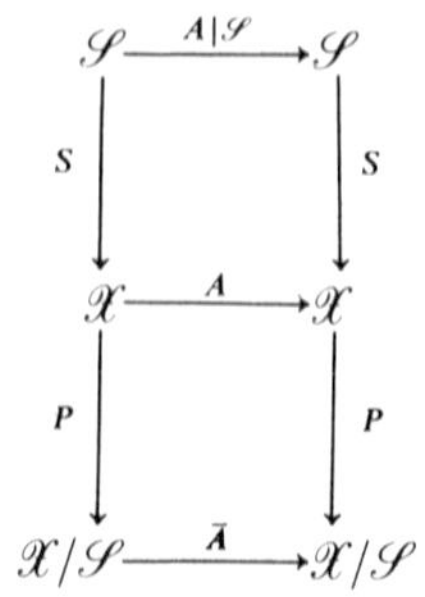

In this diagram the vertical sequence $\mathscr{S} \to \mathscr{X} \to \mathscr{X}/\mathscr{S}$ is exact at $\mathscr{X}$.

Let $A\mathscr{S} \subset \mathscr{S}$. If there exists a subspace $\mathscr{T} \subset \mathscr{X}$ such that $A\mathscr{T} \subset \mathscr{T}$ and $\mathscr{S} \oplus \mathscr{T} = \mathscr{X}$, then $\mathscr{S}$ is said to *decompose* $\mathscr{X}$ (*relative to* A). In this case, selecting $\mathscr{T}$ for the complement $\mathscr{R}$ introduced previously, we have that Mat A in (7.2) is block diagonal, and $A_2 = \text{Mat}(A \mid \mathscr{T})$.

0.8 Characteristic Polynomial. Spectrum

Let $\mathscr{X}$ be a linear space over $\mathbb{F}$, with $d(\mathscr{X}) = n$, and let $A\colon \mathscr{X} \to \mathscr{X}$ be an arbitrary endomorphism. The *characteristic polynomial* (ch.p.) of A is the nth degree monic polynomial

$$\pi(\lambda) := \det(\lambda 1 - A).$$

Here, det means determinant, the elementary properties of which we assume to be familiar. Whether $\mathbb{F}$ is $\mathbb{R}$ or $\mathbb{C}$, we define the *spectrum* of A, written $\sigma(A)$, to be the set of n complex zeros of $\pi(\lambda)$, listed according to multiplicity. The elements of $\sigma(A)$ are the *eigenvalues* of A: $\lambda \in \sigma(A)$ if and only if there exists a nonzero vector $x \in \mathscr{X}_{\mathbb{C}}$ such that $Ax = \lambda x$. Then x is an *eigenvector of A corresponding to λ.*

If $\mathbb{F} = \mathbb{R}$, the elements of Mat A are real, so $\sigma(A)$ has the general form

$$\sigma(A) = \{\alpha_1, \alpha_2, \ldots; \beta_1, \beta_1^*; \beta_2, \beta_2^*; \ldots\},$$

where $\alpha_i \in \mathbb{R}$, $\beta_j \in \mathbb{C}$ and * denotes complex conjugate. Such a set of complex numbers will be called *symmetric* (about the real axis).

As an example of spectrum calculation, we note that the block triangular structure of Mat A in (7.2) implies, via the characteristic polynomial,

$$\sigma(A) = \sigma(A \mid \mathscr{S}) \uplus \sigma(\bar{A}),$$

where $\uplus$ denotes union with any common elements repeated.

0.9 Polynomial Rings

In the sequel, certain polynomials associated with linear transformations play an important role. The set of all polynomials in a single "indeterminate" λ, and with coefficients in a field $\mathbb{F}$, has the structure of a *ring*, in particular of a *principal ideal domain*, under the usual rules of polynomial addition and multiplication. This ring is denoted by $\mathbb{F}[\lambda]$. For our purposes it is enough to recall a few of the basic facts. A polynomial is *monic* if its leading coefficient (i.e. the coefficient of its highest power of λ) is 1. Associated with any finite set of nonzero polynomials $\rho_1(\lambda), \ldots, \rho_k(\lambda) \in \mathbb{F}[\lambda]$ is their *least common multiple* (LCM), defined as the unique monic polynomial $\mu(\lambda)$ of least degree such that $\rho_i(\lambda) \mid \mu(\lambda)$ [i.e. $\rho_i(\lambda)$ divides $\mu(\lambda)$] for $i \in \mathbf{k}$. Similarly, there exists a unique monic polynomial $\delta(\lambda)$, the *greatest common divisor* (GCD) of the $\rho_i(\lambda)$, defined as the monic polynomial of greatest degree such that $\delta(\lambda) \mid \rho_i(\lambda)$, $i \in \mathbf{k}$. If

$$\delta(\lambda) = \mathrm{GCD}[\rho_1(\lambda), \ldots, \rho_k(\lambda)],$$

there exist polynomials $\sigma_1(\lambda), \ldots, \sigma_k(\lambda)$ (not unique, or necessarily monic) such that

$$\sigma_1(\lambda)\rho_1(\lambda) + \cdots + \sigma_k(\lambda)\rho_k(\lambda) = \delta(\lambda). \tag{9.1}$$

The set $\{\rho_i(\lambda), i \in \mathbf{k}\}$ is *coprime* if $\delta(\lambda) = 1$. Both δ and a suitable set σ_i can be calculated by the well known *Euclidean algorithm.*

An *irreducible element* $\pi(\lambda) \in \mathbb{F}[\lambda]$ is a polynomial which cannot be factored as a product of polynomials of lower degree. If $\mathbb{F} = \mathbb{C}$, the irreducible polynomials are those of form $c_1 \lambda + c_2$; if $\mathbb{F} = \mathbb{R}$ the irreducible polynomials are of form

$$r_1 \lambda + r_2 \quad \text{or} \quad r_3 \lambda^2 + r_4 \lambda + r_5$$

with $r_i \in \mathbb{R}$ and $r_4^2 - 4r_3 r_5 < 0$. Finally, any polynomial in $\mathbb{F}[\lambda]$ of degree ≥ 1 can be factored as a product of powers of pairwise coprime irreducible polynomials of degree ≥ 1, and such a *prime* factorization is unique up to the order of factors and the selection of (nonzero) leading coefficients in the polynomials involved.

Occasionally we need the ring of polynomials in N indeterminates, denoted by $\mathbb{F}[\boldsymbol{\lambda}]$, where $\boldsymbol{\lambda} := (\lambda_1, \ldots, \lambda_N)$. Finally, we recall that $\mathbb{F}[\boldsymbol{\lambda}]$ can be imbedded in the *fraction field* of rational expressions in $\boldsymbol{\lambda}$, denoted by $\mathbb{F}(\boldsymbol{\lambda})$.

0.10 Rational Canonical Structure

Let $\mathscr{X}$ be a linear space over $\mathbb{F}$ with $d(\mathscr{X}) = n$, and let $A: \mathscr{X} \to \mathscr{X}$ be an arbitrary endomorphism. Write $\pi(\lambda)$ for the ch.p. of A. The *Hamilton-Cayley Theorem* states that $\pi(A) = 0$. The *minimal polynomial* (m.p.) of A is the monic polynomial $\alpha(\lambda)$ of least degree such that $\alpha(A) = 0$. The m.p. of A is unique, and divides every nonzero polynomial $\beta(\lambda)$ such that $\beta(A) = 0$; in particular $\alpha(\lambda) | \pi(\lambda)$, so that $\deg \alpha \leq n$. Let $x \in \mathscr{X}$. The *minimal polynomial* of x (relative to A) is the unique monic polynomial $\xi_x(\lambda)$ of least degree such that $\xi_x(A)x = 0$. We have $\xi_x(\lambda) | \alpha(\lambda)$ for all x; furthermore

$$\alpha(\lambda) = \text{LCM}\{\xi_x(\lambda); x \in \mathscr{X}\}$$

We shall need the important

Proposition 0.1. *If $A: \mathscr{X} \to \mathscr{X}$ and the m.p. of A is $\alpha(\lambda)$, there exists $x \in \mathscr{X}$ such that $\alpha(\lambda)$ is the m.p. of x relative to A, i.e. $\xi_x(\lambda) = \alpha(\lambda)$.*

If $\alpha(\lambda) = \pi(\lambda)$, i.e. $\deg \alpha = n$, A is said to be *cyclic*, and there exists $g \in \mathscr{X}$ such that the vectors

$$g, Ag, \ldots, A^{n-1}g$$

form a basis for $\mathscr{X}$. Such a vector g is a (*cyclic*) *generator* for $\mathscr{X}$ (relative to A). If g is a generator the set of all generators coincides with the set of vectors $\gamma(A)g$, where $\gamma(\lambda) \in \mathbb{F}[\lambda]$ is coprime with $\alpha(\lambda)$.

Let A be cyclic with generator g and let

$$\alpha(\lambda) = \lambda^n - (a_1 + a_2 \lambda + \cdots + a_n \lambda^{n-1}).$$

Define auxiliary polynomials

$$\begin{aligned} \alpha^{(0)}(\lambda) &:= \alpha(\lambda) \\ \alpha^{(1)}(\lambda) &:= \lambda^{n-1} - (a_2 + a_3\lambda + \cdots + a_n\lambda^{n-2}) \\ &\vdots \\ \alpha^{(n-1)}(\lambda) &:= \lambda - a_n \\ \alpha^{(n)}(\lambda) &:= 1. \end{aligned} \tag{10.1}$$

The $\alpha^{(i)}$ satisfy the recursion relation

$$\lambda\alpha^{(i)}(\lambda) = \alpha^{(i-1)}(\lambda) + a_i\alpha^{(n)}(\lambda), \qquad i \in \mathbf{n}. \tag{10.2}$$

Now introduce the vectors

$$e_i := \alpha^{(i)}(A)g, \qquad i \in \mathbf{n},$$

with $e_0 := 0$. The e_i $(i \in \mathbf{n})$ are clearly a basis for $\mathscr{X}$. Replacing λ by A in (10.2) and operating on g, we get

$$Ae_i = e_{i-1} + a_i e_n, \qquad i \in \mathbf{n}. \tag{10.3}$$

By (10.3), in this basis

$$\text{Mat } A = \begin{bmatrix} 0 & 1 & 0 & \cdot & \cdot & \cdot & 0 \\ 0 & 0 & 1 & 0 & \cdot & \cdot & 0 \\ \cdot & \cdot & \cdot & \cdot & \cdot & \cdot & \cdot \\ 0 & \cdot & \cdot & \cdot & \cdot & 0 & 1 \\ a_1 & a_2 & \cdot & \cdot & \cdot & \cdot & a_n \end{bmatrix}.$$

This is the *companion* form of Mat A.

Now let $A: \mathscr{X} \to \mathscr{X}$ be arbitrary. A subspace $\mathscr{S} \subset \mathscr{X}$ with $A\mathscr{S} \subset \mathscr{S}$ is *A-cyclic* if $A|\mathscr{S}$ is cyclic. Let A have m.p. $\alpha(\lambda)$ and, by Proposition 0.1, choose $x \in \mathscr{X}$ with m.p. $\alpha(\lambda)$. If $\deg \alpha = m$, the vectors

$$x, Ax, \ldots, A^{m-1}x$$

span an m-dimensional A-cyclic subspace with cylic generator x. In general, we call $\mathscr{S} \subset \mathscr{X}$ *maximal cyclic* (*relative to* A) if $\mathscr{S}$ is A-cyclic and $A|\mathscr{S}$ has m.p. $\alpha(\lambda)$. Our next result states that such $\mathscr{S}$ can be "split off" from $\mathscr{X}$.

Proposition 0.2. *If $A: \mathscr{X} \to \mathscr{X}$ and $\mathscr{S} \subset \mathscr{X}$ is maximal cyclic relative to A, then $\mathscr{S}$ decomposes $\mathscr{X}$ relative to A.*

By successive applications of Propositions 0.1 and 0.2, $\mathscr{X}$ can be decomposed into a minimal number of A-cyclic direct summands, in an essentially unique way. This is the main result in linear algebra; the precise statement follows.

Theorem 0.1 (Rational Canonical Structure). *Let $A: \mathscr{X} \to \mathscr{X}$ be an endomor-*

phism of $\mathscr{X}$. There exist a positive integer k and subspaces $\mathscr{X}_i \subset \mathscr{X}$ $(i \in \mathbf{k})$ with the properties:

i. $\mathscr{X} = \mathscr{X}_1 \oplus \cdots \oplus \mathscr{X}_k$.
ii. *For $i \in \mathbf{k}$, $A\mathscr{X}_i \subset \mathscr{X}_i$ and $A \mid \mathscr{X}_i$ is cyclic.*
iii. *If $\alpha_i(\lambda)$ is the m.p. of $A \mid \mathscr{X}_i$ then α_1 is the m.p. of A, and*

$$\alpha_2 \mid \alpha_1,\ \alpha_3 \mid \alpha_2,\ \ldots,\ \alpha_k \mid \alpha_{k-1}.$$

iv. *There are exactly one integer k and one list of monic polynomials $\alpha_1, \ldots, \alpha_k$ such that a family of subspaces $\mathscr{X}_1, \ldots, \mathscr{X}_k$ exists with the properties (i)–(iii).*

We shall apply this theorem only when $\mathbb{F} = \mathbb{R}$; then, of course, the $\alpha_i \in \mathbb{R}[\lambda]$. The integer k will be called the *cyclic index* of A. By use of the Jordan decomposition (Section 0.11) it can be shown that

$$k = \max\{d[\operatorname{Ker}(A - \lambda 1)]: \lambda \in \sigma(A)\}. \tag{10.4}$$

The polynomials $\alpha_1, \ldots, \alpha_k$ are the *invariant factors* of A, and characterize Mat A to within a transformation of form $T^{-1}AT$ (i.e. a *similarity* transformation). Note that the theorem does not claim that the $\mathscr{X}_i$ themselves are unique; in fact, generally they are not.

If $A_i = A \mid \mathscr{X}_i$, a basis in $\mathscr{X}_i$ can be chosen as above such that Mat A_i is a companion matrix with ch.p. $\alpha_i(\lambda)$. Then,

$$\text{Mat } A = \operatorname{diag}[\text{Mat } A_1, \ldots, \text{Mat } A_k],$$

the *rational canonical form* of Mat A.

The following generalization of Proposition 0.2 states that an A-invariant subspace $\mathscr{S}$ decomposes $\mathscr{X}$ (relative to A) if the rational canonical structure of $A \mid \mathscr{S}$ is "maximal."

Proposition 0.3. *Let $A: \mathscr{X} \to \mathscr{X}$ and $A\mathscr{S} \subset \mathscr{S} \subset \mathscr{X}$. Suppose*

$$\mathscr{S} = \mathscr{S}_1 \oplus \cdots \oplus \mathscr{S}_j,$$

where $A\mathscr{S}_i \subset \mathscr{S}_i$ $(i \in \mathbf{j})$ and $A \mid \mathscr{S}_i$ is cyclic with m.p. equal to the ith invariant factor α_i of A. Then there exists $\mathscr{T} \subset \mathscr{X}$ such that $A\mathscr{T} \subset \mathscr{T}$, $\mathscr{S} \oplus \mathscr{T} = \mathscr{X}$, and

$$\mathscr{T} = \mathscr{T}_{j+1} \oplus \cdots \oplus \mathscr{T}_k,$$

where the $\mathscr{T}_i$ are A-invariant and A-cyclic, and the m.p. of $A \mid \mathscr{T}_i$ is α_i $(i = j + 1, \ldots, k)$.

Two maps $A: \mathscr{X} \to \mathscr{X}$ and $\hat{A}: \hat{\mathscr{X}} \to \hat{\mathscr{X}}$ are *similar* if they are related by a similarity transformation: namely, there is an isomorphism $T: \mathscr{X} \simeq \hat{\mathscr{X}}$ such that $\hat{A}T = TA$, i.e. the diagram below commutes.

$$\begin{array}{ccc} \mathscr{X} & \xrightarrow{A} & \mathscr{X} \\ {\scriptstyle T}\downarrow\wr & & \wr\downarrow{\scriptstyle T} \\ \hat{\mathscr{X}} & \xrightarrow{\hat{A}} & \hat{\mathscr{X}} \end{array}$$

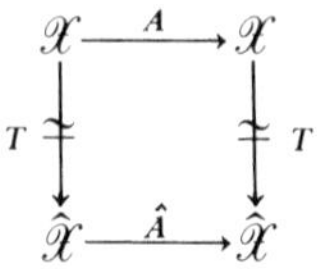

We write in this case $A \simeq \hat{A}$. It is clear that two maps are similar if and only if they have the same rational canonical structure.

0.11 Jordan Decomposition

In the notation of Section 0.10, let $\mathbb{F} = \mathbb{R}$, and

$$\alpha(\lambda) = \gamma_1(\lambda)\gamma_2(\lambda) \cdots \gamma_p(\lambda), \tag{11.1}$$

where the $\gamma_i(\lambda) \in \mathbb{R}[\lambda]$ are pairwise coprime. Define

$$\tilde{\mathscr{X}}_i := \operatorname{Ker} \gamma_i(A), \qquad i \in \mathbf{p}. \tag{11.2}$$

By use of (9.1) it is easy to check that

$$\mathscr{X} = \tilde{\mathscr{X}}_1 \oplus \cdots \oplus \tilde{\mathscr{X}}_p, \tag{11.3}$$

$$A\tilde{\mathscr{X}}_i \subset {}_i, \qquad i \in \mathbf{p}, \tag{11.4}$$

and the m.p. of $A \,|\, \tilde{\mathscr{X}}_i$ is γ_i. If (11.1) is actually a prime factorization of $\alpha(\lambda)$ over $\mathbb{R}[\lambda]$, then (11.3) provides a decomposition of $\mathscr{X}$ into generalized eigenspaces of A, which is unique. Of course, $A \,|\, \tilde{\mathscr{X}}_i$ need not be cyclic. In general, a decomposition of $\mathscr{X}$ of the form (11.3), corresponding to a partition of $\sigma(A)$ into disjoint subsets of $\mathbb{C}$, will be called a *modal decomposition* of $\mathscr{X}$.

The prime-factor modal decomposition applied to each map $A \,|\, \mathscr{X}_i$, with the $\mathscr{X}_i$ as in Theorem 0.1, yields

$$\mathscr{X}_i = \tilde{\mathscr{X}}_{i1} \oplus \cdots \oplus \tilde{\mathscr{X}}_{ip_i}, \qquad i \in \mathbf{k}.$$

Since $A \,|\, \mathscr{X}_i$ is cyclic, so is $A \,|\, \tilde{\mathscr{X}}_{ij}$, and we obtain a decomposition of $\mathscr{X}$ into cyclic subspaces on each of which the m.p. $\alpha_{ij}(\lambda)$ of A is of form $q(\lambda)^\nu$, where $q(\lambda)$ is an irreducible polynomial, of first or second degree, and ν is a positive integer. This is the (real) *Jordan decomposition* of $\mathscr{X}$. The polynomials $\alpha_{ij}(\lambda)$ ($j \in \mathbf{p}_i$, $i \in \mathbf{k}$) are the *elementary divisors* of A.

The corresponding canonical form of Mat A is obtained as follows. Let the m.p. of $A \,|\, \tilde{\mathscr{X}}_{ij}$ be $\theta(\lambda)$ (where i, j are fixed). First suppose $\theta(\lambda) = (\lambda - \mu)^\nu$, where μ is real. Let g be a generator for $\tilde{\mathscr{X}}_{ij}$, and define a basis $\{e_1, \ldots, e_\nu\}$ for $\tilde{\mathscr{X}}_{ij}$ according to

$$e_t = (A - \mu 1)^{\nu - t} g, \qquad t \in \boldsymbol{\nu}.$$

Then

$$(A - \mu 1)e_1 = 0,$$

$$(A - \mu 1)e_{t+1} = e_t, \qquad t \in \boldsymbol{\nu} - \mathbf{1},$$

so in this basis

$$\operatorname{Mat}(A \,|\, \tilde{\mathscr{X}}_{ij}) = \begin{bmatrix} \mu & 1 & 0 & \cdot & \cdot & \cdot & \cdot \\ 0 & \mu & 1 & 0 & \cdot & \cdot & \cdot \\ \cdot & \cdot & \cdot & \cdot & \cdot & \cdot & \cdot \\ 0 & \cdot & \cdot & \cdot & \cdot & \mu & 1 \\ 0 & \cdot & \cdot & \cdot & \cdot & 0 & \mu \end{bmatrix}. \tag{11.5}$$

Next, suppose $\theta(\lambda) = [(\lambda - \mu_1)^2 + \mu_2^2]^\nu$ with μ_1, μ_2 real and $\mu_2 \neq 0$. To find a convenient real basis, factor $\theta(\lambda) = \varphi(\lambda)\varphi^*(\lambda)$ over $\mathbb{C}$, where

$$\varphi(\lambda) = (\lambda - \mu_1 - i\mu_2)^\nu,$$
$$\varphi^*(\lambda) = (\lambda - \mu_1 + i\mu_2)^\nu.$$

Writing $\mathscr{X}$ for $\tilde{\mathscr{X}}_{ij}$ and A for $A \mid \tilde{\mathscr{X}}_{ij}$, let $\mathscr{X}_\mathbb{C}$ be the complexification of $\mathscr{X}$, and note that

$$\mathscr{X}_\mathbb{C} = \operatorname{Ker} \varphi(A) \oplus \operatorname{Ker} \varphi^*(A) = \mathscr{X}_0 \oplus \mathscr{X}_0^*,$$

say, where $\mathscr{X}_0$ (resp. $\mathscr{X}_0^*$) is cyclic with m.p. $\varphi(\lambda)$ (resp. $\varphi^*(\lambda)$). Let $\mathscr{X}_0$ have a cyclic generator $g = g_1 + ig_2$, where $g_1 = \mathfrak{Re}\, g$, $g_2 = \mathfrak{Im}\, g$, so that

$$\mathscr{X}_0 = \operatorname{Span}_\mathbb{C}\{(A - \mu_1 1 - i\mu_2 1)^{t-1}(g_1 + ig_2), t \in \mathbf{v}\}.$$

Define

$$e_{2\nu} = g_2, \qquad e_{2\nu-1} = g_1$$

and

$$\begin{aligned} e_{2t} &= \mathfrak{Im}(A - \mu_1 1 - i\mu_2 1)(e_{2t+1} + ie_{2t+2}) \\ e_{2t-1} &= \mathfrak{Re}(A - \mu_1 1 - i\mu_2 1)(e_{2t+1} + ie_{2t+2}) \end{aligned} \tag{11.6}$$

for $t \in \mathbf{v} - \mathbf{1}$. Then,

$$\mathscr{X}_0 = \operatorname{Span}_\mathbb{C}\{e_1 + ie_2, \ldots, e_{2\nu-1} + ie_{2\nu}\};$$

furthermore

$$(A - \mu_1 1 - i\mu_2 1)(e_1 + ie_2) = 0$$

so that

$$\begin{aligned} Ae_1 &= \mu_1 e_1 - \mu_2 e_2 \\ Ae_2 &= \mu_2 e_1 + \mu_1 e_2; \end{aligned} \tag{11.7}$$

and from (11.6)

$$\begin{aligned} Ae_{2t-1} &= \mu_1 e_{2t-1} - \mu_2 e_{2t} + e_{2t-3} \\ Ae_{2t} &= \mu_2 e_{2t-1} + \mu_1 e_{2t} + e_{2t-2} \end{aligned} \tag{11.8}$$

for $t = 2, \ldots, \nu$. Now, if $\varphi(A)x = 0$ and

$$x_1 = \mathfrak{Re}\, x, \qquad x_2 = \mathfrak{Im}\, x$$

then clearly $\varphi^*(A)x^* = 0$, where $x^* = x_1 - ix_2$; and the reverse is true. It follows that

$$\mathscr{X}_0^* = \operatorname{Span}_\mathbb{C}\{e_1 - ie_2, \ldots, e_{2\nu-1} - ie_{2\nu}\}.$$

Therefore, the 2ν vectors

$$e_1 \pm ie_2, \ldots, e_{2\nu-1} \pm ie_{2\nu}$$

are linearly independent over $\mathbb{C}$, which implies that

$$e_1, e_2, \ldots, e_{2\nu-1}, e_{2\nu} \tag{11.9}$$

are linearly independent over $\mathbb{R}$. We can now take the set (11.9) as a basis for the (real) space $\mathscr{X} = \tilde{\mathscr{X}}_{ij}$: by (11.7) and (11.8),

$$\operatorname{Mat} A = \operatorname{Mat}(A \mid \tilde{\mathscr{X}}_{ij}) = \begin{bmatrix} M & 1_2 & 0 & \cdot & \cdot & \cdot & \cdot \\ 0 & M & 1_2 & 0 & \cdot & \cdot & \cdot \\ \cdot & \cdot & \cdot & \cdot & \cdot & \cdot & \cdot \\ 0 & \cdot & \cdot & \cdot & \cdot & M & 1_2 \\ 0 & \cdot & \cdot & \cdot & \cdot & 0 & M \end{bmatrix}_{2\nu \times 2\nu}, \tag{11.10}$$

where

$$M = \begin{bmatrix} \mu_1 & \mu_2 \\ -\mu_2 & \mu_1 \end{bmatrix}, \qquad 1_2 = \begin{bmatrix} 1 & 0 \\ 0 & 1 \end{bmatrix}.$$

Thus, the complete real Jordan form of Mat A will be the appropriate diagonal array of blocks of type (11.5) and (11.10).

For the case $\mathbb{F} = \mathbb{C}$, the (complex) Jordan form is even simpler, each $\theta(\lambda)$ being of form $(\lambda - \mu)^\nu$ with $\mu \in \mathbb{C}$.

The following is a useful decomposition property of arbitrary invariant subspaces.

Proposition 0.4. *Let the m.p. of A be $\alpha = \gamma_1 \gamma_2 \cdots \gamma_p$ where the γ_i are pairwise coprime, and let $\tilde{\mathscr{X}}_i = \operatorname{Ker} \gamma_i(A)$, $i \in \mathbf{p}$. Then (as already noted)*

$$\mathscr{X} = \tilde{\mathscr{X}}_1 \oplus \cdots \oplus \tilde{\mathscr{X}}_p;$$

and if $\mathscr{R} \subset \mathscr{X}$ is A-invariant,

$$\mathscr{R} = \mathscr{R} \cap \tilde{\mathscr{X}}_1 \oplus \cdots \oplus \mathscr{R} \cap \tilde{\mathscr{X}}_p.$$

Thus a modal decomposition of $\mathscr{X}$ relative to A induces a corresponding modal decomposition of any A-invariant subspace of $\mathscr{X}$.

To conclude our discussion of canonical structure, we shall give a criterion for an invariant subspace to decompose $\mathscr{X}$, and relate this result to the solvability of Sylvester's matrix equation. We assume that a subspace $\mathscr{R} \subset \mathscr{X}$ is given, with $A\mathscr{R} \subset \mathscr{R}$. Let R: $\mathscr{R} \to \mathscr{X}$ be the insertion map of $\mathscr{R}$ in $\mathscr{X}$, $1_{\mathscr{R}}$ the identity on $\mathscr{R}$, and $A_1 = A \mid \mathscr{R}$. It is easily seen that $\mathscr{R}$ decomposes $\mathscr{X}$ relative to A if and only if there exists a map Q: $\mathscr{X} \to \mathscr{R}$ such that

$$QR = 1_{\mathscr{R}} \tag{11.11}$$

$$QA = A_1 Q. \tag{11.12}$$

Indeed if (11.11) and (11.12) hold, set $\mathscr{S} = \operatorname{Ker} Q$. Then, if $x \in \mathscr{X}$,

$$x = RQx + (1 - RQ)x;$$

since $Q(1 - RQ)x = 0$, we have $x \in \mathscr{R} + \mathscr{S}$, so that $\mathscr{R} + \mathscr{S} = \mathscr{X}$. Also, $x \in \mathscr{R} \cap \mathscr{S}$ implies $x = Rr$, say, and $Qx = 0$; thus $0 = QRr = r$, so that $x = 0$, hence $\mathscr{R} \cap \mathscr{S} = 0$. Finally, $Qx = 0$ implies $QAx = A_1 Qx = 0$, so $A\mathscr{S} \subset \mathscr{S}$. Conversely, if $\mathscr{R} \oplus \mathscr{S} = \mathscr{X}$ with $A\mathscr{S} \subset \mathscr{S}$, let Q be the natural projection $\mathscr{R} \oplus \mathscr{S} \to \mathscr{R}$.

Now let $\mathscr{R} \oplus \tilde{\mathscr{S}} = \mathscr{X}$, where $\tilde{\mathscr{S}}$ is an arbitrary complement of $\mathscr{R}$ in $\mathscr{X}$. In a compatible basis A and R have matrices

$$A = \begin{bmatrix} A_1 & A_3 \\ 0 & A_2 \end{bmatrix}, \qquad R = \begin{bmatrix} 1 \\ 0 \end{bmatrix}. \tag{11.13}$$

By (11.13), the relations (11.11) and (11.12) are equivalent to

$$Q = [1 \quad Q_2]$$

and

$$A_1 Q_2 - Q_2 A_2 - A_3 = 0. \tag{11.14}$$

Thus, to check whether $\mathscr{R}$ decomposes $\mathscr{X}$ it is enough to verify that the linear matrix equation (11.14) (*Sylvester's equation*) has a solution Q_2. This computational problem is in principle straightforward.

Of greater theoretical interest is the following result, which can be obtained from the structure theory already presented.

Proposition 0.5. *$\mathscr{R}$ decomposes $\mathscr{X}$ if and only if the elementary divisors of $A|\mathscr{R}$, with those of the induced map $\bar{A}$ in $\mathscr{X}/\mathscr{R}$, together give all the elementary divisors of A.*

In (11.13), A_1 is the matrix of $A|\mathscr{R}$ and A_2 that of $\bar{A}$. Proposition 0.5 thus solves the existence problem for (11.14) in a style which respects the role of A_1 and A_2 as endomorphisms in their own right. As a special case (and already a consequence of (11.1)–(11.4)), (11.14) has a solution which is even unique, if the spectra of A_1 and A_2 are disjoint. So, in this case the *Sylvester map*

$$S\colon \mathbb{F}^{n_1 \times n_2} \to \mathbb{F}^{n_1 \times n_2},$$

given by

$$S(Q) := A_1 Q - Q A_2,$$

is an isomorphism. This important map is discussed further in Section 0.13.

0.12 Dual Spaces

Let $\mathscr{X}$ be a linear vector space over $\mathbb{F}$. The set of all linear functionals $x'\colon \mathscr{X} \to \mathbb{F}$ is denoted by $\mathscr{X}'$. $\mathscr{X}'$ is turned into a linear vector space over $\mathbb{F}$ by the definitions

$$(x_1' + x_2')x := x_1'x + x_2'x; \qquad x_i' \in \mathscr{X}', x \in \mathscr{X}$$

$$(cx_1')x := c(x_1'x); \qquad x_1' \in \mathscr{X}', x \in \mathscr{X}, c \in \mathbb{F}.$$

If $\{x_1, \ldots, x_n\}$ is a basis for $\mathscr{X}$, the corresponding *dual basis* for $\mathscr{X}'$ is the unique set $\{x'_1, \ldots, x'_n\} \subset \mathscr{X}'$ such that $x'_i x_j = \delta_{ij}$ $(i, j \in \mathbf{n})$.

If $C: \mathscr{X} \to \mathscr{Y}$, its *dual* map $C': \mathscr{Y}' \to \mathscr{X}'$ is defined as follows. Temporarily write $C'(y')$ for the value in $\mathscr{X}'$ of C' at y'. Then, let

$$C'(y') := y'C, \qquad y' \in \mathscr{Y}', \tag{12.1}$$

where the definition makes sense because y' is a map from $\mathscr{Y}$ to $\mathbb{F}$. By choosing arbitrary bases in $\mathscr{X}$ and $\mathscr{Y}$, and their duals in $\mathscr{X}'$ and $\mathscr{Y}'$, it is easily verified that if

$$\text{Mat } C = [c_{ij}]$$

then

$$\text{Mat } C' = [c_{ji}],$$

the *transpose* of Mat C. The notation of (12.1) then matches the matrix convention that $x \in \mathscr{X}$ is represented as a column vector and $y' \in \mathscr{Y}'$ as a row vector.

A nice consequence of (12.1) is that every commutative diagram has a dual commutative diagram obtained by replacing all maps and spaces by their duals and reversing all the arrows. Under dualization exact sequences remain exact. Thus, the sequence

$$0 \longrightarrow \mathscr{X} \xrightarrow{\ C\ } \mathscr{Y},$$

expressing the fact that $C: \mathscr{X} \to \mathscr{Y}$ is monic, has the dual

$$0 \longleftarrow \mathscr{X}' \xleftarrow{\ C'\ } \mathscr{Y}'$$

which states that $C': \mathscr{Y}' \to \mathscr{X}'$ is epic. Similarly, C epic implies that C' is monic.

Let $\mathscr{S} \subset \mathscr{X}$. The *annihilator* of $\mathscr{S}$, written $\mathscr{S}^\perp$, is the set of all $x' \in \mathscr{X}'$ such that $x'\mathscr{S} = 0$. Clearly, $\mathscr{S}^\perp$ is a subspace of $\mathscr{X}'$. Thus, $0^\perp = \mathscr{X}'$, $\mathscr{X}^\perp = 0$, and, in general, $\mathscr{S}^\perp \simeq \mathscr{X}/\mathscr{S}$.

If $\mathscr{R} \subset \mathscr{X}$ and $\mathscr{S} \subset \mathscr{X}$, then

$$(\mathscr{R} + \mathscr{S})^\perp = \mathscr{R}^\perp \cap \mathscr{S}^\perp,$$

$$(\mathscr{R} \cap \mathscr{S})^\perp = \mathscr{R}^\perp + \mathscr{S}^\perp,$$

and $\mathscr{R} \subset \mathscr{S}$ implies $\mathscr{R}^\perp \supset \mathscr{S}^\perp$.

Fix $x \in \mathscr{X}$, and in $\tilde{\mathscr{X}} := (\mathscr{X}')'$ define $\tilde{x}$ by

$$\tilde{x}(x') = x'(x), \qquad x' \in \mathscr{X}'. \tag{12.2}$$

On the other hand, if $\tilde{x} \in \tilde{\mathscr{X}}$ let $\{x'_i, i \in \mathbf{n}\}$ be a basis for $\mathscr{X}'$ and define $x \in \mathscr{X}$ (uniquely) by the requirement

$$x'_i x = \tilde{x}(x'_i), \qquad i \in \mathbf{n}. \tag{12.3}$$

Equations (12.2) and (12.3) provide a natural (i.e. basis-independent) isomorphism $\tilde{\mathscr{X}} \simeq \mathscr{X}$, and from now on we identify $(\mathscr{X}')' = \mathscr{X}$. Thus, if $\mathscr{R} \subset \mathscr{X}$ then $(\mathscr{R}^\perp)^\perp = \mathscr{R}$.

If $C\colon \mathcal{X} \to \mathcal{Y}$ then

$$(\operatorname{Im} C)^{\perp} = \operatorname{Ker} C' \text{ and } (\operatorname{Ker} C)^{\perp} = \operatorname{Im} C'.$$

Finally, if $\mathcal{R} \subset \mathcal{X}$ and $\mathcal{S} \subset \mathcal{Y}$,

$$(C\mathcal{R})^{\perp} = (C')^{-1}\mathcal{R}^{\perp} \text{ and } (C^{-1}\mathcal{S})^{\perp} = C'\mathcal{S}^{\perp}.$$

0.13 Tensor Product. The Sylvester Map[1]

In the study of certain linear matrix equations (like Sylvester's) it will be convenient to use the formalism of tensor products. Let $\mathcal{X}$, $\mathcal{Y}$, $\mathcal{Z}$ be linear spaces over $\mathbb{F}$. A function $\varphi\colon \mathcal{X} \times \mathcal{Y} \to \mathcal{Z}$ is *bilinear* if it is linear in each of its arguments when the other is held fixed, i.e.

$$\varphi(c_1x_1 + c_2x_2, y_1) = c_1\varphi(x_1, y_1) + c_2\varphi(x_2, y_1)$$
$$\varphi(x_1, c_1y_1 + c_2y_2) = c_1\varphi(x_1, y_1) + c_2\varphi(x_1, y_2)$$

for all $x_i \in \mathcal{X}$, $y_i \in \mathcal{Y}$, and $c_i \in \mathbb{F}$. The *tensor product* $(\mathcal{X} \otimes \mathcal{Y}, \otimes)$ of $\mathcal{X}$ and $\mathcal{Y}$ consists of a linear space $\mathcal{X} \otimes \mathcal{Y}$ over $\mathbb{F}$, and a bilinear map

$$\otimes\colon \mathcal{X} \times \mathcal{Y} \to \mathcal{X} \otimes \mathcal{Y}\colon (x, y) \mapsto x \otimes y$$

with the following properties:

i. $\operatorname{Span}\{\operatorname{Im} \otimes\} := \operatorname{Span}\{\otimes(\mathcal{X} \times \mathcal{Y})\} = \mathcal{X} \otimes \mathcal{Y}$
ii. For every $\mathcal{Z}$ and bilinear map $\varphi\colon \mathcal{X} \times \mathcal{Y} \to \mathcal{Z}$, there exists a unique linear map $\psi\colon \mathcal{X} \otimes \mathcal{Y} \to \mathcal{Z}$, such that the following diagram commutes:

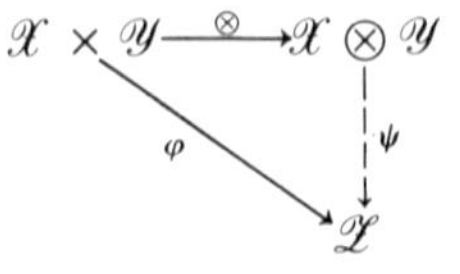

Any two tensor products of $\mathcal{X}$ and $\mathcal{Y}$ are connected by an isomorphism; i.e. tensor product is essentially unique. If $\{x_j, j \in \mathbf{n}\}$, $\{y_i, i \in \mathbf{m}\}$ are bases for $\mathcal{X}$ and $\mathcal{Y}$, respectively, then the nm products $\{x_j \otimes y_i, j \in \mathbf{n}, i \in \mathbf{m}\}$ provide a basis for $\mathcal{X} \otimes \mathcal{Y}$; thus, $d(\mathcal{X} \otimes \mathcal{Y}) = d(\mathcal{X})d(\mathcal{Y})$. Next, if $\mathcal{X}'$, $\mathcal{Y}'$ are the spaces dual to $\mathcal{X}$ and $\mathcal{Y}$, there is a natural isomorphism

$$(\mathcal{X} \otimes \mathcal{Y})' \simeq \mathcal{X}' \otimes \mathcal{Y}'.$$

From now on we identify these tensor products. Finally, if

$$\mathcal{X} = \mathcal{X}_1 + \mathcal{X}_2, \qquad \mathcal{Y} = \mathcal{Y}_1 + \mathcal{Y}_2 \tag{13.1}$$

then

$$\mathcal{X} \otimes \mathcal{Y} = \mathcal{X}_1 \otimes \mathcal{Y}_1 + \mathcal{X}_1 \otimes \mathcal{Y}_2 + \mathcal{X}_2 \otimes \mathcal{Y}_1 + \mathcal{X}_2 \otimes \mathcal{Y}_2. \tag{13.2}$$

If both sums in (13.1) are direct, so is the sum in (13.2).

[1] The material in this section is required only in Chapter 8.

Now let $C: \mathscr{X} \to \mathscr{Y}$ be a linear map. It will be shown that C can be regarded as an element c of $\mathscr{Y} \otimes \mathscr{X}'$. In fact, C determines a bilinear function $\gamma: \mathscr{Y}' \times \mathscr{X} \to \mathbb{F}$ according to the rule

$$\gamma(y', x) := y'Cx, \qquad y' \in \mathscr{Y}', x \in \mathscr{X}.$$

By property (ii) of tensor product there exists a unique linear map $c: \mathscr{Y}' \otimes \mathscr{X} \to \mathbb{F}$ such that

$$c(y' \otimes x) = \gamma(y', x), \qquad y' \in \mathscr{Y}', x \in \mathscr{X},$$

and so we identify C with $c \in (\mathscr{Y}' \otimes \mathscr{X})' = \mathscr{Y} \otimes \mathscr{X}'$. This process can be reversed, showing that

$$\mathscr{Y} \otimes \mathscr{X}' \simeq \mathbf{L}(\mathscr{X}, \mathscr{Y}), \tag{13.3}$$

where $\mathbf{L}(\mathscr{X}, \mathscr{Y})$ is the space of linear maps from $\mathscr{X}$ to $\mathscr{Y}$. Explicitly, if $\{x_j', j \in \mathbf{n}\}$ is the basis of $\mathscr{X}'$ dual to the basis $\{x_j\}$, and if

$$Cx_j = \sum_{i=1}^{m} c_{ij} y_i, \qquad j \in \mathbf{n}, \tag{13.4}$$

then

$$c = \sum_{i=1}^{m} \sum_{j=1}^{n} c_{ij} y_i \otimes x_j'. \tag{13.5}$$

Let $S: \mathscr{X} \to \mathscr{U}$ and $T: \mathscr{Y} \to \mathscr{V}$ be linear maps. Their *tensor product* is the linear map

$$S \otimes T: \mathscr{X} \otimes \mathscr{Y} \to \mathscr{U} \otimes \mathscr{V}: x \otimes y \mapsto Sx \otimes Ty. \tag{13.6}$$

Since $\mathscr{X} \otimes \mathscr{Y}$ is spanned by the elements $x \otimes y$, $S \otimes T$ is fully determined by (13.6). For compatible S_1, S_2 and T_1, T_2, we have

$$(S_1 \otimes T_1)(S_2 \otimes T_2) = S_1 S_2 \otimes T_1 T_2.$$

Now let $A: \mathscr{Y} \to \mathscr{W}$ and $B: \mathscr{U} \to \mathscr{X}$. The map

$$C \mapsto ACB \tag{13.7}$$

is linear from $\mathbf{L}(\mathscr{X}, \mathscr{Y}) \simeq \mathscr{Y} \otimes \mathscr{X}'$ to $\mathbf{L}(\mathscr{U}, \mathscr{W}) \simeq \mathscr{W} \otimes \mathscr{U}'$. By use of (13.5) and (13.6) it can be verified that the map (13.7) has the representation

$$c \mapsto (A \otimes B')c. \tag{13.8}$$

As an important application, let $A_i: \mathscr{X}_i \to \mathscr{X}_i$ ($i \in \mathbf{2}$). The *Sylvester map*

$$S: \mathbf{L}(\mathscr{X}_2, \mathscr{X}_1) \to \mathbf{L}(\mathscr{X}_2, \mathscr{X}_1): V \mapsto A_1 V - V A_2,$$

introduced in Section 0.11, has the representation

$$S: \mathscr{X}_1 \otimes \mathscr{X}_2' \to \mathscr{X}_1 \otimes \mathscr{X}_2': v \mapsto (A_1 \otimes 1_2' - 1_1 \otimes A_2')v.$$

Here, $v \in \mathscr{X}_1 \otimes \mathscr{X}_2'$ is the representation of $V \in \mathbf{L}(\mathscr{X}_2, \mathscr{X}_1)$, and $1_2'$ (resp. 1_1) is the identity on $\mathscr{X}_2'$ (resp. $\mathscr{X}_1$). Sylvester showed that S is an isomorphism (i.e.

is nonsingular) if and only if $\sigma(A_1) \cap \sigma(A_2) = \emptyset$. Later Frobenius calculated $d(\text{Ker } S)$, as follows. Let $\{\delta_i(\lambda), i \in \mathbf{k}_1\}$, $\{\epsilon_j(\lambda), j \in \mathbf{k}_2\}$ be the lists of invariant factors of A_1 and A_2, respectively. Then,

$$d(\text{Ker } S) = \sum_{i=1}^{k_1} \sum_{j=1}^{k_2} \deg \text{GCD}(\delta_i, \epsilon_j). \tag{13.9}$$

In (13.9) the elementary divisors of A_1 and A_2 could be substituted for the invariant factors without changing the result. In the latter version (13.9) is readily proved by reducing the matrices A_1, A_2 to Jordan form, then computing the contribution to Ker S from each pair of Jordan blocks (J_1, J_2) taken from A_1, A_2, respectively.

To complete this section we note the standard matrix representation of tensor product, although it will not be needed in the sequel. With reference to (13.3)–(13.5), order the basis vectors $y_i \otimes x_j'$ of $\mathscr{Y} \otimes \mathscr{X}'$ lexicographically, i.e. as

$$y_1 \otimes x_1', \ldots, y_1 \otimes x_n'; \ldots; y_m \otimes x_1', \ldots, y_m \otimes x_n'. \tag{13.10}$$

To this ordering corresponds the representation of c as the $mn \times 1$ column vector

$$c = \text{col}[c_{11} \cdots c_{1n} \cdots c_{m1} \cdots c_{mn}].$$

In (13.7), (13.8) let $d(\mathscr{U}) = p$, $d(\mathscr{W}) = q$, and take an ordered basis for $\mathscr{W} \otimes \mathscr{U}'$ by the same rule as in (13.10). If Mat $A = [a_{rs}]$ it can be checked from (13.6) that $\text{Mat}(A \otimes B')$ is the $qp \times mn$ array formed by replacing each element a_{rs} in Mat A by the $p \times n$ block $a_{rs} B'$:

$$\text{Mat}(A \otimes B') = \begin{bmatrix} a_{11} B' & \cdots & a_{1m} B' \\ \vdots & & \vdots \\ a_{q1} B' & \cdots & a_{qm} B' \end{bmatrix}. \tag{13.11}$$

The matrix (13.11) is the *Kronecker product* of the matrices A and B'. A simple computation will verify that, as matrix operations, (13.7) and (13.8) are notationally consistent.

0.14 Inner Product Spaces

It is sometimes useful to regard $\mathscr{X}$ as an inner product space and thereby identify $\mathscr{X}$ with its dual $\mathscr{X}'$. Assume $\mathbb{F} = \mathbb{C}$; the results for $\mathbb{R}$ are immediate by specialization. Let $\{x_1, \ldots, x_n\}$ be a fixed basis for $\mathscr{X}$. If $x, y \in \mathscr{X}$ with

$$x = \sum_{i=1}^{n} c_i x_i, \qquad y = \sum_{i=1}^{n} d_i x_i,$$

we define the *inner product* of x and y (with respect to the given basis) as

$$\langle x, y \rangle := \sum_{i=1}^{n} c_i d_i^*.$$

The inner product is linear in x, and antilinear (i.e. linear within conjugation of scalar multiples) in y. Such a function is sometimes called *sesquilinear.*

With the basis $\{x_i, i \in \mathbf{n}\}$ fixed, an isomorphism $\mathscr{X}' \simeq \mathscr{X}: x' \mapsto x$ is induced as follows: define x (uniquely) by the requirement

$$\langle x, x_i \rangle = x'x_i, \qquad i \in \mathbf{n}.$$

Explicitly, if $\{x_i', i \in \mathbf{n}\}$ is the dual basis in $\mathscr{X}'$, and

$$x' = c_1 x_1' + \cdots + c_n x_n'$$

then

$$x = c_1 x_1 + \cdots + c_n x_n.$$

Under this isomorphism it is often convenient to identify $\mathscr{X}'$ with $\mathscr{X}$, and write the inner product $\langle x, y \rangle$ as $x'y^*$. Here, if

$$y = d_1 x_1 + \cdots + d_n x_n, \qquad d_i \in \mathbb{C}$$

then

$$y^* := d_1^* x_1 + \cdots + d_n^* x_n.$$

The *Euclidean norm* of $x \in \mathscr{X}$, written $|x|$, is

$$|x| := +\sqrt{\langle x, x \rangle} = +\sqrt{(x^*)'x} = +\sqrt{\sum_{i=1}^{n} |c_i|^2}.$$

0.15 Hermitian and Symmetric Maps

Let $\mathscr{X}$ be an inner product space over $\mathbb{C}$; the results for $\mathbb{R}$ follow by specialization. A map $P: \mathscr{X} \to \mathscr{X}$ is *Hermitian* if $\langle x, Py \rangle = \langle Px, y \rangle$ for all $x, y \in \mathscr{X}$. Equivalently, if the inner product is related to a basis $\{x_i, i \in \mathbf{n}\}$ as in Section 0.14, we have

$$x'(Py)^* = (Px)'y^* = x'P'y^*.$$

This implies that $P' = P^*$, where P^* is defined by

$$P^* x_i := (Px_i)^*, \qquad i \in \mathbf{n}.$$

Thus P is Hermitian if and only if $P = (P')^*$: in matrix terms, P coincides with its conjugate transpose. The main result on Hermitian maps is the following.

Theorem 0.2 (Spectral Theorem). *Let $P: \mathscr{X} \to \mathscr{X}$ be Hermitian. Then the eigenvalues of P are all real. Furthermore, if the distinct eigenvalues $\lambda_1, \ldots, \lambda_k$*

occur with multiplicity n_i ($i \in \mathbf{k}$), there exist unique subspaces $\mathscr{X}_i$ ($i \in \mathbf{k}$) with the properties

i. $\mathscr{X} = \mathscr{X}_1 \oplus \cdots \oplus \mathscr{X}_k, \qquad d(\mathscr{X}_i) = n_i.$
ii. $P\mathscr{X}_i \subset \mathscr{X}_i, \qquad i \in \mathbf{k}.$
iii. $P|\mathscr{X}_i = \lambda_i 1_{\mathscr{X}_i}, \qquad i \in \mathbf{k}.$
iv. *The $\mathscr{X}_i$ are orthogonal, in the sense that $\langle x_i, x_j \rangle = 0$ for all $x_i \in \mathscr{X}_i$, $x_j \in \mathscr{X}_j$ with $j \neq i$.*

As a simple consequence, if $x^{*\prime}Px = 0$ for all x, then $P = 0$.

We shall mainly need Theorem 0.2 when $\mathscr{X}$ is defined over $\mathbb{R}$. Then "Hermitian" is to be replaced by "symmetric": P is *symmetric* if $P' = P$. In the complexification $\mathscr{X}_{\mathbb{C}}$ one has that $P^* = P$, and "symmetric" does mean "Hermitian." Keeping $\mathbb{F} = \mathbb{R}$, we call $R: \mathscr{X} \to \mathscr{X}$ *orthogonal* if R is invertible and $R^{-1} = R'$. Thus, $(Rx)'Ry = x'y$ for all $x, y \in \mathscr{X}$. In matrices, Theorem 0.2 states that, for suitable orthogonal R,

$$R'PR = \operatorname{diag}[\lambda_1 1_{n_1}, \ldots, \lambda_k 1_{n_k}].$$

A Hermitian map P is *positive definite*, written $P > 0$ (or *positive semidefinite*, written $P \geq 0$) if $\langle x, Px \rangle > 0$ (or ≥ 0) for all nonzero $x \in \mathscr{X}$. By Theorem 0.2, $P \geq 0$ and $\langle x, Px \rangle = 0$ implies $x \in \operatorname{Ker} P$. With Q also Hermitian, write $P \geq Q$ if $P - Q \geq 0$. Observe that $P \geq Q$ and $Q \geq P$ imply $P = Q$. Thus, the class of Hermitian maps on $\mathscr{X}$ is partially ordered by inequality ($\geq$), although if $d(\mathscr{X}) > 1$ it does not form a lattice.

The *norm* of P is the number

$$\begin{aligned} |P| &:= \max\{|\langle x, Px \rangle| : |x| = 1\} \\ &= \max\{|x^{*\prime}Px| : |x| = 1\} \\ &= \max\{|\lambda| : \lambda \in \sigma(P)\}. \end{aligned}$$

In the remainder of this section, we take $\mathbb{F} = \mathbb{R}$. In the sense of the partial ordering of symmetric maps, we may speak of *monotone nondecreasing* sequences $\{P_k\}$, written $P_k \uparrow$ (or *nonincreasing*, written $P_k \downarrow$), such that $P_{k+1} \geq P_k$ (or $P_{k+1} \leq P_k$). We have

Proposition 0.6. *If P_k, Q are symmetric maps such that $P_k \geq Q$ ($k = 1, 2, \ldots$) and $P_k \downarrow$, then*

$$P := \lim P_k, \qquad k \to \infty,$$

exists.

Here the limit means

$$y'Px = \lim y'P_k x, \qquad k \to \infty,$$

for all $x, y \in \mathscr{X}$; in matrix terms, the limits exist elementwise. A similar result holds for monotone nondecreasing sequences which are bounded above.

0.16 Well-Posedness and Genericity

Let A, B, ... be matrices with elements in $\mathbb{R}$ and suppose $\Pi(A, B, \ldots)$ is some property which may be asserted about them. In applications where A, B, ... represent the data of a physical problem, it is often important to know various topological features of Π. For instance, if Π is true at a nominal parameter set $\mathbf{p} = (A_0, B_0, \ldots)$ it may be desirable or natural that Π be true at points $\mathbf{p}$ in a neighborhood of $\mathbf{p}_0$, corresponding to small deviations of the parameters from their nominal values.

Most of the properties of interest to us will turn out to hold true for all sets of parameter values except possibly those which correspond to points $\mathbf{p}$ which lie on some algebraic hypersurface in a suitable parameter space, and which are thus, in an intuitive sense, atypical. To make this idea precise, we borrow some terminology from algebraic geometry. Let

$$\mathbf{p} = (p_1, \ldots, p_N) \in \mathbb{R}^N,$$

and consider polynomials $\varphi(\lambda_1, \ldots, \lambda_N)$ with coefficients in $\mathbb{R}$. A *variety* $\mathbf{V} \subset \mathbb{R}^N$ is defined to be the locus of common zeros of a finite number of polynomials $\varphi_1, \ldots, \varphi_k$:

$$\mathbf{V} = \{\mathbf{p}\colon \varphi_i(p_1, \ldots, p_N) = 0,\ i \in \mathbf{k}\}.$$

$\mathbf{V}$ is *proper* if $\mathbf{V} \neq \mathbb{R}^N$ and *nontrivial* if $\mathbf{V} \neq \varnothing$. A *property* Π is merely a function $\Pi\colon \mathbb{R}^N \to \{0, 1\}$, where $\Pi(\mathbf{p}) = 1$ (or 0) means Π holds (or fails) at $\mathbf{p}$. Let $\mathbf{V}$ be a proper variety. We shall say that Π is *generic relative to* $\mathbf{V}$ provided $\Pi(\mathbf{p}) = 0$ only for points $\mathbf{p} \in \mathbf{V}$; and that Π is *generic* provided such a $\mathbf{V}$ exists. If Π is generic, we sometimes write

$$\Pi = 1(g).$$

Assign to $\mathbb{R}^N$ the usual Euclidean topology. In general, a property Π is said to be *well-posed at* $\mathbf{p}$ if Π holds throughout some neighborhood of $\mathbf{p}$ in $\mathbb{R}^N$. By extension a "problem" that is parametrized by data in $\mathbb{R}^N$ will be called well-posed at the data point $\mathbf{p}$ if it is solvable for all data points $\mathbf{p}'$ in some neighborhood of $\mathbf{p}$. If $\mathbf{V}$ is any variety in $\mathbb{R}^N$ it is clear from the continuity of its defining polynomials that $\mathbf{V}$ is a closed subset of $\mathbb{R}^N$. Thus, if Π is generic relative to $\mathbf{V}$ (so that $\mathbf{V}$ is proper) then Π is well-posed at every point in the complement $\mathbf{V}^c$.

Let $\mathbf{p}_0 \in \mathbf{V}$, with $\mathbf{V}$ nontrivial and proper. It is clear that every neighborhood of $\mathbf{p}_0$ contains points $\mathbf{p} \in \mathbf{V}^c$; otherwise, each defining polynomial φ of $\mathbf{V}$ vanishes identically in some neighborhood of $\mathbf{p}_0$, hence vanishes on $\mathbb{R}^N$, and therefore, $\mathbf{V} = \mathbb{R}^N$, in contradiction to the assumption that $\mathbf{V}$ is proper. Thus, if Π is generic relative to $\mathbf{V}$ and if Π fails at $\mathbf{p}_0$, Π can be made to hold if $\mathbf{p}_0$ is shifted by a suitable perturbation, which can be chosen arbitrarily small. We conclude that the set of points $\mathbf{p}$ where a generic property is well-posed, is both open and dense in $\mathbb{R}^N$; furthermore, it can be shown that its complement has zero Lebesgue measure.

We shall sometimes use terms like "almost surely" or "almost all" to indicate genericity in the sense defined. Thus a well-posed property holds almost surely at $\mathbf{p}$ if $\mathbf{p}$ is selected "randomly."

As a primitive illustration of these ideas, let $C \in \mathbb{R}^{m \times n}$, $y \in \mathbb{R}^{m \times 1}$ and consider the assertion: there exists $x \in \mathbb{R}^{n \times 1}$ such that $Cx = y$. Say that $\mathbf{p} := (C, y)$ has property Π (i.e. $\Pi(\mathbf{p}) = 1$) if and only if our assertion is true. By listing the elements of C and y in some arbitrary order, regard $\mathbf{p}$ as a data point in $\mathbb{R}^N$, $N = mn + m$. Now $\Pi(\mathbf{p}) = 1$ if and only if $y \in \operatorname{Im} C$, i.e.

$$\operatorname{Rank}[C, y] = \operatorname{Rank} C. \tag{16.1}$$

It follows easily that Π is well-posed at $\mathbf{p}$ if and only if Rank $C = m$, and Π is generic if and only if $m \leq n$.

To verify these statements note first that (16.1) fails only if

$$\operatorname{Rank} C = d(\operatorname{Im} C) < d(\mathscr{Y}) = m. \tag{16.2}$$

But (16.2) implies that all $m \times m$ minors of C vanish: let $\mathbf{V} \subset \mathbb{R}^N$ be the variety so determined. If $m \leq n$, $\mathbf{V}$ is clearly proper, hence Π is generic, as claimed. On the other hand, if $m \geq n + 1$, (16.1) holds only if all $(n + 1) \times (n + 1)$ minors of $[C, y]$ vanish. The variety $\mathbf{W}$ so defined is proper, and $\Pi(\mathbf{p}) = 0$ for $\mathbf{p} \in \mathbf{W}^c$, hence Π cannot be generic. Finally, if Rank $C = m$ at $\mathbf{p}$ then (equivalently) at least one $m \times m$ minor of C is nonzero at $\mathbf{p}$, hence nonzero in a neighborhood of $\mathbf{p}$, so Π is well-posed at $\mathbf{p}$. Conversely, if Rank $C < m$ at $\mathbf{p}$ then a suitable $\tilde{y}$, with $|\tilde{y} - y|$ arbitrarily small, will make

$$\operatorname{Rank}[C, \tilde{y}] = \operatorname{Rank} C + 1;$$

namely, if $\tilde{\mathbf{p}} := (C, \tilde{y})$, then $\Pi(\tilde{\mathbf{p}}) = 0$, hence Π is not well-posed at $\mathbf{p}$.

As a second illustration we consider the intersection of two subspaces "in general position." More precisely if $\mathscr{R}, \mathscr{S} \subset \mathscr{X}$ then $\mathscr{R}$ and $\mathscr{S}$ are said to be *transverse*, or to *intersect transversely*, if $d(\mathscr{R} + \mathscr{S})$ is a maximum (equivalently $d(\mathscr{R} \cap \mathscr{S})$ is a minimum) compatible with the dimensions of $\mathscr{R}$, $\mathscr{S}$ and $\mathscr{X}$; namely

$$d(\mathscr{R} + \mathscr{S}) = \min\{d(\mathscr{R}) + d(\mathscr{S}), d(\mathscr{X})\}.$$

If $R\colon n \times r$, $S\colon n \times s$ are the matrices of insertion maps for $\mathscr{R}$ and $\mathscr{S}$, then $\mathscr{R}$, $\mathscr{S}$ are transverse if and only if

$$\operatorname{Rank}[R, S] = \min[r + s, n].$$

From a consideration of minors it is clear that transversality is a well-posed property at any data point $\mathbf{p} = (R, S) \in \mathbb{R}^{nr+ns}$ where the property holds. Furthermore, in the space of such data points, transversality is generic. Intuitively, two (or more) subspaces selected "at random" will "almost surely" intersect "in general position."

0.17 Linear Systems

We consider mainly finite-dimensional, constant-parameter (i.e. time-invariant) linear systems, modeled by equations of form

$$\begin{aligned}\dot{x}(t) &= Ax(t) + Bu(t)\\ y(t) &= Cx(t)\end{aligned} \tag{17.1}$$

for $t \geq 0$. The vectors x, y, u belong to real linear vector spaces $\mathscr{X}$, $\mathscr{Y}$, $\mathscr{U}$, respectively, with

$$d(\mathscr{X}) = n, \quad d(\mathscr{Y}) = p, \quad d(\mathscr{U}) = m.$$

Here $\mathscr{X}$ is the *state space*, $\mathscr{Y}$ the *output space*, and $\mathscr{U}$ the *input space*. For our purposes it is sufficient to assume that $u(\cdot)$ is piecewise continuous.

In some applications the output equation may appear in the form

$$y(t) = C_1 x(t) + C_2 u(t),$$

involving direct control feedthrough. We shall later indicate how this situation can be reduced algebraically to the standard case, in the context of the various specific problems of system synthesis treated in the chapters to follow.

Virtually all the theoretical developments considered in this book apply without change to the discrete-time system

$$x(t + 1) = Ax(t) + Bu(t),$$

$t = 0, 1, 2, \ldots$. An exception is the theory of quadratic optimization (Chapters 12 and 13), where the Lyapunov and matrix quadratic equations require superficial modification, while the final results are essentially the same as for the continuous version presented in the text. These modifications are left to the exercises.

If $x(0) = x_0$ then (17.1) implies

$$x(t) = e^{tA}x_0 + \int_0^t e^{(t-\tau)A}Bu(\tau)\, d\tau, \qquad t \geq 0,$$

or more generally

$$x(t) = e^{(t-t_0)A}x(t_0) + \int_{t_0}^t e^{(t-\tau)A}Bu(\tau)\, d\tau$$

for $t_0 \geq 0$, $t \geq 0$.

It is sometimes convenient to know e^{tA} explicitly. For this let $\pi(\lambda)$ be the ch.p. of A:

$$\pi(\lambda) = \lambda^n - (p_1 + p_2\lambda + \cdots + p_n\lambda^{n-1}).$$

Define auxiliary polynomials (cf. (10.1))

$$\pi^{(r)}(\lambda) = \lambda^{n-r} - (p_{r+1} + p_{r+2}\lambda + \cdots + p_n\lambda^{n-r-1})$$

for $r \in \mathbf{n}$. A short calculation verifies that

$$\pi(\lambda)(\lambda 1 - A)^{-1} = \sum_{r=1}^{n} \pi^{(r)}(\lambda) A^{r-1}. \tag{17.2}$$

Then if $\mathfrak{G}$ is any simple closed contour enclosing $\sigma(A)$ in the complex plane, we have by Cauchy's theorem

$$e^{tA} = \frac{1}{2\pi i} \oint_{\mathfrak{G}} (z1 - A)^{-1} e^{tz}\, dz = \sum_{r=1}^{n} \psi_r(t) A^{r-1}, \tag{17.3}$$

where

$$\psi_r(t) := \frac{1}{2\pi i} \oint_{\mathfrak{G}} \frac{\pi^{(r)}(z)}{\pi(z)} e^{tz}\, dz \qquad r \in \mathbf{n}.$$

Instead of the characteristic polynomial π, any multiple of the m.p. of A could be used in this calculation, if the auxiliary polynomials are defined accordingly.

0.18 Transfer Matrices. Signal Flow Graphs

If in (17.1), $x(0) = 0$, and $|u(\cdot)|$ grows at most exponentially fast as $t \uparrow \infty$, then $y(\cdot)$ has the Laplace transform

$$\begin{aligned}
\hat{y}(s) &:= \int_0^\infty e^{-st} y(t)\, dt \\
&= \int_0^\infty e^{-st} \left[C \int_0^t e^{(t-\tau)A} Bu(\tau)\, d\tau \right] dt \\
&= C \int_0^\infty e^{-st} \left[\int_0^t e^{(t-\tau)A} Bu(\tau)\, d\tau \right] dt \\
&= C(s1 - A)^{-1} B\hat{u}(s),
\end{aligned}$$

defined for $\Re e\, s$ sufficiently large. The matrix

$$H(s) := C(s1 - A)^{-1} B$$

is the *transfer matrix* of the triple of matrices (C, A, B), and is defined for $s \in \mathbb{C} - \sigma(A)$. For such s, $H(s)$ can be viewed as the matrix of a map $\mathscr{U} \to \mathscr{Y}$, taken as linear spaces over $\mathbb{C}$.

While transfer matrices play no role in the synthesis procedures of this book, they will be useful for casual descriptive purposes. In this regard, a composite system comprising interconnected subsystems can be represented by its *signal flow graph*. Informally, this is a directed graph in which the nodes represent variables (like u, x, y, ...) and the branches represent the transfer matrices relating them. A node variable is computed as the weighted sum of node variables at the tail of each entering branch, the weights being the branch transfer matrices; if no branch enters a node, its variable is an "input"; if no branch leaves a node, it is an "output." The signal flow graph is drawn after taking formal Laplace transforms of everything in sight. Thus, the system equations

$$\dot{x} = Ax + Bu, \qquad x(0) = x_0$$
$$u = Fx + v, \qquad y = Cx$$

yield

$$\hat{x}(s) = (s1 - A)^{-1}(x_0 + B\hat{u}(s)),$$
$$\hat{u}(s) = F\hat{x}(s) + \hat{v}(s), \qquad \hat{y}(s) = C\hat{x}(s),$$

whence the graph shown in Fig. 0.1. From now on we drop the caret (ˆ) on node variables in graphs; and frequently leave off the initial values x_0, with their corresponding branches.

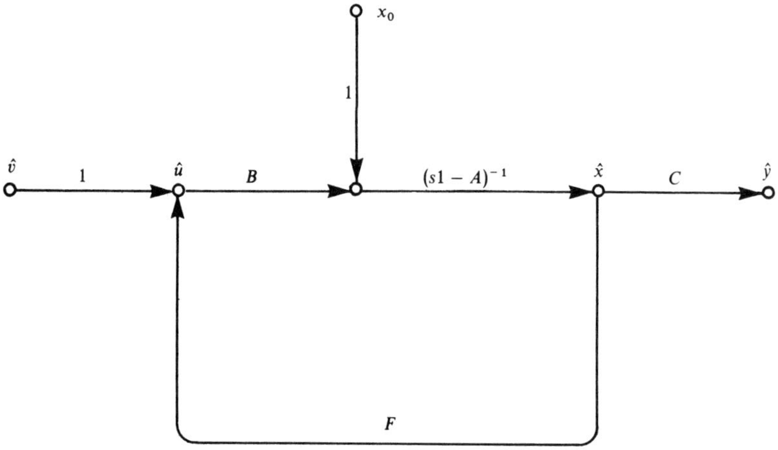

Figure 0.1 Signal Flow Graph for the System:
$\hat{x}(s) = (s1 - A)^{-1}[x_0 + B\hat{u}(s)]$
$\hat{u}(s) = F\hat{x}(s) + \hat{v}(s)$
$\hat{y}(s) = C\hat{x}(s)$

0.19 Rouché's Theorem

This well known result from complex-function theory will find application to "root loci."

Theorem 0.3. *If $f(z)$ and $g(z)$ are analytic inside and on a simple closed contour $\mathfrak{G}$, and $|g(z)| < |f(z)|$ for z on $\mathfrak{G}$, then $f(z)$ and $f(z) + g(z)$ have the same number of zeros inside $\mathfrak{G}$.*

0.20 Exercises

0.1. Prove, or look up the proofs of, all the unproved assertions in this chapter. The exercises to follow provide hints for the easier items; the hard structural results of Section 0.10 are amply covered in the textbooks cited below in Section 0.21.

0.2. Prove the modular distributive rule (3.1). Hint: This and the remaining identities among subspaces in this chapter are all provable by the standard technique of establishing the inclusions LHS $\subset$ RHS and RHS $\subset$ LHS by direct computation. Thus, for (3.1), note that $x \in \mathscr{R} \cap \mathscr{S} + \mathscr{R} \cap \mathscr{T}$ means $x = r_1 + r_2$, say, where $r_1 \in \mathscr{R} \cap \mathscr{S}$ and $r_2 \in \mathscr{R} \cap \mathscr{T}$. Thus, $r_1, r_2 \in \mathscr{R}$, hence $r_1 + r_2 \in \mathscr{R}$ (since $\mathscr{R}$ is a subspace); similarly, $r_1 + r_2 \in \mathscr{S} + \mathscr{T}$, by definition of subspace addition; so $r_1 + r_2 \in \mathscr{R} \cap (\mathscr{S} + \mathscr{T})$, as claimed. For the reverse inclusion, $x \in \mathscr{R} \cap (\mathscr{S} + \mathscr{T})$ means, in obvious notation, $x = r = s + t$, say; but $\mathscr{R} \supset \mathscr{S}$ implies $s \in \mathscr{R}$, hence $s \in \mathscr{R} \cap \mathscr{S}$; then $t = r - s \in \mathscr{R}$ implies $t \in \mathscr{R} \cap \mathscr{T}$; therefore,

$$x = s + t \in \mathscr{R} \cap \mathscr{S} + \mathscr{R} \cap \mathscr{T}$$

as claimed, and the proof is complete.

0.3. Prove (3.4). Hint: Note that if $\mathscr{T} \subset \mathscr{R}$, one can always write $\mathscr{R} = \hat{\mathscr{R}} \oplus \mathscr{T}$ for suitable $\hat{\mathscr{R}} \subset \mathscr{R}$: simply take a basis $\{t_1, \ldots, t_k\}$ for $\mathscr{T}$, extend it to a basis $\{t_1, \ldots, t_k, \hat{r}_1, \ldots, \hat{r}_l\}$ for $\mathscr{R}$, and set $\hat{\mathscr{R}} := \mathrm{Span}\{\hat{r}_1, \ldots, \hat{r}_l\}$. Of course, $\hat{\mathscr{R}}$ is not unique, as one sees by simple pictures in $\mathbb{R}^2$ or $\mathbb{R}^3$.

0.4. Given $C: \mathscr{X} \to \mathscr{Y}$ epic, prove the existence of a right inverse $\check{C}_r$ as in (4.4). Hint: The technique is to define a map $\check{C}_r$ by specifying its action on a basis. Let $\{y_i, i \in \mathbf{p}\}$ be a basis for $\mathscr{Y}$. C being epic, there are $x_i \in \mathscr{X}$ $(i \in \mathbf{p})$ such that $Cx_i = y_i$ $(i \in \mathbf{p})$, so define $\check{C}_r y_i := x_i$ $(i \in \mathbf{p})$. In general, the x_i are not unique, hence $\check{C}_r$ is not unique either.

0.5. The following miscellaneous facts are sometimes useful; the proofs are straightforward.
   i. $C(C^{-1}\mathscr{S}) = \mathscr{S} \cap \mathrm{Im}\ C$.
   ii. $C^{-1}(C\mathscr{R}) = \mathscr{R} + \mathrm{Ker}\ C$.
   iii. $C\mathscr{R} \subset \mathscr{S}$ if and only if $\mathscr{R} \subset C^{-1}\mathscr{S}$.
   iv. In general, $C^{-1}\mathscr{S} \subset \mathscr{R}$ does not imply, and is not implied by, $\mathscr{S} \subset C\mathscr{R}$.
   v. If $A: \mathscr{X} \to \mathscr{X}$ and $\mathscr{B}, \mathscr{R}, \mathscr{S} \subset \mathscr{X}$, then $A\mathscr{R} \subset A\mathscr{S} + \mathscr{B}$ if and only if $\mathscr{R} \subset \mathscr{S} + A^{-1}\mathscr{B}$.
   vi. If $\mathscr{R} \subset \mathscr{X}$ and $A: \mathscr{X} \to \mathscr{X}$ then for $j = 0, 1, 2, \ldots$, define

$$A^{-j}\mathscr{R} := A^{-1}(\cdots (A^{-1}(A^{-1}\mathscr{R})) \cdots) \qquad (j\text{-fold});$$

   and prove:

$$((A')^j \mathscr{R}^\perp)^\perp = A^{-j}\mathscr{R} = (A^j)^{-1}\mathscr{R}.$$

   vii. If A, B, C are endomorphisms of $\mathscr{X}$, then

$$d[\mathrm{Im}(AB)] + d[\mathrm{Im}(BC)] \le d(\mathrm{Im}\ B) + d[\mathrm{Im}(ABC)].$$

   viii. If $A: \mathscr{X} \to \mathscr{X}$ and $\mathscr{B}, \mathscr{R} \subset \mathscr{X}$, then

$$\frac{A\mathscr{R} + \mathscr{B}}{\mathscr{B}} \simeq \frac{\mathscr{R}}{\mathscr{R} \cap A^{-1}\mathscr{B}}.$$

ix. If $\mathscr{R}, \mathscr{S} \subset \mathscr{X}$ and $C: \mathscr{X} \to \mathscr{Y}$, then

$$\frac{C\mathscr{R} \cap C\mathscr{S}}{C(\mathscr{R} \cap \mathscr{S})} \simeq \frac{(\mathscr{R} + \mathscr{S}) \cap \operatorname{Ker} C}{\mathscr{R} \cap \operatorname{Ker} C + \mathscr{S} \cap \operatorname{Ker} C}.$$

x. If $\mathscr{R}, \mathscr{S} \subset \mathscr{Y}$ and $C: \mathscr{X} \to \mathscr{Y}$, then

$$\frac{C^{-1}(\mathscr{R} + \mathscr{S})}{C^{-1}\mathscr{R} + C^{-1}\mathscr{S}} \simeq \frac{(\mathscr{R} + \mathscr{S}) \cap \operatorname{Im} C}{R \cap \operatorname{Im} C + \mathscr{S} \cap \operatorname{Im} C}.$$

0.6. Develop matrix algorithms for the computation of $\mathscr{R} + \mathscr{S}$, $\mathscr{R} \cap \mathscr{S}$, and $A^{-1}\mathscr{R}$. Hint: If R and S are insertion maps for $\mathscr{R}$ and $\mathscr{S}$, consider the corresponding matrix $[R, S]$. The span of its columns is $\mathscr{R} + \mathscr{S}$. To get an insertion map for $\mathscr{R} + \mathscr{S}$ simply eliminate redundant columns: e.g. working from left to right, eliminate columns which are linearly dependent on their predecessors. For the intersection, represent the elements of the dual space $\mathscr{X}'$ as row vectors, and let $R^{\perp}: \mathscr{R}^{\perp} \to \mathscr{X}'$ be an insertion map for $\mathscr{R}^{\perp}$, with a similar definition for $S^{\perp}$. Thus, $R^{\perp}$ can be any matrix with independent rows, such that $R^{\perp}x = 0$ if and only if $x \in \mathscr{R}$. Noting that $(\mathscr{R} \cap \mathscr{S})^{\perp} = \mathscr{R}^{\perp} + \mathscr{S}^{\perp}$, conclude that

$$\mathscr{R} \cap \mathscr{S} = \operatorname{Ker} \begin{bmatrix} R^{\perp} \\ S^{\perp} \end{bmatrix}.$$

Elimination of redundant rows will give an insertion map for $(\mathscr{R} \cap \mathscr{S})^{\perp}$. As an immediate result of the definitions, one now has

$$A^{-1}\mathscr{R} = \operatorname{Ker}[R^{\perp}A].$$

0.7. Let $A: \mathscr{X} \to \mathscr{X}$. Show that the family of A-invariant subspaces of $\mathscr{X}$ is a lattice, relative to $\subset$, $+$, and $\cap$, hence is a sublattice of the lattice of all subspaces of $\mathscr{X}$. Hint: It is enough to show that if $\mathscr{R}$ and $\mathscr{S}$ are A-invariant, so are $\mathscr{R} + \mathscr{S}$ and $\mathscr{R} \cap \mathscr{S}$.

0.8. Let $A: \mathscr{X} \to \mathscr{X}$, $A\mathscr{N} \subset \mathscr{N}$, $A\mathscr{S} \subset \mathscr{S}$, and $\mathscr{S} \supset \mathscr{N}$. Let $P_1: \mathscr{X} \to \mathscr{X}/\mathscr{N}$ and $Q: \mathscr{X} \to \mathscr{X}/\mathscr{S}$ be the respective canonical projections, and let $\bar{A}$, $\bar{\bar{A}}$ be the maps induced by A in respectively $\mathscr{X}/\mathscr{N}$ and $\mathscr{X}/\mathscr{S}$. Prove the existence of a map $P_2: \mathscr{X}/\mathscr{N} \to \mathscr{X}/\mathscr{S}$ such that the diagram below commutes.

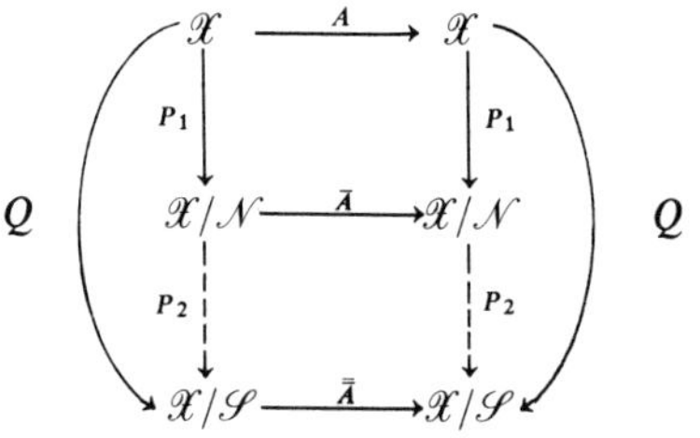

Hint: Note that $\mathscr{X}/\mathscr{S} \simeq (\mathscr{X}/\mathscr{N})/(\mathscr{S}/\mathscr{N})$ in a natural way, and consider the canonical projection from $\mathscr{X}/\mathscr{N}$ to its indicated factor space.

0.9. Prove (11.3). Hint: First suppose $p = 2$. By coprimeness (cf. (9.1)), one has

$$1 = \sigma_1(\lambda)\gamma_1(\lambda) + \sigma_2(\lambda)\gamma_2(\lambda)$$

for suitable $\sigma_i(\lambda)$ ($i \in \mathbf{p}$). Replacing λ by A and operating on $x \in \mathscr{X}$ yield the representation

$$x = \sigma_1(A)\gamma_1(A)x + \sigma_2(A)\gamma_2(A)x,$$

which is clearly of the form required. The proof is finished by induction on p.

0.10. Prove the second statement of Proposition 0.4. Hint: Factor the m.p. of $A \mid \mathscr{R}$.

0.11. Let $\mathscr{V} \subset \mathscr{X}$. Prove that there are natural (i.e. basis-independent) isomorphisms

$$\mathscr{V}' \simeq \frac{\mathscr{X}'}{\mathscr{V}^{\perp}}, \qquad \left(\frac{\mathscr{X}}{\mathscr{V}}\right)' \simeq \mathscr{V}^{\perp}.$$

From this show that if $P: \mathscr{X} \to \mathscr{X}/\mathscr{V}$ is the canonical projection then $P': (\mathscr{X}/\mathscr{V})' \to \mathscr{X}'$ is (i.e. can be identified with) the insertion $\mathscr{V}^{\perp} \to \mathscr{X}'$. Dually, if $V: \mathscr{V} \to \mathscr{X}$ is the insertion then $V': \mathscr{X}' \to \mathscr{X}'/\mathscr{V}^{\perp}$ is the canonical projection.

0.12. Prove Proposition 0.6. Hint: Use the polarization identity

$$2x'Py = (x + y)'P(x + y) - x'Px - y'Py,$$

plus the fact that the numerical sequences $\{x'P_k x\}$ are monotone and bounded.

0.13. Let $P \geq 0$ be a symmetric map on $\mathscr{X}$ (over $\mathbb{R}$). Show that P has a unique, nonnegative, symmetric square root: i.e. there exists $Q \geq 0$ symmetric, with $Q^2 = P$, and these properties determine Q uniquely. Hint: First prove the assertion for $P = 0$ and $P = 1$, then exploit Theorem 0.2.

0.21 Notes and References

The material in this chapter is standard, although not all of it is readily accessible in any one source. For coverage of linear algebra at the theoretical level required, see Gantmacher [1], Greub [1], Jacobson [1] or MacLane and Birkhoff [1]. Of these, and for our requirements, the most useful all-round text is probably Gantmacher's. Linear algebra from a numerical viewpoint is treated by Noble [1] and Strang [1]. For tensor products consult Greub [2] or Marcus [1]. An introduction to algebraic geometry is given in Chapter 16 of Van der Waerden [1].

The term "well-posed" is borrowed from partial differential equations, where it was introduced by Hadamard [1] to signify the continuity of a solution with respect to initial or boundary data. The results needed on linear differential equations are amply covered by Gantmacher [1], Lefschetz [1] or Hale [1]. Rouché's Theorem is proved in Titchmarsh [1]. For the general background in systems theory desirable as a prerequisite for this book, see especially Desoer [1]; also helpful are Porter [1] and Chapter 2 of Kalman, Falb and Arbib [1].

1 Introduction to Controllability

It is natural to say that a dynamic system is "controllable" if, by suitable manipulation of its inputs, the system outputs can be made to behave in some desirable way. In this chapter one version of this concept will be made precise, and some of its implications explored, for the system of Section 0.17:

$$\dot{x}(t) = Ax(t) + Bu(t), \qquad t \geq 0. \tag{0.1}$$

We start by examining those states which, roughly speaking, the control $u(\cdot)$ in (0.1) is able to influence.

1.1 Reachability

Let **U** denote the linear space of piecewise continuous controls $t \mapsto u(t) \in \mathscr{U}$, defined for $t \geq 0$; and denote by $\varphi(t; x_0, u)$ the corresponding solution of (0.1) with $x(0) = x_0$; i.e.

$$\varphi(t; x_0, u) = e^{tA}x_0 + \int_0^t e^{(t-s)A}Bu(s)\, ds. \tag{1.1}$$

A state $x \in \mathscr{X}$ is *reachable from* x_0 if there exist t and u, with $0 < t < \infty$ and $u \in \mathbf{U}$, such that $\varphi(t; x_0, u) = x$. Let $\mathscr{R}_0$ be the set of states reachable from $x_0 = 0$. It is readily checked from (1.1), and the admissibility of piecewise continuous controls, that $\mathscr{R}_0$ is a linear subspace of $\mathscr{X}$. We now describe $\mathscr{R}_0$ directly in terms of A and B. For this, let $\mathscr{B} := \operatorname{Im} B$ and

$$\langle A \,|\, \mathscr{B} \rangle := \mathscr{B} + A\mathscr{B} + \cdots + A^{n-1}\mathscr{B}. \tag{1.2}$$

Theorem 1.1

$$\mathscr{R}_0 = \langle A \,|\, \mathscr{B} \rangle.$$

PROOF. If $x \in \mathscr{R}_0$ then for suitable t and $u(\cdot)$,

$$\begin{aligned} x &= \int_0^t e^{(t-s)A} Bu(s)\, ds \\ &= \sum_{i=1}^n A^{i-1} B \int_0^t \psi_i(t-s) u(s)\, ds \in \langle A \,|\, \mathscr{B} \rangle, \end{aligned}$$

by (0.17.3). For the reverse inclusion, we show first that

$$\langle A \,|\, \mathscr{B} \rangle = \operatorname{Im} W_t, \qquad t > 0, \tag{1.3}$$

where

$$W_t := \int_0^t e^{sA} BB' e^{sA'}\, ds.$$

As in Section 0.14, we here identify $\mathscr{X}' = \mathscr{X}$, $\mathscr{U}' = \mathscr{U}$, and regard W_t as a map in $\mathscr{X}$. As W_t is symmetric, (1.3) is equivalent to

$$\langle A \,|\, \mathscr{B} \rangle^\perp = \operatorname{Ker} W_t, \qquad t > 0.$$

If $x \in \operatorname{Ker} W_t$, then $x' W_t x = 0$, i.e.

$$\int_0^t |B' e^{sA'} x|^2\, ds = 0,$$

and so

$$B' e^{sA'} x = 0, \qquad 0 \le s \le t.$$

Repeated differentiation at $s = 0$ yields

$$B' A'^{i-1} x = 0, \qquad i \in \mathbf{n},$$

so that

$$\begin{aligned} x \in \bigcap_{i=1}^n \operatorname{Ker}(B' A'^{i-1}) &= \bigcap_{i=1}^n [\operatorname{Im}(A^{i-1}B)]^\perp \\ &= \left[\sum_{i=1}^n \operatorname{Im}(A^{i-1}B) \right]^\perp = \langle A \,|\, \mathscr{B} \rangle^\perp. \end{aligned}$$

If $x \in \langle A \,|\, \mathscr{B} \rangle^\perp$, reversing the steps of the argument yields $x' W_t x = 0$, and $W_t \ge 0$ implies that $x \in \operatorname{Ker} W_t$.

Now let $x \in \langle A \,|\, \mathscr{B} \rangle$ and fix $t > 0$. Then $x = W_t z$ for some $z \in \mathscr{X}$. Setting

$$u(s) = B' e^{(t-s)A'} z, \qquad 0 \le s \le t, \tag{1.4}$$

we see that

$$W_t z = \varphi(t; 0, u) \in \mathscr{R}_0. \qquad \square$$

By the construction used in the proof of Theorem 1.1, $x \in \mathscr{R}_0$ implies that for every $t > 0$ there exists $u \in \mathbf{U}$ such that $x = \varphi(t; 0, u)$. From (1.1) it now follows that $x \in \mathscr{X}$ is reachable from x_0 if and only if $x - e^{tA}x_0 \in \mathscr{R}_0$ for some t, $0 < t < \infty$. Equivalently (as will be clear from Section 1.2 below), if $\bar{\mathscr{X}} := \mathscr{X}/\mathscr{R}_0$ and $\bar{A}$ is the map induced by A on $\bar{\mathscr{X}}$, then x is reachable from x_0 if and only if $\bar{x} = e^{t\bar{A}}\bar{x}_0$ for some t, $0 < t < \infty$.

1.2 Controllability

The subspace $\mathscr{R}_0 = \langle A \mid \mathscr{B} \rangle \subset \mathscr{X}$ is the *controllable subspace* of the pair (A, B). From (1.2) (and the Hamilton-Cayley theorem) it is clear that $A\mathscr{R}_0 \subset \mathscr{R}_0$, i.e. $\mathscr{R}_0$ is A-invariant. It is easy to see that $\mathscr{R}_0$ is, in fact, the smallest A-invariant subspace containing $\mathscr{B}$. Now let $\bar{\mathscr{X}} = \mathscr{X}/\mathscr{R}_0$, $P: \mathscr{X} \to \bar{\mathscr{X}}$ be the canonical projection, $\bar{A}$ the map induced in $\bar{\mathscr{X}}$ by A; and write $\bar{x} = Px$. Since $PB = 0$, we have from (0.1),

$$\dot{\bar{x}}(t) = \bar{A}\bar{x}(t).$$

Thus the control $u(\cdot)$ has no influence on the coset of x mod $\mathscr{R}_0$. In this notation, Theorem 1.1 says that all states can be reached from 0 when $\bar{\mathscr{X}} = 0$, i.e. $\mathscr{R}_0 = \mathscr{X}$. Thus we are led to the definition: the pair (A, B) is *controllable* if its controllable subspace is the whole space, i.e. $\langle A \mid \mathscr{B} \rangle = \mathscr{X}$.

With (A, B) controllable, we have that

$$W_t = \int_0^t e^{sA}BB'e^{sA'}\, ds$$

is positive definite for every $t > 0$. With $t > 0$ fixed, set

$$u(s) := B'e^{(t-s)A'}W_t^{-1}(x - e^{tA}x_0), \qquad 0 \le s \le t.$$

Then it is clear that $\varphi(t; x_0, u) = x$. That is, every state x can be reached from any state x_0 in a time interval of arbitrary positive length.

Next, we note that controllability of (A, B) is preserved under arbitrary automorphisms of $\mathscr{X}$ and $\mathscr{U}$; indeed, this is virtually obvious from the basis-independent style of the foregoing discussion.

Proposition 1.1. *Let $T: \mathscr{X} \simeq \mathscr{X}$ and $G: \mathscr{U} \simeq \mathscr{U}$, and let (A, B) be controllable. Then, $(T^{-1}AT, T^{-1}BG)$ is controllable.*

PROOF

$$\sum_{i=1}^{n} (T^{-1}AT)^{i-1}\,\mathrm{Im}(T^{-1}BG) = \sum_{i=1}^{n} (T^{-1}AT)^{i-1}T^{-1}\,\mathrm{Im}(BG)$$

$$= T^{-1}\sum_{i=1}^{n} A^{i-1}\mathscr{B} = T^{-1}\mathscr{X} = \mathscr{X}. \qquad \square$$

The next two propositions state that controllability of (A, B) implies controllability in factor spaces, and in subspaces which decompose A.

Proposition 1.2. *Let* $\langle A \mid \mathscr{B} \rangle = \mathscr{X}$ *and* $A\mathscr{V} \subset \mathscr{V} \subset \mathscr{X}$. *Write* $\bar{\mathscr{X}} = \mathscr{X}/\mathscr{V}$, $\bar{\mathscr{B}} = (\mathscr{B} + \mathscr{V})/\mathscr{V}$ *and let* $\bar{A}$ *be the map induced by* A *in* $\bar{\mathscr{X}}$. *Then,*

$$\langle \bar{A} \mid \bar{\mathscr{B}} \rangle = \bar{\mathscr{X}}.$$

PROOF. Let $P: \mathscr{X} \to \bar{\mathscr{X}}$ be the canonical projection; thus, $\bar{\mathscr{B}} = P\mathscr{B}$ and $\bar{A}P = PA$. Then,

$$\begin{aligned} \bar{\mathscr{X}} &= P\langle A \mid \mathscr{B} \rangle = P(\mathscr{B} + A\mathscr{B} + \cdots + A^{n-1}\mathscr{B}) \\ &= \bar{\mathscr{B}} + \bar{A}\bar{\mathscr{B}} + \cdots + \bar{A}^{n-1}\bar{\mathscr{B}} = \langle \bar{A} \mid \bar{\mathscr{B}} \rangle. \end{aligned}$$ □

The geometric relationships in Proposition 1.2 are exhibited in the commutative diagram below:

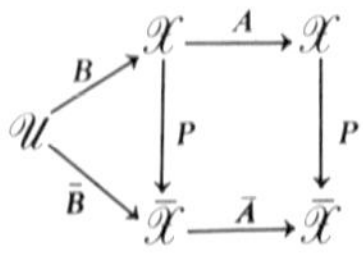

The pair $(\bar{A}, \bar{B})$ constitutes a "lower-order model" of the pair (A, B) from which the structure of $A \mid \mathscr{V}$ has been erased. The proposition states that the model is controllable if the original pair (A, B) is.

Proposition 1.3. *Let* $\langle A \mid \mathscr{B} \rangle = \mathscr{X}$ *and let* $\mathscr{R} \subset \mathscr{X}$ *decompose* A: *i.e.* $A\mathscr{R} \subset \mathscr{R}$ *and there exists* $\mathscr{S} \subset \mathscr{X}$ *with* $A\mathscr{S} \subset \mathscr{S}$ *and* $\mathscr{R} \oplus \mathscr{S} = \mathscr{X}$. *If* Q *is the projection on* $\mathscr{R}$ *along* $\mathscr{S}$ *then*

$$\mathscr{R} = \langle A \mid Q\mathscr{B} \rangle.$$

PROOF. Clearly, $QA = AQ$. Therefore,

$$\begin{aligned} \mathscr{R} = Q\mathscr{X} &= Q(\mathscr{B} + A\mathscr{B} + \cdots + A^{n-1}\mathscr{B}) \\ &= Q\mathscr{B} + AQ\mathscr{B} + \cdots + A^{n-1}Q\mathscr{B} = \langle A \mid Q\mathscr{B} \rangle. \end{aligned}$$ □

In matrix terms Proposition 1.2 states that if (A, B) is controllable, with

$$A = \begin{bmatrix} A_1 & A_3 \\ 0 & A_2 \end{bmatrix}, \qquad B = \begin{bmatrix} B_1 \\ B_2 \end{bmatrix},$$

then (A_2, B_2) is controllable. In general, (A_1, B_1) is not controllable, as shown by the example

$$A = \begin{bmatrix} 0 & 1 \\ 0 & 0 \end{bmatrix}, \qquad B = \begin{bmatrix} 0 \\ 1 \end{bmatrix}.$$

In this notation Proposition 1.3 states that (A_1, B_1) is controllable if $A_3 = 0$; the extension of Proposition 1.3 to the case where $A = \text{diag}[A_1 \cdots A_k]$ is left to the reader.

Finally, it may be of interest to point out that if (A, B) is controllable and $A\mathscr{R} \subset \mathscr{R}$ then there always exists $\mathscr{S} \subset \mathscr{X}$ (not necessarily A-invariant) such that $\mathscr{R} \oplus \mathscr{S} = \mathscr{X}$ and, if Q is the projection on $\mathscr{R}$ along $\mathscr{S}$, then $\langle A \mid Q\mathscr{B} \rangle = \mathscr{R}$. In matrix terms it is always possible to complete a basis for $\mathscr{R}$ to a basis for $\mathscr{X}$ in such a way that, in the notation above, (A_1, B_1) is controllable; in fact, "almost any" complementary basis will do. For our example, with $\mathscr{R} = \text{Im}[\begin{smallmatrix}1\\0\end{smallmatrix}]$, let $\mathscr{S} = \text{Im}[\begin{smallmatrix}a\\1\end{smallmatrix}]$ $(a \neq 0)$. In the basis $\{[\begin{smallmatrix}1\\0\end{smallmatrix}], [\begin{smallmatrix}a\\1\end{smallmatrix}]\}$, we have

$$A_{\text{new}} = \begin{bmatrix} 0 & 1 \\ 0 & 0 \end{bmatrix}, \qquad B_{\text{new}} = \begin{bmatrix} -a \\ 1 \end{bmatrix}$$

and $(A_1, B_1) = (0, -a)$ is now controllable. A proof of this remark, that exploits the properties of feedback, is deferred to Exercise 2.5.

1.3 Single-Input Systems

Let $B = b \neq 0$, that is, $\mathscr{B} = \text{Span}\{b\} = \mathscr{b}$ for some $b \in \mathscr{X}$. The corresponding system equation is $\dot{x} = Ax + bu$, where $u(\cdot)$ is scalar-valued, i.e. the system has a single control input. Suppose (A, b) is controllable. Since $\langle A \mid \mathscr{b} \rangle = \mathscr{X}$ it follows that the vectors $\{b, Ab, \ldots, A^{n-1}b\}$ form a basis for $\mathscr{X}$; thus, A is cyclic, and b is a generator.

Let the minimal polynomial (m.p.) of A be

$$\alpha(\lambda) = \lambda^n - (a_1 + a_2\lambda + \cdots + a_n\lambda^{n-1}).$$

Introduce the auxiliary polynomials $\alpha^{(i)}(\lambda)$ defined in (0.10.1) and the corresponding basis

$$e_i = \alpha^{(i)}(A)b, \qquad i \in \mathbf{n}. \tag{3.1}$$

Then, $b = e_n$, and the matrices of A and b are

$$A = \begin{bmatrix} 0 & 1 & 0 & 0 & \cdots & 0 & 0 \\ 0 & 0 & 1 & 0 & \cdots & 0 & 0 \\ \cdot & \cdot & \cdot & \cdot & \cdots & \cdot & \cdot \\ 0 & \cdot & \cdot & \cdot & \cdots & 0 & 1 \\ a_1 & a_2 & \cdot & \cdot & \cdots & \cdot & a_n \end{bmatrix}, \qquad b = \begin{bmatrix} 0 \\ \vdots \\ 0 \\ 1 \end{bmatrix}. \tag{3.2}$$

We refer to (3.2) as the *standard canonical* form of the controllable (matrix) pair (A, b).

Call (A, b) and (A_1, b_1) *similar* if for some $T: \mathscr{X} \simeq \mathscr{X}$,

$$A_1 = T^{-1}AT, \qquad b_1 = T^{-1}b.$$

Similarity is an equivalence relation on (A, b) pairs. Since similarity leaves invariant the characteristic polynomial of A, hence the a_i ($i \in \mathbf{n}$), our discussion shows that every controllable pair is similar to exactly one pair of form (3.2). It is even true that the basis in which the matrices of (A, b) have standard canonical form is unique: for if $T^{-1}AT = A$ and $Tb = b$ then by (3.1), $Te_i = e_i$ ($i \in \mathbf{n}$) and therefore $T = 1$.

1.4 Multi-Input Systems

Let $\langle A | \mathscr{B} \rangle = \mathscr{X}$, with $d(\mathscr{B}) = m$. Since each vector $b \in \mathscr{B}$ generates a cyclic subspace $\langle A | \mathscr{b} \rangle$ it is plausible that we must have $m \geq k$, where k is the cyclic index of A. This is true, and more: it is possible to select independent vectors $b_i \in \mathscr{B}$ ($i \in \mathbf{k}$) such that the subspaces $\langle A | \mathscr{b}_i \rangle$ span $\mathscr{X}$; unfortunately, however, these subspaces cannot, in general, be chosen to be independent.

Theorem 1.2. *Let $\langle A | \mathscr{B} \rangle = \mathscr{X}$, with $d(\mathscr{B}) = m$; and let k be the cyclic index of A. Then $m \geq k$. Let the invariant factors of A be $\alpha_1, \ldots, \alpha_k$. There exist A-invariant subspaces $\mathscr{X}_i \subset \mathscr{X}$, and vectors $b_i \in \mathscr{B}$ ($i \in \mathbf{k}$), with the properties:*

i. $\mathscr{X} = \mathscr{X}_1 \oplus \cdots \oplus \mathscr{X}_k$.
ii. $A | \mathscr{X}_i$ *is cyclic with m.p.* α_i, $i \in \mathbf{k}$.
iii. $\langle A | \mathscr{b}_1 + \cdots + \mathscr{b}_i \rangle = \mathscr{X}_1 \oplus \cdots \oplus \mathscr{X}_i$, $i \in \mathbf{k}$.

Briefly, the theorem says that $\mathscr{X}$ has a rational canonical decomposition having the special property (iii) with respect to $\mathscr{B}$. The proof depends on simple properties of polynomials in $\mathbb{R}[\lambda]$ and on controllability of subspaces.

Lemma 1.1. *Let $\alpha, \gamma_1, \ldots, \gamma_m$ belong to $\mathbb{R}[\lambda]$, with α monic. Then*

$$\operatorname*{LCM}_{i \in \mathbf{m}} \left[\frac{\alpha}{\text{GCD}(\alpha, \gamma_i)} \right] = \frac{\alpha}{\text{GCD}(\alpha, \gamma_1, \ldots, \gamma_m)}.$$

PROOF. If π^μ is a prime factor of α and $\text{GCD}(\alpha, \gamma_i)$ has the corresponding factor π^{μ_i} then

$$\operatorname*{LCM}_{i \in \mathbf{m}} \left(\frac{\pi^\mu}{\pi^{\mu_i}} \right) = \pi^{\max_i(\mu - \mu_i)},$$

and

$$\frac{\pi^\mu}{\text{GCD}(\pi^\mu, \pi^{\mu_1}, \ldots, \pi^{\mu_m})} = \frac{\pi^\mu}{\pi^{\min_i \mu_i}} = \pi^{\mu - \min_i \mu_i}$$

Since $\max_i(\mu - \mu_i) = \mu - \min_i \mu_i$, the assertion follows.

Lemma 1.2. *Let $\mathscr{B} \subset \mathscr{X}$ be an arbitrary subspace of $\mathscr{X}$, having minimal polynomial β with respect to A. There exists an element $b \in \mathscr{B}$ with minimal polynomial β.*

Note that $\mathscr{B}$ is not assumed to be A-invariant.

PROOF. Let

$$\mathscr{X} = \bigoplus_{i=1}^{k} \mathscr{X}_i,$$

be any decomposition of $\mathscr{X}$ such that $A \mid \mathscr{X}_i$ is cyclic with minimal polynomial α_i. Let $\mathscr{X}_i = \langle A \mid \mathscr{g}_i \rangle$ $(i \in \mathbf{k})$ and let $\mathscr{B} = \operatorname{Span}\{b_1, \ldots, b_m\}$. We have

$$b_i = \gamma_{i1}(A)g_1 + \cdots + \gamma_{ik}(A)g_k, \qquad i \in \mathbf{m},$$

for suitable $\gamma_{ij} \in \mathbb{R}[\lambda]$; and we may arrange that $\deg \gamma_{ij} < \deg \alpha_j$. Let β_i $(i \in \mathbf{m})$ be the m.p. of b_i; thus,

$$\beta_i = \operatorname*{LCM}_{j \in \mathbf{k}} \left[\frac{\alpha_j}{\operatorname{GCD}(\alpha_j, \gamma_{ij})} \right].$$

Then

$$\begin{aligned} \beta = \operatorname{LCM}(\beta_1, \ldots, \beta_m) &= \operatorname*{LCM}_{j \in \mathbf{k}} \operatorname*{LCM}_{i \in \mathbf{m}} \left[\frac{\alpha_j}{\operatorname{GCD}(\alpha_j, \gamma_{ij})} \right] \\ &= \operatorname*{LCM}_{j \in \mathbf{k}} \left[\frac{\alpha_j}{\operatorname{GCD}(\alpha_j, \gamma_{1j}, \ldots, \gamma_{mj})} \right], \end{aligned}$$

by application of Lemma 1.1. Now define

$$b = r_1 b_1 + \cdots + r_m b_m,$$

where the $r_i \in \mathbb{R}$ $(i \in \mathbf{m})$ are to be determined. Then,

$$b = \sum_{j=1}^{k} \sum_{i=1}^{m} r_i \gamma_{ij}(A) g_j,$$

and if β_0 is the m.p. of b,

$$\beta_0 = \operatorname*{LCM}_{j \in \mathbf{k}} \left[\frac{\alpha_j}{\operatorname{GCD}\left(\alpha_j, \sum_{i=1}^{m} r_i \gamma_{ij}\right)} \right].$$

We wish to choose the r_i so that $\beta_0 = \beta$, and for this it is clearly sufficient that

$$\operatorname{GCD}\left(\alpha_j, \sum_{i=1}^{m} r_i \gamma_{ij}\right) = \operatorname{GCD}(\alpha_j, \gamma_{1j}, \ldots, \gamma_{mj}), \qquad j \in \mathbf{k}.$$

Denote the GCD on the right by γ_j and let $\gamma_{ij} = \hat{\gamma}_{ij}\gamma_j$ $(i \in \mathbf{m}, j \in \mathbf{k})$. It is enough to choose the r_i so that

$$\mathrm{GCD}\left(\alpha_j, \sum_{i=1}^{m} r_i \hat{\gamma}_{ij}\right) = 1, \qquad j \in \mathbf{k}. \tag{4.1}$$

For this, let $\lambda_{i1}, \ldots, \lambda_{i\nu(i)}$ be the roots of $\alpha_i(\lambda)$ over $\mathbb{C}$. Then (4.1) holds provided

$$\sum_{i=1}^{m} r_i \hat{\gamma}_{ij}(\lambda_{j\mu}) \neq 0; \qquad \mu \in \mathbf{v}_j, j \in \mathbf{k}. \tag{4.2}$$

Observe that not all the terms $\hat{\gamma}_{ij}(\lambda_{j\mu})$ can vanish in any sum; otherwise, for some j, μ and all $i \in \mathbf{m}$, $\hat{\gamma}_{ij}$ has the factor $\lambda - \lambda_{j\mu}$ in common with α_j, a contradiction. It follows that $r_i \in \mathbb{R}$ exist such that (4.2) is true: indeed, for each μ, j either the real or imaginary part of the sum in (4.2) is required not to vanish identically; writing $r' = (r_1, \ldots, r_m)$ we see that (4.2) is equivalent to $r'v_\nu \neq 0$ for a finite number of real, nonzero m-vectors v_ν; and so it is required merely to choose a point $r \in \mathbb{R}^m$ in the complement of a finite union of $(m-1)$-dimensional hyperplanes. □

The proof shows that, roughly speaking, a vector $b \in \mathscr{B}$ is almost certain to have the property required, if it is chosen at random. Indeed, the result depends on the existence of plenty of vectors in the space, and is false if the underlying field is finite (Exercise 1.7).

An alternative, shorter proof of Lemma 1.2 using simple ideas from algebraic geometry is presented in Exercise 1.8.

Corollary 1.1. *If A is cyclic and (A, B) is controllable, there exists a vector $b \in \mathscr{B}$ such that (A, b) is controllable.*

Proof (of Theorem 1.2). Observe that the m.p. of a vector $b \in \mathscr{X}$ coincides with the m.p. of $\langle A | \mathscr{b} \rangle$. Since $\langle A | \mathscr{B} \rangle = \mathscr{X}$, Lemma 1.2 provides a vector $b_1 \in \mathscr{B}$ whose m.p. coincides with the m.p. of A, which we denote by α_1. Define $\mathscr{X}_1 = \langle A | \mathscr{b}_1 \rangle$. Since $\mathscr{X}_1 \subset \mathscr{X}$ is maximal cyclic, there exists, by Proposition 0.2, a subspace $\mathscr{Y} \subset \mathscr{X}$, such that

$$\langle A | \mathscr{b}_1 \rangle \oplus \mathscr{Y} = \mathscr{X}$$

and

$$A\mathscr{Y} \subset \mathscr{Y}.$$

Let Q be the projection on $\mathscr{Y}$ along $\mathscr{X}_1$. By Proposition 1.3,

$$\langle A | Q\mathscr{B} \rangle = \mathscr{Y}.$$

If $\hat{\alpha}_2$ is the m.p. of $A | \mathscr{Y}$ there exists, again by Lemma 1.2, a vector $b_2 \in \mathscr{B}$ such that the m.p. of Qb_2 is $\hat{\alpha}_2$. Define $\mathscr{X}_2 = \langle A | Qb_2 \rangle$. Then, $b_2 \in \mathscr{X}_1 \oplus \mathscr{X}_2$ and

$$\langle A | \mathscr{b}_1 + \mathscr{b}_2 \rangle = \mathscr{X}_1 \oplus \mathscr{X}_2.$$

Continuing in this way, we obtain eventually, for some r,

$$\mathscr{X} = \mathscr{X}_1 \oplus \cdots \oplus \mathscr{X}_r,$$

with $A \mid \mathscr{X}_i$ cyclic with m.p. $\hat{\alpha}_i$. Since the subspace $\mathscr{X}_i$ split off at the ith stage is maximal cyclic,

$$\hat{\alpha}_2 \mid \alpha_1, \ldots, \hat{\alpha}_r \mid \hat{\alpha}_{r-1};$$

and by the uniqueness of the rational canonical decomposition (Theorem 0.1) it follows that $r = k$ and $\hat{\alpha}_i = \alpha_i$ ($i \in \mathbf{k}$). Again, by the construction, we have

$$\langle A \mid \mathscr{b}_1 + \cdots + \mathscr{b}_i \rangle = \mathscr{X}_1 \oplus \cdots \oplus \mathscr{X}_i, \qquad i \in \mathbf{k},$$

as required. □

Theorem 1.2 implies that every controllable pair (A, B) admits a matrix representation

$$A = \operatorname{diag}[A_1 \cdots A_k]$$

$$B = \begin{bmatrix} b_{11} & \cdots & b_{1k} & * \\ & \ddots & \vdots & \\ \bigcirc & & b_{kk} & * \end{bmatrix}.$$

Here, A is in rational canonical form and each pair (A_i, b_{ii}) is controllable.

1.5 Controllability is Generic

On the basis of the discussion of genericity in Section 0.16, we can easily show the following.

Theorem 1.3. *Let (A, B) be a matrix pair with $A \in \mathbb{R}^{n \times n}$, $B \in \mathbb{R}^{n \times m}$. The property that (A, B) be controllable is generic, and is well-posed at every point (A, B) where it holds.*

Proof. By listing the entries of A and B, we regard $\mathbf{p} = (A, B)$ as a point in $\mathbb{R}^N$, where $N = n^2 + nm$. It is easily seen (Exercise 1.2) that (A, B) is controllable if and only if the $n \times nm$ matrix

$$R := [B, AB, \ldots, A^{n-1}B]$$

has rank n. Write $x_1, \ldots, x_{nm}$ for the columns of R. Then controllability of (A, B) fails if and only if every determinant formed by selecting n columns x_i vanishes; that is,

$$\varphi_i(\mathbf{p}) = \det[x_{i_1} \cdots x_{i_n}] = 0,$$

where i ranges over all multi-indices $i = (i_1, \ldots, i_n)$ with $1 \leq i_1 < i_2 < \cdots < i_n \leq nm$. Let $\mathbf{V} \subset \mathbb{R}^N$ be the set of common zeros of the φ_i. Clearly, $\mathbf{V}$ is a variety in $\mathbb{R}^N$. Also, choosing an (A, B) pair with A and the first column of B in the standard canonical form (3.2), we see that controllable pairs exist, hence $\mathbf{V}$ is proper. It is now obvious that controllability is generic relative to $\mathbf{V}$, and well-posedness at $\mathbf{p} \in \mathbf{V}^c$ is clear. □

The foregoing discussion suggests that it is "easy" for a pair (A, B) to be controllable. However, it should be borne in mind that we have defined controllability in a purely qualitative, algebraic way. In practice, it could well turn out that the controllability matrix R introduced in the proof is poorly conditioned (typically, if $n/m \gg 1$). But apart from pointing out their existence, we shall not attempt to discuss these important numerical problems here.

1.6 Exercises

1.1. Discuss the behavior of $u(\cdot)$ defined by (1.4) as $t \downarrow 0$. Hint: Consider controls of the form $u(t) = u_0\,\delta(t) + u_1\,\delta'(t) + \cdots$.

1.2. Show that (A, B) is a controllable matrix pair if and only if

$$\operatorname{Rank}[B, AB, \ldots, A^{n-1}B] = n.$$

This $n \times nm$ block matrix is the *controllability matrix* of (A, B).

1.3. Regard $\mathscr{U}$, $\mathscr{X}$ as linear vector spaces over $\mathbb{C}$. Show that (A, B) is controllable if and only if the $n \times (n + m)$ matrix $[A - \lambda 1, B]$ has rank n for every eigenvalue λ of A (and hence for every $\lambda \in \mathbb{C}$). Hint: Consider

$$\operatorname{Ker}\begin{bmatrix} A' - \lambda 1' \\ B' \end{bmatrix},$$

where the matrix is regarded as a map from $\mathscr{X}'$ to $\mathscr{X}' \oplus \mathscr{U}'$.

1.4. Show that there is a basis for $\mathscr{X}$ in which A and B have matrices of form

$$A = \begin{bmatrix} A_1 & A_3 \\ 0 & A_2 \end{bmatrix}, \qquad B = \begin{bmatrix} B_1 \\ 0 \end{bmatrix},$$

where (A_1, B_1) is controllable. Describe $\langle A \,|\, \mathscr{B} \rangle$ in this basis.

1.5. Suppose $\mathscr{X} = \mathscr{X}_1 \oplus \cdots \oplus \mathscr{X}_l$ and $A\mathscr{X}_i \subset \mathscr{X}_i$, $i \in \mathbf{l}$. Write $A_i = A \,|\, \mathscr{X}_i$, P_i: $\mathscr{X} \to \mathscr{X}_i$ for the natural projection on $\mathscr{X}_i$, and $B_i = P_i B$, $i \in \mathbf{l}$. Assuming the m.p. of the A_i are coprime in pairs, show that (A, B) is controllable if and only if (A_i, B_i) is controllable for each $i \in \mathbf{l}$. This result shows that "controllability is preserved under modal decomposition." Hint: Apply Proposition 0.4.

1.6. Show by an example that the subspaces $\langle A \,|\, \mathscr{b}_i \rangle$ in Theorem 1.2 cannot, in general, be chosen to be independent.

1.7. Lemma 1.2 is false, in general, if A and B are matrices over a finite field. For a counterexample consider the binary field $GF(2)$, i.e. the integers (mod 2), let A

be cyclic with m.p. $\alpha(\lambda)$ and generator g, and let $B = [b_1, b_2]$ with $b_i = \alpha_i(A)g$. Show that (A, B) is controllable, yet (A, b) is not controllable for any $b \in \mathscr{B}$, if and only if

$$\begin{aligned} \mathrm{GCD}(\alpha_1, \alpha_2, \alpha) &= 1, \\ \mathrm{GCD}(\alpha_1 + \alpha_2, \alpha) &\neq 1, \\ \mathrm{GCD}(\alpha_i, \alpha) &\neq 1, \qquad i = 1, 2. \end{aligned}$$

Specifically choose, for instance,

$$\begin{aligned} \alpha_1(\lambda) = \lambda, \qquad \alpha_2(\lambda) &= \lambda^2 + 1 \\ \alpha(\lambda) &= \alpha_1(\lambda)\alpha_2(\lambda)(\alpha_1(\lambda) + \alpha_2(\lambda)) \\ &= \lambda^5 + \lambda^4 + \lambda^2 + \lambda, \end{aligned}$$

and verify directly that (A, B) provides the counterexample required. Here, you may take

$$A = \begin{bmatrix} 0 & 1 & 0 & 0 & 0 \\ 0 & 0 & 1 & 0 & 0 \\ 0 & 0 & 0 & 1 & 0 \\ 0 & 0 & 0 & 0 & 1 \\ 0 & 1 & 1 & 0 & 1 \end{bmatrix}, \qquad B = \begin{bmatrix} 0 & 0 \\ 0 & 0 \\ 0 & 1 \\ 1 & 1 \\ 1 & 0 \end{bmatrix}, \qquad g = \begin{bmatrix} 0 \\ 0 \\ 0 \\ 0 \\ 1 \end{bmatrix}.$$

1.8. An alternative and short proof of Lemma 1.2 can be based on the following *Lemma*: *Let* $\mathscr{B}$, $\mathscr{S}_1, \ldots, \mathscr{S}_N$ *be subspaces of* $\mathscr{X}$ *with* $\mathscr{B} \subset \mathscr{S}_1 \cup \cdots \cup \mathscr{S}_N$, *the set-theoretic union. Then* $\mathscr{B} \subset \mathscr{S}_i$ *for some* $i \in N$. Hint: First prove the *Sublemma. If* $\mathbf{V}_1, \ldots, \mathbf{V}_N$ *are algebraic varieties in* $\mathbb{R}^m$, *and if*

$$\mathbf{V}_1 \cup \cdots \cup \mathbf{V}_N = \mathbb{R}^m, \tag{6.1}$$

then $\mathbf{V}_i = \mathbb{R}^m$ *for some* $i \in \mathbf{N}$. {Proof: Let

$$\mathbf{V}_i = \{x \colon x \in \mathbb{R}^m, \varphi_{ij}(x) = 0, j \in k_i\}$$

for polynomials $\varphi_{ij}(\lambda) \in \mathbb{R}[\lambda_1, \ldots, \lambda_m]$. Then

$$\mathbf{V}_i = \{x \colon x \in \mathbb{R}^m, \varphi_i(x) = 0\},$$

where

$$\varphi_i(\lambda) := \sum_{j=1}^{k_i} \varphi_{ij}(\lambda)^2;$$

and so

$$\mathbf{V}_1 \cup \cdots \cup \mathbf{V}_N = \{x \colon x \in \mathbb{R}^m, \varphi_1(x) \cdots \varphi_N(x) = 0\}.$$

If (6.1) is true then $\varphi_1 \cdots \varphi_N = 0$ in $\mathbb{R}[\lambda]$, and because $\mathbb{R}[\lambda]$ is an integral domain, $\varphi_i = 0$ for some $i \in \mathbf{N}$.} The lemma is an immediate consequence. {For $\mathbb{R}^m$ take

$$\mathscr{B} = (\mathscr{B} \cap \mathscr{S}_1) \cup \cdots \cup (\mathscr{B} \cap \mathscr{S}_N)$$

and note that $\mathscr{B} \cap \mathscr{S}_i$ is an algebraic variety.} To reprove Lemma 1.2, let $\beta(\lambda)$ be the m.p. of $\mathscr{B}$ and let $\beta_1(\lambda), \ldots, \beta_N(\lambda)$ be a list of the distinct m.p. of vectors b obtained as b runs through $\mathscr{B}$. The list is finite because each such m.p. must divide the m.p. of A. Thus, $\mathscr{B} \subset \bigcup_{i=1}^{N} \operatorname{Ker} \beta_i(A)$, and so by the lemma, $\mathscr{B} \subset \operatorname{Ker} \beta_j(A)$ for some $j \in \mathbf{N}$. Therefore, $\beta | \beta_j$; at the same time, since $\beta = \mathrm{LCM}\{\beta_i: i \in \mathbf{N}\}$, we have that $\beta_j | \beta$. So $\beta_j = \beta$, and for b we may take any vector in $\mathscr{B}$ with m.p. β_j. □

1.7 Notes and References

The definition of controllability used here, and the identification of this concept as one of fundamental importance, are due to Kalman [1], [2]; see also Gilbert [1]. The standard canonical form for a single-input controllable pair was discovered independently by several workers around 1961, but apparently first published by Popov [1]. Corollary 1.1 is due to Wonham [1] and its subsequent generalization in the form of Theorem 1.2 to Heymann [2]; see also Guidorzi [1]. The short proof of Lemma 1.2 given in Exercise 1.8 was suggested by Jan C. Willems [private communication]; for the sublemma see also Section 16.1 of Van der Waerden [1]. The fact that the set of controllable pairs is open and dense was pointed out by Lee and Markus [1]. The result in Exercise 1.3 is due to Hautus [1]. Controllability of systems defined over arbitrary fields (cf. Exercise 1.7) is discussed by Mitter and Foulkes [1].

2 Controllability, Feedback and Pole Assignment

Consider as usual the system

$$\dot{x}(t) = Ax(t) + Bu(t), \qquad t \geq 0. \tag{0.1}$$

Suppose we are free to modify (0.1) by setting

$$u(t) = Fx(t) + v(t), \qquad t \geq 0, \tag{0.2}$$

where $v(\cdot)$ is a new external input, and $F: \mathscr{X} \to \mathscr{U}$ is an arbitrary map. We refer to F as the *state feedback*. The obvious result of introducing state feedback is to change the pair (A, B) in (0.1) into the pair $(A + BF, B)$. We shall explore the effect of such a transformation of pairs on controllability and on the spectrum of $A + BF$. Our main result is that if (A, B) is controllable then $\sigma(A + BF)$ can be assigned arbitrarily by suitable choice of F, and this property in turn implies controllability.

2.1 Controllability and Feedback

We first prove the simple and gratifying result that controllability is not affected by state feedback. The following statement goes a little further.

Lemma 2.1. *For any state feedback $F: \mathscr{X} \to \mathscr{U}$,*

$$\langle A + BF \,|\, \mathscr{B} \rangle = \langle A \,|\, \mathscr{B} \rangle.$$

In particular, if (A, B) is controllable, so is $(A + BF, B)$.

PROOF. Observe that

$$\mathscr{B} + (A + BF)\mathscr{R} = \mathscr{B} + A\mathscr{R}$$

for all $\mathscr{R} \subset \mathscr{X}$ and $F: \mathscr{X} \to \mathscr{U}$. Writing $\hat{A} = A + BF$, we then have

$$\begin{aligned}\langle A + BF \,|\, \mathscr{B} \rangle &= \mathscr{B} + \hat{A}\mathscr{B} + \cdots + \hat{A}^{n-1}\mathscr{B} \\ &= \mathscr{B} + \hat{A}(\mathscr{B} + \hat{A}(\cdots(\mathscr{B} + \hat{A}\mathscr{B}))\cdots) \\ &= \mathscr{B} + A\mathscr{B} + \cdots + A^{n-1}\mathscr{B} \\ &= \langle A \,|\, \mathscr{B} \rangle.\end{aligned}$$

□

Our next observation is that state feedback can be used to replace a multi-input controllable system ($d(\mathscr{B}) > 1$) by a single-input controllable system. Furthermore, the single controlling input can enter via any nonzero vector $b \in \mathscr{B}$, if feedback is chosen accordingly.

Lemma 2.2. *Let $0 \neq b \in \mathscr{B}$. If (A, B) is controllable, there exists $F: \mathscr{X} \to \mathscr{U}$ such that $(A + BF, b)$ is controllable.*

PROOF. Let $b_1 = b$, and let $n_1 = d(\langle A \,|\, \mathscr{b}_1 \rangle)$. Put $x_1 = b_1$ and $x_j = Ax_{j-1} + b_1$ ($j = 2, \ldots, n_1$). Then the x_j ($j \in \mathbf{n}_1$) are a basis for $\langle A \,|\, \mathscr{b}_1 \rangle$. If $n_1 < n$ choose $b_2 \in \mathscr{B}$ such that $b_2 \notin \langle A \,|\, \mathscr{b}_1 \rangle$; such a b_2 exists by controllability. Let n_2 be the dimension of $\langle A \,|\, \mathscr{b}_2 \rangle$ mod$\langle A \,|\, \mathscr{b}_1 \rangle$, i.e. the largest integer such that the vectors

$$x_1, \ldots, x_{n(1)}, b_2, Ab_2, \ldots, A^{n_2 - 1}b_2$$

are independent; and define

$$x_{n(1)+i} = Ax_{n(1)+i-1} + b_2, \qquad i \in \mathbf{n}_2.$$

Then $\{x_1, \ldots, x_{n(1)+n(2)}\}$ is a basis for $\langle A \,|\, \mathscr{b}_1 + \mathscr{b}_2 \rangle$. Continuing thus, we obtain eventually $x_1, \ldots, x_n$ independent, and x_{i+1} has the form

$$x_{i+1} = Ax_i + \bar{b}_i, \qquad i \in \mathbf{n - 1},$$

where $\bar{b}_i \in \mathscr{B}$. Choose F, such that

$$BFx_i = \bar{b}_i, \qquad i \in \mathbf{n},$$

where $\bar{b}_n \in \mathscr{B}$ is arbitrary: since $\bar{b}_i = Bu_i$ for suitable $u_i \in \mathscr{U}$, and the x_i are independent, F certainly exists. Then,

$$(A + BF)x_i = x_{i+1}, \qquad i \in \mathbf{n - 1},$$

so that

$$x_i = (A + BF)^{i-1}b, \qquad i \in \mathbf{n},$$

and, therefore, $\mathscr{X} = \langle A + BF \,|\, \mathscr{b} \rangle$. □

2.2 Pole Assignment

In applications state feedback is introduced to change the dynamic behavior of the free, uncontrollable system $\dot{x} = Ax$ in some desirable way: to achieve stability, say, or to speed up response. Such criteria can sometimes be expressed as conditions on the spectrum of the modified system matrix $A + BF$. Thus,

$$\max\{\mathfrak{Re}\ \lambda : \lambda \in \sigma(A + BF)\} < 0$$

for stability; and, with suitable $\alpha > 0$, $\beta \geq 0$,

$$\max\{\mathfrak{Re}\ \lambda : \lambda \in \sigma(A + BF)\} \leq -\alpha$$

$$\max\{|\mathfrak{Im}\ \lambda| : \lambda \in \sigma(A + BF)\} \leq \beta$$

for rapid response with limited frequency of oscillation. It is an important fact that any spectral criterion can be met by state feedback, provided (A, B) is controllable. Conversely, this property of (A, B) characterizes controllability. For a single-input system the result is virtually obvious by inspection of the standard canonical form (1.3.2), and we exploit this observation in the proof.

Theorem 2.1. *The pair* (A, B) *is controllable if and only if, for every symmetric set* Λ *of* n *complex numbers, there exists a map* $F: \mathscr{X} \to \mathscr{U}$ *such that* $\sigma(A + BF) = \Lambda$.

PROOF. (Only if) First suppose $d(\mathscr{B}) = 1$, $B = b$. It was shown in Section 1.3 that there is a basis for $\mathscr{X}$ in which A, b have the standard canonical matrices (1.3.2); there A has the characteristic polynomial

$$\lambda^n - (a_1 + a_2\lambda + \cdots + a_n\lambda^{n-1}).$$

Let $\Lambda = \{\lambda_1, \ldots, \lambda_n\}$ and write

$$(\lambda - \lambda_1)\cdots(\lambda - \lambda_n) = \lambda^n - (\hat{a}_1 + \hat{a}_2\lambda + \cdots + \hat{a}_n\lambda^{n-1}).$$

On the assumption that (A, b) is in standard canonical form, let f' be the row vector

$$f' = (\hat{a}_1 - a_1, \ldots, \hat{a}_n - a_n).$$

Then it is clear that the matrix $A + bf'$ is again of form (1.3.2), with a_i replaced by $\hat{a}_i$ $(i \in \mathbf{n})$. This completes the proof when $d(\mathscr{B}) = 1$.

For the general case choose, by Lemma 2.2, any vector $b = Bu \in \mathscr{B}$ and a map $F_1: \mathscr{X} \to \mathscr{U}$ such that $(A + BF_1, b)$ is controllable. Regard b as a map $\mathbb{R} \to \mathscr{X}$. We have just shown the existence of $f': \mathscr{X} \to \mathbb{R}$ such that $\sigma(A + BF_1 + bf') = \Lambda$. Then,

$$F = F_1 + uf'$$

is a map with the property required.

(If) Let λ_i $(i \in \mathbf{n})$ be real and distinct, with $\lambda_i \notin \sigma(A)$ $(i \in \mathbf{n})$. Choose F so that $\sigma(A + BF) = \{\lambda_1, \ldots, \lambda_n\}$. Let $x_i \in \mathscr{X}$ $(i \in \mathbf{n})$ be the corresponding eigenvectors: that is,

$$(A + BF)x_i = \lambda_i x_i, \qquad i \in \mathbf{n},$$

so that

$$x_i = (\lambda_i 1 - A)^{-1} BF x_i, \qquad i \in \mathbf{n}.$$

Now by (0.17.2)

$$(\lambda 1 - A)^{-1} = \sum_{j=1}^{n} \rho_j(\lambda) A^{j-1}$$

for suitable rational functions $\rho_j(\lambda)$, defined in $\mathbb{C} - \sigma(A)$. So,

$$x_i = \sum_{j=1}^{n} \rho_j(\lambda_i) A^{j-1} BF x_i \in \langle A \mid \mathscr{B} \rangle, \qquad i \in \mathbf{n}.$$

Since the x_i span $\mathscr{X}$, $\langle A \mid \mathscr{B} \rangle = \mathscr{X}$ as claimed. □

Remark 1. The result just proved is sometimes called the "pole assignment" theorem, in reference to the fact that the eigenvalues of $A + BF$ are the poles of the closed-loop system transfer matrix

$$(s1 - A - BF)^{-1} B. \tag{2.1}$$

Here the system output is taken to be the state. One method of computing $F = F(\Lambda)$ is suggested in Exercise 2.1.

Remark 2. In practice the assignment of $\sigma(A + BF)$ would only partially meet typical design requirements for the closed loop transfer matrix (2.1). An interesting problem that merits further exploration (cf. Exercise 2.2) is how to utilize the remaining freedom of choice of F (in case $m \geq 2$, when such freedom exists) to achieve additional desirable properties. These could relate, for instance, to overshoot in step response or to parameter sensitivity.

2.3 Incomplete Controllability and Pole Shifting

Suppose (A, B) is not controllable. Write $\mathscr{R} := \langle A \mid \mathscr{B} \rangle$, with $d(\mathscr{R}) = \rho < n$. By Lemma 2.1,

$$\langle A + BF \mid \mathscr{B} \rangle = \mathscr{R} \tag{3.1}$$

for all $F: \mathscr{X} \to \mathscr{U}$. Let $P: \mathscr{X} \to \mathscr{X}/\mathscr{R}$ be the canonical projection and denote by a bar the map induced in $\mathscr{X}/\mathscr{R}$ by a map in $\mathscr{X}$. By (3.1), $(A + BF)\mathscr{R} \subset \mathscr{R}$. Since $\overline{A + BF}$ is defined uniquely by the relation

$$(\overline{A + BF})P = P(A + BF),$$

and since by (3.1) $PB = 0$, we have $\overline{A + BF} = \bar{A}$, where $\bar{A}$ is the induced map determined by $\bar{A}P = PA$. There follows

$$\begin{aligned}\sigma(A + BF) &= \sigma[(A + BF)|\mathscr{R}] \uplus \sigma(\overline{A + BF}) \\ &= \sigma[(A + BF)|\mathscr{R}] \uplus \sigma(\bar{A}).\end{aligned}$$

Thus, the $n - \rho$ eigenvalues $\sigma(\bar{A}) \subset \sigma(A)$ are invariant under all transformations of A by state feedback, whereas Theorem 2.1 shows that the remaining ρ eigenvalues, corresponding to "modes" in $\mathscr{R}$, can be assigned arbitrarily by suitable choice of F.

In applications we may wish to distinguish between "good" (e.g. stable) eigenvalues and "bad" eigenvalues. For this let the complex plane be partitioned as

$$\mathbb{C} = \mathbb{C}_g \cup \mathbb{C}_b, \qquad \mathbb{C}_g \cap \mathbb{C}_b = \varnothing. \tag{3.2}$$

[In place of (3.2) we often write $\mathbb{C} = \mathbb{C}_g \uplus \mathbb{C}_b$.] Choose $\mathbb{C}_g$, so that

$$\mathbb{C}_g^* = \mathbb{C}_g \quad \text{and} \quad \mathbb{C}_g \cap \mathbb{R} \neq \varnothing; \tag{3.3}$$

i.e. if $s \in \mathbb{C}_g$ then $s^* \in \mathbb{C}_g$, and $\mathbb{C}_g$ includes at least one point on the real axis. A partition (3.2) with properties (3.3) will be called *symmetric*. Now let the m.p. of A be $\alpha(\lambda)$, and factor $\alpha(\lambda)$ in the form

$$\alpha(\lambda) = \alpha_g(\lambda)\alpha_b(\lambda),$$

where the zeros in $\mathbb{C}$ of α_g (resp. α_b) belong to $\mathbb{C}_g$ (resp. $\mathbb{C}_b$). Since α_g and α_b are coprime, we have by (0.11.3)

$$\mathscr{X} = \mathscr{X}_g(A) \oplus \mathscr{X}_b(A),$$

where

$$\mathscr{X}_g(A) := \operatorname{Ker} \alpha_g(A), \qquad \mathscr{X}_b(A) := \operatorname{Ker} \alpha_b(A).$$

The subspaces $\mathscr{X}_g(A)$ and $\mathscr{X}_b(A)$ can be thought of as the good and bad modal subspaces of A, respectively. The criterion of Theorem 2.2, below, says that the bad eigenvalues of A can be converted to good ones by state feedback, if and only if the bad modes of A are controllable.

Lemma 2.3. *If* $\mathbb{C} = \mathbb{C}_g \uplus \mathbb{C}_b$ *and* $T: \mathscr{X} \to \mathscr{X}$ *has m.p.* $\tau(\lambda) = \tau_g(\lambda)\tau_b(\lambda)$, *then* $\operatorname{Ker} \tau_b(T) \neq 0$ *only if* $\sigma(T) \cap \mathbb{C}_b \neq \varnothing$.

PROOF. Suppose $\sigma(T) \cap \mathbb{C}_b = \varnothing$. Then $\tau(\lambda)$ can have no zeros in $\mathbb{C}_b$, i.e. $\tau_b(\lambda) \equiv 1$. Thus, $\tau_b(T) = 1_{\mathscr{X}}$ and $\operatorname{Ker} \tau_b(T) = 0$. □

Theorem 2.2. *Let* $\mathbb{C} = \mathbb{C}_g \uplus \mathbb{C}_b$ *be a symmetric partition of* $\mathbb{C}$. *There exists* $F: \mathscr{X} \to \mathscr{U}$ *such that*

$$\sigma(A + BF) \subset \mathbb{C}_g$$

if and only if

$$\mathscr{X}_b(A) \subset \langle A|\mathscr{B}\rangle.$$

PROOF. In the proof we write $\mathscr{X}_g, \mathscr{X}_b$ for $\mathscr{X}_g(A), \mathscr{X}_b(A)$.

(If) The argument is summarized in the commutative diagram below.

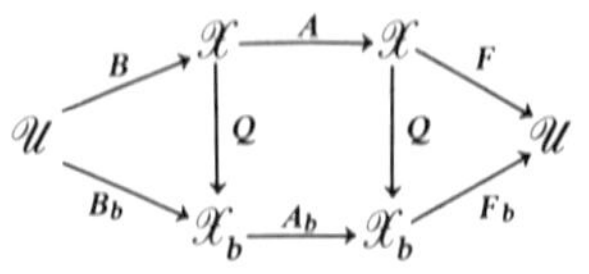

Let $Q: \mathscr{X}_b \oplus \mathscr{X}_g \to \mathscr{X}_b$ be the natural projection. By (3.4)

$$\mathscr{X}_b = Q\langle A \mid \mathscr{B} \rangle = \langle A_b \mid \mathscr{B}_b \rangle,$$

where

$$A_b = A \mid \mathscr{X}_b, \qquad B_b = QB.$$

Since (A_b, B_b) is controllable, there exists $F_b: \mathscr{X}_b \to \mathscr{U}$ such that[1]

$$\sigma(A_b + B_b F_b) \subset \mathbb{C}_g.$$

Define

$$F = F_b Q, \qquad A_g = A \mid \mathscr{X}_g.$$

Then

$$(A + BF) \mid \mathscr{X}_g = A_g$$

and

$$Q(A + BF) = (A_b + B_b F_b)Q.$$

Thus the map induced by $A + BF$ on $\mathscr{X}/\mathscr{X}_g \simeq \mathscr{X}_b$ is similar to $A_b + B_b F_b$, and so

$$\sigma(A + BF) = \sigma(A_g) \cup \sigma(A_b + B_b F_b) \subset \mathbb{C}_g.$$

(Only if) We use the notation introduced at the beginning of this section. Let $\bar{\alpha}$ be the m.p. of $\bar{A}$. Clearly, $\bar{\alpha} = \bar{\alpha}_g \bar{\alpha}_b$, where $\bar{\alpha}_g \mid \alpha_g$ and $\bar{\alpha}_b \mid \alpha_b$. Since $\bar{\alpha}_g, \bar{\alpha}_b$ must be coprime,

$$\bar{\mathscr{X}} = \operatorname{Ker} \bar{\alpha}_g(\bar{A}) \oplus \operatorname{Ker} \bar{\alpha}_b(\bar{A}). \tag{3.5}$$

Now suppose $\mathscr{X}_b \not\subset \mathscr{R} := \langle A \mid \mathscr{B} \rangle$. There exists $x \in \mathscr{X}$ with

$$\alpha_b(A)x = 0, \qquad x \notin \mathscr{R}.$$

Writing $\bar{x} = Px$, we have

$$\alpha_b(\bar{A})\bar{x} = 0, \qquad \bar{x} \neq 0.$$

In accordance with (3.5) write $\bar{x} = \bar{x}_g + \bar{x}_b$. Then

$$0 = \alpha_b(\bar{A})\bar{x} = \alpha_b(\bar{A})\bar{x}_g.$$

[1] If $d(\mathscr{X}_b)$ is odd we must here use the hypothesis $\mathbb{C}_g \cap \mathbb{R} \neq \varnothing$.

Therefore,

$$\bar{x}_g \in \text{Ker } \alpha_b(\bar{A}) \cap \text{Ker } \bar{\alpha}_g(\bar{A}) = 0,$$

since $\alpha_b(\lambda)$ and $\bar{\alpha}_g(\lambda)$ are coprime. So, $\bar{x} = \bar{x}_b \in \text{Ker } \bar{\alpha}_b(\bar{A})$ and, therefore,

$$\text{Ker } \bar{\alpha}_b(\bar{A}) \neq 0.$$

But $\overline{A + BF} = \bar{A}$ for all F, so Lemma 2.3 implies

$$\sigma(\overline{A + BF}) \cap \mathbb{C}_b \neq \emptyset.$$ □

2.4 Stabilizability

As an application of the foregoing ideas, define the stability and instability regions

$$\mathbb{C}^- := \{s: \Re e\ s < 0\}, \qquad \mathbb{C}^+ := \{s: \Re e\ s \geq 0\}.$$

We say that (A, B) is *stabilizable* if there exists $F: \mathscr{X} \to \mathscr{U}$, such that

$$\sigma(A + BF) \subset \mathbb{C}^-.$$

Let the m.p. of A be α and factor $\alpha = \alpha^- \alpha^+$, where the complex zeros of α^- (resp. α^+) belong to $\mathbb{C}^-$ (resp. $\mathbb{C}^+$). The subspace Ker $\alpha^+(A) \subset \mathscr{X}$ is the subspace of "unstable modes" of A, and we conclude from Theorem 2.2 that (A, B) is stabilizable if and only if the unstable modes of A are controllable. This special case is important enough to state separately.

Theorem 2.3. *(A, B) is stabilizable if and only if*

$$\text{Ker } \alpha^+(A) \subset \langle A \mid \mathscr{B} \rangle.$$

2.5 Exercises

2.1. Let

$$A = \begin{bmatrix} 1 & 0 & 0 & 0 \\ 0 & 0 & 1 & 0 \\ 0 & 0 & 0 & 0 \\ 1 & 0 & 0 & 0 \end{bmatrix}, \qquad B = \begin{bmatrix} 1 & 0 \\ 1 & 0 \\ 0 & 1 \\ 0 & 0 \end{bmatrix}.$$

Find F, such that

$$\sigma(A + BF) = \{-1, -1, -1 + i, -1 - i\}.$$

Describe as completely as you can the class of F for which the given spectrum is assigned. Hint: Step 1. Check that (A, B) is controllable. Step 2. Take a random linear combination $b := Bu$ of the columns of B and choose a random F_0: then $A_0 := A + BF_0$ almost surely has distinct eigenvalues, hence is cyclic, so by Corollary 1.1 the pair (A_0, b) is almost surely controllable: check it. Step 3.

Follow the constructive proof of Theorem 2.1 to get f' such that $A_0 + bf'$ has the required spectrum. Step 4. Set $F = F_0 + uf'$. With F selected, write out the ch.p. of $A + BF + B\tilde{F}$ to determine the $\tilde{F}$ for which $\sigma(A + BF + B\tilde{F}) = \sigma(A + BF)$.

2.2. Let $\{\lambda_1, \ldots, \lambda_n\}$ be a symmetric set of n distinct complex numbers, and let $\{v_1, \ldots, v_n\}$ be nonzero vectors in $\mathscr{X}_{\mathbb{C}}$ such that $v_i \in \mathscr{X}$ if λ_i is real and $v_i = v_j^*$ if $\lambda_i = \lambda_j^*$. Show that there exists $F: \mathscr{X} \to \mathscr{U}$ such that $(A + BF)v_i = \lambda_i v_i$ ($i \in \mathbf{n}$) if and only if the v_i are independent and

$$v_i \in \mathscr{X}(\lambda_i) := (\lambda_i 1 - A)^{-1}\mathscr{B}.$$

Furthermore, F is unique just when Ker $B = 0$. Show also that $\mathscr{X}(\lambda_i)$ is the projection on $\mathscr{X}_{\mathbb{C}}$ along $\mathscr{U}_{\mathbb{C}}$ of Ker$[\lambda_i 1 - A, B]$. Using Exercise 1.3, relate these results to Theorem 2.1.

2.3. Give a matrix proof of Theorem 2.2 based on the representation of Exercise 1.4.

2.4. Consider the system pair

$$A = \begin{bmatrix} 0 & 1 & 0 & 0 & 0 \\ 0 & 0 & 0 & 0 & 0 \\ 0 & 0 & -2 & 0 & 0 \\ 0 & 0 & 0 & 0 & 1 \\ 0 & 0 & 0 & 0 & 0 \end{bmatrix}, \qquad B = \begin{bmatrix} 0 & 0 \\ 1 & 1 \\ 0 & 0 \\ 0 & 0 \\ 1 & -1 \end{bmatrix}.$$

Say an eigenvalue λ is "good" if and only if $\Re e\ \lambda \le -1$. Check convertibility of bad eigenvalues by state feedback, using the criterion of Theorem 2.2.

2.5. Verify the remark at the end of Section 1.2. Hint: Let $P: \mathscr{X} \to \bar{\mathscr{X}} := \mathscr{X}/\mathscr{R}$. Then $\langle \bar{A} \mid \bar{\mathscr{B}} \rangle = \bar{\mathscr{X}}$ and you can choose $\bar{F}: \bar{\mathscr{X}} \to \mathscr{U}$ such that

$$\sigma(\bar{A} + \bar{B}\bar{F}) \cap \sigma(A \mid \mathscr{R}) = \varnothing.$$

Let $F := \bar{F}P$, $\hat{A} := A + BF$. Select $\mathscr{S} \subset \mathscr{X}$, $\mathscr{X} = \mathscr{R} \oplus \mathscr{S}$, $\hat{A}\mathscr{S} \subset \mathscr{S}$; this can be done in just one way (how?), and you may check that $\mathscr{S}$ does the job.

2.6. Assume (A, B) controllable and construct a basis for $\mathscr{X}$ as follows. Let B have column matrix $[b_1, \ldots, b_m]$. Let n_i be the dimension of $\langle A \mid \mathscr{b}_i \rangle$ mod$\langle A \mid \mathscr{b}_1 + \cdots + \mathscr{b}_{i-1} \rangle$ ($i \in \mathbf{m}$, $\mathscr{b}_0 = 0$). By reordering the b_i if necessary we can arrange that $n_i \ne 0$ ($i \in \mathbf{l}$) and $n_1 + \cdots + n_l = n$ for some $l \le m$. Show that in the basis

$$\{b_1, \ldots, A^{n_1 - 1}b_1, b_2, \ldots, A^{n_2 - 1}b_2, \ldots, b_l, \ldots, A^{n_l - 1}b_l\}$$

the matrix of A is in block upper-triangular form. Then show that F_1 can be chosen so that $A + BF_1$ has distinct eigenvalues, hence is cyclic. Using Corollary 1.1 complete the proof of the "only if" half of Theorem 2.1.

2.6 Notes and References

Lemma 2.1 has been noted independently by various authors. Lemma 2.2 is due to Heymann [1]; the proof here follows Wonham and Morse [1]. Theorem 2.1 was proved, for complex A, B, F and arbitrary complex Λ, by Langenhop [1] and by Popov [2]. The result for real A, B, F and symmetric Λ is due to Wonham [1], who used the argument outlined in Exercise 2.6. Theorem 2.2 generalizes the criterion for stabilizability given by Wonham [1] and repeated here as Theorem 2.3. A geometric criterion for stabilizability, expressed in terms of the differential equation but equivalent to the criterion of Theorem 2.3, was introduced earlier by Krasovskii [1]; cf. also Krasovskii [2].

A computational algorithm for pole assignment along the lines of Exercise 2.1 is described by Davison and Chow [1]. The alternative approach of Exercise 2.2 is due to Moore [1]; for an extension to multiple eigenvalues see Klein and Moore [1]. Related design methods, including pole assignment using various optimization techniques, are presented by Porter and Crossley [1] and by Lee and Jordan [1].

3 Observability and Dynamic Observers

Observability is a property of a dynamic system together with its observable inputs and outputs, according to which the latter alone suffice to determine exactly the state of the system. A data processor which performs state determination is called an "observer." In an intuitive sense observability is a property dual to controllability: a system is controllable if any state can be reached by suitable choice of input; it is observable if (when the input is known) its state can be computed by suitable processing of the output. For linear time-invariant systems this intuitive duality translates into a precise algebraic duality.

In this chapter we discuss observability, and observers of various kinds, for our standard system. As in Chapter 1 we start with a problem description in systemic terms, but move quickly to the underlying algebraic questions.

3.1 Observability

Consider the system

$$\dot{x}(t) = Ax(t) + v(t), \qquad t \geq 0, \tag{1.1}$$

$$y(t) = Cx(t), \qquad t \geq 0, \tag{1.2}$$

$$x(0) = x_0. \tag{1.3}$$

Our point of view is the following. The system map A and output map C are known, as are the input $v(s)$ and output $y(s)$ on some interval $t - \alpha \leq s \leq t$. The state $x(t - \alpha)$ is unknown, so $x(t)$ cannot *in general* be computed from the data listed. The situation of interest is just when such a computation is possible.

To formalize this idea, let **V** (resp. **Y**) be the set of piecewise-continuous functions $[0, \infty) \to \mathscr{X}$ (resp. $[0, \infty) \to \mathscr{Y}$); and for $t \geq \alpha > 0$ let

$$\begin{aligned} \omega_\alpha(t, s) &= 1, \qquad t - \alpha \leq s \leq t, \\ &= 0, \qquad \text{otherwise.} \end{aligned}$$

We define the system (1.1), (1.2) to be *observable* if for some $\alpha > 0$ there exists a function

$$\Omega\colon [\alpha, \infty) \times \mathbf{V} \times \mathbf{Y} \to \mathscr{X} \tag{1.4}$$

with the property

$$\Omega[t, \omega_\alpha(t, \cdot)v(\cdot), \omega_\alpha(t, \cdot)y(\cdot)] = x(t), \qquad t \geq \alpha$$

for all solutions $x(\cdot)$, $y(\cdot)$ of (1.1) and (1.2), with $v \in \mathbf{V}$ and $x_0 \in \mathscr{X}$.

The choice of function sets **V** and **Y** appearing in the definition is certainly not crucial, nor is the restriction that the processing interval $[t - \alpha, t]$ be of fixed length. Actually, for linear time-invariant systems all plausible definitions of observability turn out to be equivalent to the following algebraic condition:

Let $d(\mathscr{X}) = n$. The pair of maps (C, A) is *observable* if

$$\bigcap_{i=1}^{n} \operatorname{Ker}(CA^{i-1}) = 0. \tag{1.5}$$

Theorem 3.1. *The system* (1.1), (1.2) *is observable if and only if the pair* (C, A) *is observable.*

The proof depends on the "dual" of a construction already used in proving Theorem 1.1.

Lemma 3.1. *The pair of maps* (C, A) *is observable if and only if the symmetric map*

$$W_\alpha = \int_0^\alpha e^{-\sigma A'} C' C e^{-\sigma A}\, d\sigma \tag{1.6}$$

is positive definite for every $\alpha > 0$.

The simple proof is omitted.

Proof (of Theorem 3.1). ("*If*" *statement*) Applying Lemma 3.1, define

$$\Omega[t, \omega_\alpha(t, \cdot)v(\cdot), \omega_\alpha(t, \cdot)y(\cdot)]$$

$$:= W_\alpha^{-1} \int_0^\alpha e^{-\sigma A'} C' \left[y(t - \sigma) + C \int_0^\sigma e^{-\tau A} v(t - \sigma + \tau)\, d\tau \right] d\sigma \tag{1.7}$$

It is enough to check that the right side of (1.7) reduces to $x(t)$ for $t \geq \alpha$. For this, note from (1.1) and (1.2) that

$$y(s) = Cx(s) = C\left[e^{-(t-s)A}x(t) - \int_s^t e^{-(\tau-s)A}v(\tau)\,d\tau\right] \tag{1.8}$$

for $t - \alpha \leq s \leq t$. Multiply both sides of (1.8) by $e^{-(t-s)A'}C'$, integrate over $[t-\alpha, t]$ and use (1.6) to obtain the desired result. □

The proof of necessity in Theorem 3.1 is deferred to the next section.

3.2 Unobservable Subspace

The definition (1.5) suggests that the subspace $\mathcal{N} \subset \mathcal{X}$, defined as

$$\mathcal{N} := \bigcap_{i=1}^{n} \operatorname{Ker}(CA^{i-1}),$$

plays a significant role. We call $\mathcal{N}$ the *unobservable subspace* of (C, A). Clearly, $A\mathcal{N} \subset \mathcal{N}$; in fact, $\mathcal{N}$ is the largest A-invariant subspace contained in Ker C. Let $\bar{\mathcal{X}} = \mathcal{X}/\mathcal{N}$, $P: \mathcal{X} \to \bar{\mathcal{X}}$ be the canonical projection and $\bar{A}: \bar{\mathcal{X}} \to \bar{\mathcal{X}}$ the map induced in $\bar{\mathcal{X}}$ by A. Since Ker $C \supset \mathcal{N}$, there exists a map $\bar{C}: \bar{\mathcal{X}} \to \mathcal{Y}$ such that $\bar{C}P = C$, as shown below.

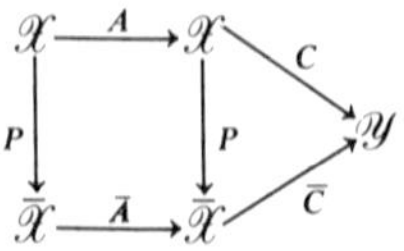

(2.1)

Lemma 3.2. *The pair $(\bar{C}, \bar{A})$ is observable.*

Proof. Since $n = d(\mathcal{X}) \geq d(\bar{\mathcal{X}})$ it is enough to show that

$$\bar{\mathcal{N}} = \bigcap_{i=1}^{n} \operatorname{Ker}(\bar{C}\bar{A}^{i-1}) = 0.$$

If $\bar{x} = Px \in \bar{\mathcal{N}}$ then $\bar{C}\bar{A}^{i-1}Px = 0$ $(i \in \mathbf{n})$. From the commutative diagram there results $CA^{i-1}x = 0$ $(i \in \mathbf{n})$, i.e. $x \in \mathcal{N}$, so $\bar{x} = Px = 0$. □

Since $(\bar{C}, \bar{A})$ is observable, it is possible to construct an observer for the "factor system"

$$\dot{\bar{x}} = \bar{A}\bar{x} + \bar{v}, \qquad y = \bar{C}\bar{x}$$

just as described in Section 3.1: details of coordinatization are suggested in Exercise 3.2. Thus it is always possible to identify the coset of the system state modulo the unobservable subspace. Our next result states that this is the best one can do.

Lemma 3.3. *Let $x_1(\cdot)$, $x_2(\cdot)$ be solutions of* (1.1)–(1.3) *for the same input $v(\cdot)$ but possibly different initial states x_{10}, x_{20}. If for some $t \geq 0$*

$$x_1(t) - x_2(t) \in \mathcal{N}$$

then

$$y_1(s) = y_2(s), \qquad s \geq 0.$$

PROOF. For all $s \geq 0$, and $i \in \mathbf{2}$,

$$y_i(s) = C\left[e^{(s-t)A}x_i(t) + \int_t^s e^{(s-\tau)A}v(\tau)\,d\tau\right].$$

Using the representation (0.17.3) for e^{tA}, we get

$$\begin{aligned} y_1(s) - y_2(s) &= Ce^{(s-t)A}[x_1(t) - x_2(t)] \\ &= \sum_{r=1}^{n} \psi_r(s-t)CA^{r-1}[x_1(t) - x_2(t)] \\ &= 0, \qquad s \geq 0, \end{aligned}$$

by definition of $\mathcal{N}$. □

We can now complete the proof of Theorem 3.1.

PROOF (of Theorem 3.1). (*"Only if" statement*) If (C, A) is not observable, i.e. $\mathcal{N} \neq 0$, let $0 \neq x_{10} - x_{20} \in \mathcal{N}$. With $v(\cdot)$ arbitrary, the corresponding solutions $x_i(\cdot)$ of (1.1) and (1.3) satisfy

$$x_1(t) - x_2(t) = e^{tA}(x_{10} - x_{20})$$

and, therefore, $0 \neq x_1(t) - x_2(t) \in \mathcal{N}$ for $t \geq 0$. By Lemma 3.3, $y_1(s) = y_2(s)$ for $s \geq 0$, and therefore every function Ω of the type (1.4) yields

$$\Omega[t, v(\cdot), y_1(\cdot)] = \Omega[t, v(\cdot), y_2(\cdot)], \qquad t \geq 0.$$ □

3.3 Full Order Dynamic Observer

The observer of Section 3.1 computes a weighted moving average of the data over a time interval of fixed length. A dynamic structure better matched to our theoretical setup is that of a linear differential equation: for this, the averaging interval is $[0, t]$ and in the observable case the observer error tends to zero exponentially fast as $t \to \infty$. To obtain satisfactory convergence in practice, the observer's dynamic response must be rapid, and this possibility depends on the pole assignment property of *observable* pairs. For the latter we need only verify that observability and controllability are algebraically dual.

Lemma 3.4. *Let $C: \mathscr{X} \to \mathscr{Y}$ and $A: \mathscr{X} \to \mathscr{X}$ be maps with duals $C': \mathscr{Y}' \to \mathscr{X}'$ and $A': \mathscr{X}' \to \mathscr{X}'$. Then (C, A) is observable if and only if (A', C') is controllable.*

PROOF. We have

$$\mathscr{N}^{\perp} = \left[\bigcap_{i=1}^{n} \operatorname{Ker}(CA^{i-1})\right]^{\perp} = \sum_{i=1}^{n} [\operatorname{Ker}(CA^{i-1})]^{\perp}$$

$$= \sum_{i=1}^{n} \operatorname{Im}(A'^{i-1}C') = \langle A' | \operatorname{Im} C' \rangle,$$

and therefore $\mathscr{N} = 0$ if and only if $\langle A' | \operatorname{Im} C' \rangle = \mathscr{X}'$. □

From Lemma 3.4 and Theorem 2.1 there follows immediately

Theorem 3.2. *The pair (C, A) is observable if and only if, for every symmetric set Λ of n complex numbers, there exists a map, $K: \mathscr{Y} \to \mathscr{X}$ such that*

$$\sigma(A + KC) = \Lambda.$$

We now seek an observer in the form of a differential equation

$$\dot{z}(t) = Jz(t) + Ky(t) + v(t), \qquad t \geq 0, \tag{3.1}$$

$$z(0) = z_0,$$

where $z(t) \in \mathscr{X}$, $y(\cdot)$ and $v(\cdot)$ are as in (1.1) and (1.2), and $J: \mathscr{X} \to \mathscr{X}$ and $K: \mathscr{Y} \to \mathscr{X}$ are to be determined. Write

$$e(t) = x(t) - z(t), \qquad t \geq 0. \tag{3.2}$$

We wish to arrange that $e(t) \to 0$ as $t \to \infty$. Applying Theorem 3.2, select K such that

$$\sigma(A - KC) = \Lambda \subset \mathbb{C}^-$$

and then set $J = A - KC$. From (1.1), (3.1), and (3.2) the result is

$$\dot{e}(t) = Je(t), \qquad t \geq 0,$$

and so $e(t) \to 0$ for every pair of initial states x_0, z_0. In practice, Λ is chosen in such a way that convergence is rapid compared to the response of the system (1.1) which is being observed.

3.4 Minimal Order Dynamic Observer

The dynamic order of the observer (3.1) is n, the same as that of the observed system (1.1). Yet n is unnecessarily large: for if the output map C has rank p then from $y(t)$ alone we can at once compute the coset of $x(t)$ in the p-

dimensional quotient space $\mathscr{X}/\text{Ker } C$. In this section we show what is now plausible: a dynamic observer can be constructed having the same form as (3.1), but of dynamic order $n - p$, to yield exactly the missing component of $x(t)$ in the $(n - p)$-dimensional subspace Ker C. Because $v(\cdot) \in \mathbf{V}$ is assumed unrestricted it is easily seen from (1.1) that every state $x \in \mathscr{X}$ is reachable from x_0. Thus, no observer of the general form (3.1) could have order less than $n - p$, if $z(t)$ is to yield, with $y(t)$, an asymptotic identification of $x(t)$ in the limit $t \to \infty$. In this sense an $(n - p)$th order observer is "minimal."

To construct a minimal-order observer, we need a rather special preliminary result on controllability which afterwards will be dualized for the application at hand.

Lemma 3.5. *Let (A, B) be controllable, and $d(\mathscr{B}) = m$. Let Λ be a symmetric set of $n - m$ complex numbers. There exist an $(n - m)$-dimensional subspace $\mathscr{V} \subset \mathscr{X}$ and a map $F: \mathscr{X} \to \mathscr{U}$, such that*

$$\mathscr{B} \oplus \mathscr{V} = \mathscr{X},$$

$$(A + BF)\mathscr{V} \subset \mathscr{V},$$

and

$$\sigma[(A + BF)|\mathscr{V}] = \Lambda.$$

We emphasize that the subspace $\mathscr{V}$, in general, depends on Λ.

Proof. Choose $\mathscr{D}$ arbitrarily such that $\mathscr{B} \oplus \mathscr{D} = \mathscr{X}$ and let $P: \mathscr{X} \to \mathscr{X}$ be the projection on $\mathscr{D}$ along $\mathscr{B}$. We show first that

$$\langle PA | PA\mathscr{B} \rangle = \mathscr{D}. \tag{4.1}$$

For this it is enough to verify that

$$x'(\mathscr{B} \oplus \langle PA | PA\mathscr{B} \rangle) = 0$$

implies $x' = 0$, for all $x' \in \mathscr{X}'$. Now $x'\mathscr{B} = 0$ implies $x'(1 - P) = 0$, or $x'P = x'$. Then, $x'PA\mathscr{B} = 0$ yields $x'A\mathscr{B} = 0$. Similarly, $x'PAPA\mathscr{B} = 0$ implies $x'APA\mathscr{B} = 0$; $x'A\mathscr{B} = 0$ implies $x'AP = x'A$; and so $x'A^2\mathscr{B} = 0$. Induction on i yields

$$x'A^{i-1}\mathscr{B} = 0, \qquad i \in \mathbf{n},$$

i.e. $x'\langle A | \mathscr{B} \rangle = 0$, hence $x' = 0$, as claimed.

By (4.1) and the pole assignment property there exists $F_0: \mathscr{X} \to \mathscr{U}$ such that

$$\sigma[(PA + PABF_0)|\mathscr{D}] = \Lambda.$$

Let

$$\mathscr{V} = (P + BF_0)\mathscr{D}.$$

Since $\mathscr{D} \cap \mathscr{B} = 0$ it is clear that $\mathscr{V} \cap \mathscr{B} = 0$ and $\mathscr{V} \simeq \mathscr{D}$, so $\mathscr{B} \oplus \mathscr{V} = \mathscr{X}$. Define $F: \mathscr{X} \to \mathscr{U}$ such that

$$BF = (BF_0 P - 1 + P)A;$$

F certainly exists, since

$$\operatorname{Im}(BF_0 P - 1 + P) \subset \mathscr{B}.$$

A direct computation now verifies that the diagram below commutes:

$$\begin{array}{ccc} \mathscr{D} & \xrightarrow{PA + PABF_0} & \mathscr{D} \\ {\scriptstyle P+BF_0}\downarrow\simeq & & \simeq\downarrow{\scriptstyle P+BF_0} \\ \mathscr{V} & \xrightarrow{A+BF} & \mathscr{V} \end{array}$$

Thus, $(A + BF)\mathscr{V} \subset \mathscr{V}$. Since $(P + BF_0)|\mathscr{D}$ is an isomorphism $\mathscr{D} \simeq \mathscr{V}$, we have that

$$\sigma[A + BF\,|\,\mathscr{V}] = \sigma[(PA + PABF_0)|\mathscr{D}] = \Lambda. \qquad \square$$

Now assume (C, A) is observable and apply Lemma 3.5 to the controllable pair (A', C'). Write $\mathscr{C}' = \operatorname{Im} C'$: $d(\mathscr{C}') = p$. Having chosen a symmetric set $\Lambda \subset \mathbb{C}$ with $|\Lambda| = n - p$, we can find an $(n - p)$-dimensional subspace $\mathscr{V}' \subset \mathscr{X}'$, and a map $K: \mathscr{Y} \to \mathscr{X}$ such that

$$\mathscr{C}' \oplus \mathscr{V}' = \mathscr{X}', \tag{4.2}$$

$$(A - KC)'\mathscr{V}' \subset \mathscr{V}',$$

and

$$\sigma[(A - KC)'\,|\,\mathscr{V}'] = \Lambda.$$

Let $V': \mathscr{V}' \to \mathscr{X}'$ be the insertion map and let $T' = (A - KC)'\,|\,\mathscr{V}'$. Then,

$$(A - KC)'V' = V'T'$$

and

$$\sigma(T') = \Lambda.$$

Thus,

$$V(A - KC) = TV, \qquad \sigma(T) = \Lambda.$$

Next, taking annihilators in (4.2) yields

$$\begin{aligned} \operatorname{Ker} C \cap \operatorname{Ker} V &= (\operatorname{Im} C')^\perp \cap (\operatorname{Im} V')^\perp \\ &= (\mathscr{C}')^\perp \cap (\mathscr{V}')^\perp = 0. \end{aligned} \tag{4.3}$$

Let $\mathscr{Y} \oplus \mathscr{V}$ be the external direct sum of $\mathscr{Y}$ and $\mathscr{V}$. Then, (4.3) implies that the map

$$Q: \mathscr{X} \to \mathscr{Y} \oplus \mathscr{V}, \qquad x \mapsto Cx \oplus Vx$$

is monic and, as $\mathscr{Y} \oplus \mathscr{V} \simeq \mathscr{X}$, Q is also epic, hence an isomorphism. Summarizing, we have

Theorem 3.3. *Let (C, A) be observable, with $A: \mathscr{X} \to \mathscr{X}$, $d(\mathscr{X}) = n$, $C: \mathscr{X} \to \mathscr{Y}$, $d(\mathscr{Y}) = p$, and C epic. Let $\Lambda \subset \mathbb{C}$ be symmetric with $|\Lambda| = n - p$. There exists a subspace $\mathscr{V} \subset \mathscr{X}$ with $d(\mathscr{V}) = n - p$, and maps $K: \mathscr{Y} \to \mathscr{X}$, $T: \mathscr{V} \to \mathscr{V}$ and $V: \mathscr{X} \to \mathscr{V}$, such that*

$$V(A - KC) = TV, \qquad \sigma(T) = \Lambda. \tag{4.4}$$

Furthermore, the map

$$Q: \mathscr{X} \to \mathscr{Y} \oplus \mathscr{V}, \qquad x \mapsto Cx \oplus Vx \tag{4.5}$$

is an isomorphism.

We are now in a position to construct a minimal-order dynamic observer for the system (1.1), (1.2). Assuming observability, consider the differential equation

$$\dot{z}(t) = Tz(t) + VKy(t) + Vv(t), \qquad t \geq 0, \tag{4.6}$$

where T, V and K are given by Theorem 3.3. Write

$$e(t) = Vx(t) - z(t), \qquad t \geq 0.$$

Computing $\dot{e}$ from (1.1) and (4.6), and using (4.4), we find

$$\dot{e} = Te.$$

Thus, if $\Lambda \subset \mathbb{C}^-$ we have that

$$z(t) = Vx(t) - e(t), \qquad t \geq 0,$$

where $e(t) \to 0$ exponentially fast. By (1.2) and (4.5)

$$\begin{aligned} x(t) &= Q^{-1}[y(t) \oplus Vx(t)] \\ &\doteq Q^{-1}[y(t) \oplus z(t)], \end{aligned} \tag{4.7}$$

with error exponentially small as $t \to \infty$. In practice Λ is chosen such that the identification error in (4.7) vanishes rapidly compared to the response time of the observed system (1.1).

A procedure for the computation of the matrices of a minimal observer is developed in Exercise 3.5.

3.5 Observers and Pole Shifting

In (1.1) set $v(t) = Bu(t)$, to obtain

$$\dot{x} = Ax + Bu. \tag{5.1}$$

Suppose it is desired to realize dynamic behavior corresponding to a control $u = Fx$. If the directly measured variable is not x but $y = Cx$, we must synthesize control by means of an observer. For the observer (3.1), we have

$$z(t) - x(t) \to 0, \qquad t \to \infty,$$

and this suggests that we put

$$u(t) = Fz(t), \qquad t \geq 0. \tag{5.2}$$

The combined system (5.1), (1.2), (3.1) is now

$$\dot{x} = Ax + Bu \tag{5.3a}$$

$$\dot{z} = Jz + Ky + Bu, \tag{5.3b}$$

where $J = A - KC$. Setting $e = x - z$, and using (1.2) and (5.2), we get

$$\dot{x} = (A + BF)x - BFe \tag{5.4a}$$

$$\dot{e} = Je. \tag{5.4b}$$

Thus, the spectrum of the combined system matrix in (5.3) coincides with that of (5.4), namely

$$\sigma(A + BF) \uplus \sigma(J). \tag{5.5}$$

It is clear from (5.5) that, for instance, a stable combined system can be synthesized provided (A, B) is stabilizable and (C, A) is observable.

The signal flow graph corresponding to (5.2) and (5.3) is given in Fig. 3.1.

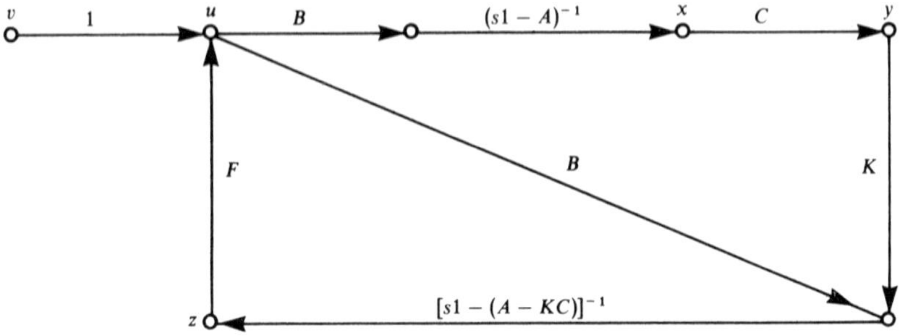

Figure 3.1 Signal Flow Graph for Observer-Compensator.

In practice it may be convenient to synthesize the combination of observer with z-feedback as a single "compensator" with $m \times m$ transfer matrix $T(s)$, such that $\hat{u}(s) = T(s)\hat{y}(s)$. By inspection of Fig. 3.1 (with $v = 0$),

$$\hat{u} = F[s1 - (A - KC)]^{-1}(B\hat{u} + K\hat{y}),$$

and so

$$T(s) = [1 - F(s1 - (A - KC))^{-1}B]^{-1}F[s1 - (A - KC)]^{-1}K$$
$$= F[s1 - (A + BF - KC)]^{-1}K.$$

However, there is no guarantee that $A + BF - KC$ is stable, even though $A + BF$ and $A - KC$ are stable individually.

The principle of stabilization via an observer also applies to an observer of minimal order. With the observer defined by (4.4)–(4.6) set, in (5.1),

$$u = FQ^{-1}(y \oplus z).$$

The combined system is

$$\dot{x} = Ax + Bu, \tag{5.6a}$$

$$\dot{z} = Tz + VKy + VBu. \tag{5.6b}$$

Set $z = Vx - e$; recall that

$$y = Cx, \qquad V(A - KC) = TV;$$

and note from (4.5) that

$$Q^{-1}(C \oplus V) = 1,$$

the identity on $\mathscr{Y} \oplus \mathscr{V}$. Then, (5.6) yields

$$\begin{aligned} \dot{x} &= Ax + BFQ^{-1}(y \oplus z) \\ &= Ax + BFQ^{-1}[Cx \oplus (Vx - e)] \\ &= Ax + BFQ^{-1}[(C \oplus V)x - (0 \oplus e)] \\ &= (A + BF)x - BFQ^{-1}(0 \oplus e) \end{aligned} \tag{5.7a}$$

and by a short computation

$$\dot{e} = Te. \tag{5.7b}$$

The spectrum of the combined system matrix in (5.7) is, therefore,

$$\sigma(A + BF) \cup \sigma(T). \tag{5.8}$$

From Theorem 3.3 and (5.8) we may draw the same conclusion as before: the combined system can be stabilized if (A, B) is stabilizable and (C, A) is observable.

Actually, in both cases the requirement on (C, A) of observability may be weakened to "detectability," as described in the section to follow.

3.6 Detectability

A property weaker than observability, but fundamental to the quadratic optimization problem considered in Chapter 12, is that (at least) the subspace of unstable modes of A be observable. This property is the dual for (C, A) of the property of stabilizability for (A, B) introduced in Section 2.4. As in that section, factor the m.p. $\alpha(\lambda)$ of A as a product

$$\alpha(\lambda) = \alpha^{+}(\lambda)\alpha^{-}(\lambda),$$

where the zeros of α^+ (resp. α^-) over $\mathbb{C}$ belong to the closed right (resp. open left) half plane, and write

$$\mathscr{X}^+(A) := \text{Ker } \alpha^+(A), \qquad \mathscr{X}^-(A) := \text{Ker } \alpha^-(A).$$

We say that (C, A) is *detectable* if

$$\bigcap_{i=1}^{n} \text{Ker}(CA^{i-1}) \subset \mathscr{X}^-(A); \tag{6.1}$$

i.e. A is stable on the unobservable subspace of (C, A).

Proposition 3.1. *The pair* (C, A) *is detectable if and only if* (A', C') *is stabilizable.*

PROOF. Note that the m.p. of A' coincides with that of A; it is then immediately verified that

$$\text{Im } \alpha^-(A') = \text{Ker } \alpha^+(A').$$

Taking annihilators in (6.1),

$$\sum_{i=1}^{n} \text{Im}(A'^{i-1}C') \supset [\text{Ker } \alpha^-(A)]^\perp = \text{Im } \alpha^-(A') = \text{Ker } \alpha^+(A')$$

and the result follows by Theorem 2.3. □

Our choice of the term "detectable" was prompted by the following observation.

Proposition 3.2. *Regard* $\mathscr{X}$ *and* $\mathscr{Y}$ *as inner product spaces over* $\mathbb{C}$, *and let* (C, A) *be detectable.* A *is stable if and only if the map*

$$W(t) := \int_0^t e^{sA'}C'Ce^{sA}\, ds$$

is a norm-bounded function of t *as* $t \uparrow \infty$.

PROOF. It is clear that $W(\cdot)$ is bounded if A is stable. For the converse assume A is unstable, and let μ be an eigenvalue of A with $\Re e\ \mu \geq 0$ and eigenvector $x \in \mathscr{X}_{\mathbb{C}}$. Then,

$$x^*W(t)x = \int_0^t e^{2s\, \Re e\, \mu}|Cx|^2\, ds.$$

Suppose the integral is bounded. Then, $Cx = 0$, so that

$$CA^{i-1}x = \mu^{i-1}Cx = 0, \qquad i \in \mathbf{n}. \tag{6.2}$$

By (6.1) and (6.2)

$$x \in \bigcap_{i=1}^{n} \text{Ker}(CA^{i-1}) \subset \mathscr{X}_{\mathbb{C}}^-(A),$$

and therefore,

$$x \in \mathscr{X}_{\mathbb{C}}^{-}(A) \cap \mathscr{X}_{\mathbb{C}}^{+}(A) = 0,$$

in contradiction to the assumption that x is an eigenvector. □

We conclude this section with the obvious remark that (C, A) is detectable if and only if there exists $K: \mathscr{Y} \to \mathscr{X}$ such that $A + KC$ is stable.

3.7 Detectors and Pole Shifting

The dynamic observer of Section 3.5 enabled us to identify the complete state x. But if our ultimate purpose is only to stabilize the system, it is enough to identify x modulo the subspace $\mathscr{X}^{-}(A)$ of stable modes of A. In general, an observer for this purpose can be constructed with dynamic order smaller than $d(\operatorname{Ker} C)$, the dimension of the minimal-order observer for x. It is also intuitively clear that such a restricted observer, or *detector*, exists only if the pair (C, A) is detectable. Loosely stated, our problem is to derive a lower-order model of the system, with observable pair $(\bar{C}, \bar{A})$ which preserves the structure of A on its unstable modal subspace $\mathscr{X}^{+}(A)$.

To formulate the correct algebraic problem consider again (1.1) and (1.2), setting $v = 0$ without loss of generality. Introduce a "model" of (1.1), (1.2) having dynamic equations

$$\dot{\bar{x}} = \bar{A}\bar{x}, \tag{7.1a}$$

$$\bar{y} = \bar{C}\bar{x}. \tag{7.1b}$$

Denote the corresponding state and output spaces by $\bar{\mathscr{X}}$ and $\bar{\mathscr{Y}}$. Since (7.1a) is to model the behavior of (1.1) on (at least) the invariant subspace

$$\mathscr{X}^{+}(A) \simeq \frac{\mathscr{X}}{\mathscr{X}^{-}(A)},$$

we identify $\bar{\mathscr{X}}$ as some factor space $\mathscr{X}/\mathscr{S}$, where

$$A\mathscr{S} \subset \mathscr{S} \subset \mathscr{X}^{-}(A).$$

Next, we must guarantee that the model output $\bar{y}$ carries no more information than does the physically available output y: realizability demands the existence of a map $D: \mathscr{Y} \to \bar{\mathscr{Y}}$ such that

$$\bar{y} = Dy. \tag{7.2}$$

Clearly, (7.2) justifies the further identification $\bar{\mathscr{Y}} = \mathscr{Y}$. In this way, we are led to the algebraic setup depicted in the diagram below.

$$\begin{array}{ccccc}
\mathscr{X} & \xrightarrow{A} & \mathscr{X} & \xrightarrow{C} & \mathscr{Y} \\
\downarrow{\scriptstyle P} & & \downarrow{\scriptstyle P} & & \downarrow{\scriptstyle D} \\
\mathscr{X}/\mathscr{S} & \xrightarrow{\bar{A}} & \mathscr{X}/\mathscr{S} & \xrightarrow{\bar{C}} & \mathscr{Y} \\
\downarrow{\scriptstyle Q} & & \downarrow{\scriptstyle Q} & & \\
\mathscr{X}^{+}(A) & \xrightarrow{A^{+}} & \mathscr{X}^{+}(A) & &
\end{array} \tag{7.3}$$

Here we may as well arrange that the pair $(\bar{C}, \bar{A})$ be observable. For if it is not, factor out its unobservable kernel in accordance with the diagram (2.1) and Lemma 3.2, then note that the corresponding canonical projection can be composed with the projection P in (7.3).

The structure displayed in the diagram is summarized in

Theorem 3.4. *Let $A: \mathscr{X} \to \mathscr{X}$ and $C: \mathscr{X} \to \mathscr{Y}$ with (C, A) detectable. Let $\mathscr{K} \subset \mathscr{X}$, $\mathscr{S} \subset \mathscr{X}$ be subspaces with the properties*

$$\mathscr{K} \supset \operatorname{Ker} C + \mathscr{S}, \qquad \bigcap_{i=1}^{n} A^{-i+1}\mathscr{K} \subset \mathscr{X}^-(A) \tag{7.4}$$

and

$$A\mathscr{S} \subset \mathscr{S} \subset \mathscr{X}^-(A).$$

Let $\bar{A}: \mathscr{X}/\mathscr{S} \to \mathscr{X}/\mathscr{S}$ be the map induced by A in $\mathscr{X}/\mathscr{S}$ and write $P: \mathscr{X} \to \mathscr{X}/\mathscr{S}$ for the canonical projection. Then,

i. *there exists a map $D: \mathscr{Y} \to \mathscr{Y}$ such that*

$$\operatorname{Ker} DC = \mathscr{K}$$

and (DC, A) is detectable;

ii. *there exists a map $\bar{C}: \mathscr{X}/\mathscr{S} \to \mathscr{Y}$ such that*

$$\bar{C}P = DC; \tag{7.5}$$

iii. *$(\bar{C}, \bar{A})$ is observable if and only if*

$$\mathscr{S} = \bigcap_{i=1}^{n} A^{-i+1}\mathscr{K},$$

i.e. $\mathscr{S}$ is the largest A-invariant subspace of $\mathscr{K}$;

iv. *if $A^+ = A \mid \mathscr{X}^+(A)$, there exists an epimorphism $Q: \mathscr{X}/\mathscr{S} \to \mathscr{X}^+(A)$ such that*

$$A^+Q = Q\bar{A}.$$

PROOF

i. By (7.4) $\mathscr{K} \supset \operatorname{Ker} C$, and the first assertion follows easily. As for detectability

$$\bigcap_{i=1}^{n} \operatorname{Ker}(DCA^{i-1}) = \bigcap_{i=1}^{n} A^{-i+1}\mathscr{K} \subset \mathscr{X}^-(A).$$

ii. The existence of $\bar{C}$ is immediate from

$$\operatorname{Ker} P = \mathscr{S} \subset \mathscr{K} = \operatorname{Ker}(DC).$$

iii. Write

$$\mathscr{S}^* = \bigcap_{i=1}^{n} A^{-i+1}\mathscr{K}.$$

A routine application of (7.5) and the definitions verifies that

$$\bigcap_{i=1}^{n} \operatorname{Ker}(\bar{C}\bar{A}^{i-1}) = \mathscr{S}^*/\mathscr{S}$$

whence the assertion follows.

iv. Write

$$x = x^+ + x^-, \qquad x \in \mathscr{X},\ x^\pm \in \mathscr{X}^\pm(A),$$

and for $\bar{x} = Px \in \mathscr{X}/\mathscr{S}$ let

$$Q\bar{x} = x^+.$$

It is straightforward to check that Q is well defined and has the stated properties. □

The foregoing discussion has reduced the problem of identifying $x \bmod \mathscr{X}^-(A)$ to that of constructing an observer (which we shall choose to be minimal in the sense of Section 3.4) for an observable model $(\bar{C}, \bar{A})$ related to (C, A) as in (7.3). Observability implies that $\mathscr{S} = \mathscr{S}^*$, hence the possible models are completely determined by the choice of $\mathscr{K}$, and the corresponding observer has dynamic order

$$d(\operatorname{Ker} \bar{C}) = d(\mathscr{K}/\mathscr{S}^*).$$

After these preliminaries, we can state the purely algebraic

Minimal Detector Problem (MDP). *Given $A: \mathscr{X} \to \mathscr{X}$ and $C: \mathscr{X} \to \mathscr{Y}$ with (C, A) detectable, find $\mathscr{K} \subset \mathscr{X}$, such that*

$$\mathscr{K} \supset \operatorname{Ker} C \tag{7.6}$$

$$\bigcap_{i=1}^{n} A^{-i+1}\mathscr{K} \subset \mathscr{X}^-(A) \tag{7.7}$$

and

$$d(\mathscr{K}) - d\left(\bigcap_{i=1}^{n} A^{-i+1}\mathscr{K}\right) = \text{minimum}. \tag{7.8}$$

An effective procedure for solving MDP is not currently available. However, a solution always exists, as (7.6) and (7.7) are satisfied in particular by $\mathscr{K} = \operatorname{Ker} C$. In general, however, $\operatorname{Ker} C$ is not minimal in the sense of (7.8). A lower bound for this minimum is easily derived:

$$d\left(\frac{\mathscr{K}}{\mathscr{S}^*}\right) \geq d\left[\frac{\operatorname{Ker} C + \mathscr{S}^*}{\mathscr{S}^*}\right] = d(\operatorname{Ker} C) - d(\mathscr{S}^* \cap \operatorname{Ker} C)$$

$$\geq d(\operatorname{Ker} C) - d[\mathscr{X}^-(A) \cap \operatorname{Ker} C]. \tag{7.9}$$

On the basis solely of (7.6) and (7.7), the bound (7.9) is the best possible, although it cannot always be attained, as the second of the following examples shows.

EXAMPLE 1. Suppose

$$\text{Ker } C = \mathscr{G}^+ \oplus \mathscr{G}^-,$$

where $\mathscr{G}^\pm \subset \mathscr{X}^\pm(A)$. Let

$$\mathscr{K} = \mathscr{X}^-(A) \oplus \mathscr{G}^+.$$

By detectability

$$\mathscr{S}^* = \bigcap_{i=1}^{n} A^{-i+1}[\mathscr{X}^-(A) \oplus \mathscr{G}^+] = \mathscr{X}^-(A);$$

so that

$$d\left(\frac{\mathscr{K}}{\mathscr{S}^*}\right) = d(\mathscr{G}^+),$$

which is minimal by (7.9).

EXAMPLE 2. Consider the detectable pair

$$A = \left[\begin{array}{ccc|cc} -1 & 1 & 0 & & \\ 0 & -1 & 1 & & \bigcirc \\ 0 & 0 & -1 & & \\ \hline & & & 0 & 1 \\ & \bigcirc & & 0 & 0 \end{array}\right], \quad C = \begin{bmatrix} 0 & 0 & 0 & 0 & 1 \\ 0 & 1 & 1 & 1 & 0 \end{bmatrix}. \tag{7.10}$$

Denoting the unit vectors by e_i ($i \in \mathbf{5}$) and setting $\mathscr{K} = \text{Ker } C$, we find

$$\text{Ker } C = \text{Span}\{e_1, e_2 - e_3, e_3 - e_4\},$$

$$\mathscr{S}^* = \text{Span}\{e_1\},$$

$$d\left(\frac{\mathscr{K}}{\mathscr{S}^*}\right) = 2$$

and

$$d\left[\frac{\text{Ker } C}{\mathscr{X}^-(A) \cap \text{Ker } C}\right] = 1.$$

It is easy to see that no other choice of $\mathscr{K}$, say $\tilde{\mathscr{K}}$, will lead to a lower value of $d(\tilde{\mathscr{K}}/\tilde{\mathscr{S}}^*)$. Indeed as

$$\tilde{\mathscr{S}}^* \subset \mathscr{X}^-(A) = \text{Span}\{e_1, e_2, e_3\}$$

and $\tilde{\mathscr{K}} \supset \text{Ker } C = \mathscr{K}$, it follows from (7.10) that $\tilde{\mathscr{S}}^*$ must be one of the subspaces

$$\mathscr{S}^*, \text{Span}\{e_1, e_2\}, \mathscr{X}^-(A).$$

If $\tilde{\mathscr{S}}^* = \mathscr{S}^*$, clearly

$$d\left(\frac{\tilde{\mathscr{K}}}{\tilde{\mathscr{S}}^*}\right) \geq d\left(\frac{\mathscr{K}}{\mathscr{S}^*}\right).$$

If $\mathscr{S}^* = \mathscr{X}^-(A)$, then $\tilde{\mathscr{K}} \supset \text{Ker } C + \mathscr{X}^-(A)$; but

$$e_4 \in [\text{Ker } C + \mathscr{X}^-(A)] \cap \mathscr{X}^+(A)$$

and $Ae_4 = 0$, so that $e_4 \in \tilde{\mathscr{S}}^*$, a contradiction. Finally, if $\tilde{\mathscr{S}}^* = \text{Span}\{e_1, e_2\}$, then

$$d(\tilde{\mathscr{K}}) \geq d(\mathscr{K}) + 1$$

and

$$d\left(\frac{\tilde{\mathscr{K}}}{\tilde{\mathscr{S}}^*}\right) \geq d(\mathscr{K}) + 1 - 2 = 2$$

as before.

We conclude this section with the obvious remark that all our results remain valid for a general symmetric partition

$$\mathbb{C} = \mathbb{C}_g \cup \mathbb{C}_b$$

and corresponding modal decomposition

$$\mathscr{X} = \mathscr{X}_g(A) \oplus \mathscr{X}_b(A),$$

where $\mathscr{X}^+(A)$, $\mathscr{X}^-(A)$ are replaced by $\mathscr{X}_b(A)$, $\mathscr{X}_g(A)$, respectively.

3.8 Pole Shifting by Dynamic Compensation

In this section we adopt an approach to pole shifting which, unlike that of Section 3.5, makes no explicit reliance on the observer action of the auxiliary dynamic element to be coupled to the original system. Of course, it will still be true that the observability property of the given system must be postulated if complete freedom of pole assignability is required.

Consider as usual

$$\dot{x} = Ax + Bu, \qquad y = Cx. \tag{8.1}$$

We shall say that (C, A, B) is *complete* if (C, A) is observable and (A, B) is controllable. When (C, A) is observable, we define the *observability index* κ_0 of (C, A) according to

$$\kappa_0 := \min\left\{j\colon 1 \leq j \leq n, \ \bigcap_{i=1}^{j} \text{Ker}(CA^{i-1}) = 0\right\}.$$

Clearly, κ_0 exists and $1 \leq \kappa_0 \leq n$.

Write $\nu := \kappa_0 - 1$, and introduce an auxiliary state space $\mathscr{W}$ with $d(\mathscr{W}) = \nu$. Our objective is to find an auxiliary dynamic system

$$\dot{w} = Ww + v \tag{8.2}$$

which, when coupled to (8.1) according to

$$u = Hy + Gw$$
$$v = Ky, \tag{8.3}$$

will assign to the composite system a desired spectrum Λ with $|\Lambda| = n + \nu$. The system (8.2) is the (dynamic) *compensator*.

That our problem is solvable is claimed by

Theorem 3.5. *Let* (C, A, B) *be complete and let the observability index of* (C, A) *be* $\nu + 1$. *Introduce* $\mathscr{W}$, *independent of* $\mathscr{X}$, *with* $d(\mathscr{W}) = \nu$. *Then for every symmetric set* Λ *of* $n + \nu$ *complex numbers, there exist maps*

$$G: \mathscr{W} \to \mathscr{U},\ H: \mathscr{Y} \to \mathscr{U},\ K: \mathscr{Y} \to \mathscr{W} \text{ and } W: \mathscr{W} \to \mathscr{W},$$

such that (in a basis adapted to $\mathscr{X} \oplus \mathscr{W}$*)*

$$\sigma\left(\begin{bmatrix} A + BHC & BG \\ KC & W \end{bmatrix}\right) = \Lambda.$$

Remark 1. In the composite system (8.1)–(8.3) only the "measurements" (y, w) are made directly accessible to the "controls" (u, v).

Remark 2. If a minimal order dynamic observer were used as compensator, as in Section 3.5, its generic order, relative to the space $\mathbb{R}^{pn+n^2}$ of all pairs (C, A), would be, obviously,

$$d(\text{Ker } C) = n - p(g).$$

One would assign the spectrum of the composite system as $\sigma = \sigma_1 \cup \sigma_2$ with $|\sigma_1| = n$, $|\sigma_2| = n - p$. On the other hand,

$$\kappa_0 = \min\{j: jp \geq n\}(g),$$

so generically, the compensator of Theorem 3.5 has dynamic order roughly n/p, and this figure is much less than $n - p$ in "typical" cases where n is large and p is relatively small.

Remark 3. Theorem 3.5 can be dualized in obvious fashion, to yield a dynamic compensator of order one less than the *controllability index* κ_c of (A, B), where

$$\kappa_c := \min\left\{j: 1 \leq j \leq n,\ \sum_{i=1}^{j} A^{i-1}\mathscr{B} = \mathscr{X}\right\}.$$

The "better" of the two results would be used in applications where reduction of compensator order is important.

To prove Theorem 3.5 we need several preliminary results, of some interest in their own right. The key step is achieved by Lemma 3.7, below.

Lemma 3.6. *Let (C, A, B) be complete and suppose that the degree of the m.p. of A is $k < n := d(\mathscr{X})$. There exist $b \in \mathscr{B}$ and $c' \in \operatorname{Im} C'$ such that $(C, A + bc', B)$ is complete and has m.p. of degree at least $k + 1$.*

PROOF. Since (C, A, B) is complete, we know (by Lemma 2.1 and its dual) that $(C, A + bc', B)$ is complete for any $b \in \mathscr{B}$ and $c' \in \operatorname{Im} C'$. By Theorem 1.2, there exists $b \in \mathscr{B}$ such that the m.p. β of b (with respect to A) coincides with the m.p. of A. Then, $\mathscr{R} := \langle A \mid \ell \rangle$ is maximal cyclic, and so by Proposition 0.2 there exists an A-invariant subspace $\mathscr{S} \subset \mathscr{X}$ such that $\mathscr{X} = \mathscr{R} \oplus \mathscr{S}$. Let $g \in \mathscr{S}$ be any vector such that $Cg \neq 0$; g exists because (C, A) is observable. Let $y' \in \mathscr{Y}'$ be any functional with $y'Cg = 1$ and set $c' = y'C$. It will be shown that g has m.p. with respect to $A + bc'$ of degree at least $k + 1$. Indeed, suppose that

$$\sum_{i=1}^{k} \gamma_i (A + bc')^{i-1} g = 0$$

for some scalars γ_i ($i \in \mathbf{k}$). Then,

$$\sum_{i=1}^{k} \gamma_i A^{i-1} g = -\gamma_k (c'g) A^{k-1} b - \cdots,$$

the remainder denoting terms in $A^{j-1}b$ $(1 \leq j \leq k - 1)$. Since the right side is a vector in $\mathscr{R}$ while the left is a vector in $\mathscr{S}$, both are zero; and as $c'g = 1$ and $\deg \beta = k$ there results $\gamma_k = 0$. Repetition of the argument yields in turn $\gamma_{k-1} = \cdots = \gamma_1 = 0$; that is, the vectors $(A + bc')^{i-1} g$ $(i \in \mathbf{k})$ are linearly independent, as claimed. □

Lemma 3.7. *If (C, A, B) is complete there exists $H: \mathscr{Y} \to \mathscr{U}$ such that $A + BHC$ is cyclic.*

PROOF. If A has m.p. of degree $k < n$, it suffices to apply Lemma 3.6 l times, for some $l \leq n - k$, to get that

$$A + b_1 c_1' + \cdots + b_l c_l'$$

has m.p. of degree n (hence, is cyclic) for suitable $b_i \in \mathscr{B}$, $c_1' \in \operatorname{Im} C'$. Now, $b_i = Bu_i$ and $c_i' = y_i' C$ for some $u_i \in \mathscr{U}$ and $y_i' \in \mathscr{Y}'$, so we need only set

$$H := \sum_{i=1}^{l} u_i y_i'. \qquad \square$$

Remark. Fix (C, A, B) complete and choose H as in Lemma 3.7. Let e be a cyclic generator for $A + BHC$. Then,

$$\det[e, (A + BHC)e, \ldots, (A + BHC)^{n-1} e] \neq 0,$$

and this inequality remains true for all $H \in \mathbb{R}^{m \times p}$ except those H which, as points in $\mathbb{R}^{mp}$, belong to a proper variety $\mathbf{V} \subset \mathbb{R}^{mp}$. Thus, $A + BHC$ is cyclic in the complement $\mathbf{V}^c$, namely for "almost all" H.

It is now clear that in proving Theorem 3.5 we can assume *a priori* that A is cyclic: otherwise, a preliminary transformation $A \mapsto A + BH_0 C$ will make it so.

The next, rather special result will provide an isomorphism needed later.

Lemma 3.8. *Let (c', A, B) be complete; let $\mathscr{T}$ be independent of $\mathscr{X}$ with $d(\mathscr{T}) = \tau \leq n - 1$; let $T: \mathscr{T} \to \mathscr{T}$ be cyclic with generator $g \in \mathscr{T}$; let $h' \in \mathscr{T}'$ be determined by*

$$h'g = \cdots = h'T^{\tau-2}g = 0, \qquad h'T^{\tau-1}g = 1; \tag{8.4}$$

let $l \in \mathscr{X}$ be a vector such that

$$l \in \mathscr{B} + A\mathscr{B} + \cdots + A^{\tau}\mathscr{B}; \tag{8.5}$$

define $\mathscr{X}_e := \mathscr{X} \oplus \mathscr{T}$; and in $\mathscr{X}_e$ let

$$A_e := \begin{bmatrix} A & -lh' \\ 0 & T \end{bmatrix}, \qquad B_e := \begin{bmatrix} B \\ 0 \end{bmatrix}.$$

Then there exist a subspace $\mathscr{S} \subset \mathscr{X}_e$ and a map $F_e: \mathscr{X}_e \to \mathscr{U}$, such that

$$\mathscr{X} \oplus \mathscr{S} = \mathscr{X}_e,$$

$$(A_e + B_e F_e)\mathscr{S} \subset \mathscr{S}, \tag{8.6}$$

and

$$\operatorname{Ker} F_e \cap \mathscr{X} \supset \operatorname{Ker} c'. \tag{8.7}$$

Under these conditions

$$(A_e + B_e F_e)|\mathscr{S} \simeq T. \tag{8.8}$$

For a systemic interpretation of the lemma, see Exercise 3.11.

PROOF. We shall construct

$$\mathscr{S} = \operatorname{Im} \begin{bmatrix} R \\ 1 \end{bmatrix} \tag{8.9}$$

for a suitable map $R: \mathscr{T} \to \mathscr{X}$. Now, by (8.5) there exists $b \in \mathscr{B}$, such that

$$l - A^{\tau}b = b_1 + Ab_2 + \cdots + A^{\tau-1}b_{\tau} \tag{8.10}$$

for suitable $b_i \in \mathscr{B}$ $(i \in \boldsymbol{\tau})$. Let $b = Bv$ $(v \in \mathscr{U})$ and set $F_e = [vc', G]$ for some $G: \mathscr{T} \to \mathscr{U}$, to be determined. Clearly, F_e satisfies (8.7), and (8.8) will be automatic if (8.6) is true. In view of (8.9), (8.6) and (8.8) are equivalent to

$$(A + bc')R + BG - lh' = RT. \tag{8.11}$$

To determine R subject to (8.11), we set

$$Rg = b \tag{8.12a}$$

and

$$\begin{aligned} R(T^i g) &= (A + bc')R(T^{i-1}g) + (BG - lh')T^{i-1}g \\ &= (A + bc')R(T^{i-1}g) + BGT^{i-1}g \end{aligned} \tag{8.12b}$$

for $i \in \boldsymbol{\tau - 1}$. It remains only to define G on the basis $\{g, Tg, \ldots, T^{\tau-1}g\}$ so as to ensure that

$$\begin{aligned} R(T^\tau g) &= (A + bc')R(T^{\tau-1}g) + (BG - lh')T^{\tau-1}g \\ &= (A + bc')R(T^{\tau-1}g) + BGT^{\tau-1}g - l \\ &\equiv R(\theta_1 1 + \theta_2 T + \cdots + \theta_\tau T^{\tau-1})g, \end{aligned} \tag{8.13}$$

$\theta_i \in \mathbb{R}$ being the coefficients of the m.p. of T. For this, use (8.10) and (8.12) to eliminate R and l from the identity (8.13). Then select the vectors Gg, $G(Tg), \ldots, G(T^{\tau-1}g)$ in turn, so that vectors formally in $A^{i-1}\mathscr{B}$ ($i \in \boldsymbol{\tau}$) are matched on both sides of the equation which results. □

Corollary 3.1. *Under the conditions of Lemma 3.8 and with b, G, R as in the proof, there follows*

$$\begin{bmatrix} 1 & R \\ 0 & 1 \end{bmatrix} \begin{bmatrix} A & lh' \\ gc' & T \end{bmatrix} = \begin{bmatrix} A + bc' & BG \\ gc' & S \end{bmatrix} \begin{bmatrix} 1 & R \\ 0 & 1 \end{bmatrix},$$

where $S := T - gc'R$.

By dualization of Corollary 3.1, we get

Corollary 3.2. *Let* (C, A, b) *be complete in* $\mathscr{X}$ *and* (h', T, g) *complete in* $\mathscr{T}$, *with* h', g *conjugate in the sense of* (8.4). *Let* $d(\mathscr{T}) = \tau \le n - 1$, $l' \in \mathscr{X}'$, *and*

$$\operatorname{Ker} l' \supset \bigcap_{i=1}^{\tau+1} \operatorname{Ker}(CA^{i-1}).$$

Then there exist maps $c' \in \operatorname{Im} C'$, $K: \mathscr{Y} \to \mathscr{T}$ *and* $S: \mathscr{T} \to \mathscr{T}$ *such that, in* $\mathscr{X} \oplus \mathscr{T}$,

$$\begin{bmatrix} A & bh' \\ gl' & T \end{bmatrix} \simeq \begin{bmatrix} A + bc' & bh' \\ KC & S \end{bmatrix}.$$

We can now deliver the *coup de grâce.*

Proof (of Theorem 3.5). As already noted, we may assume that A is cyclic. There is then $b \in \mathscr{B}$, such that (C, A, b) is complete. Choose $W_0: \mathscr{W} \to \mathscr{W}$ cyclic, then $g \in \mathscr{W}$ and $h' \in \mathscr{W}'$, such that (h', W_0, g) is complete, with

$$h'W_0^{i-1}g = 0, \qquad i \in \mathbf{v - 1}; \qquad h'W_0^{v-1}g = 1.$$

It is now simple to check that the pair

$$\begin{bmatrix} A & bh' \\ 0 & W_0 \end{bmatrix}, \quad \begin{bmatrix} 0 \\ g \end{bmatrix}$$

is controllable in $\mathscr{X} \oplus \mathscr{W}$, hence there exist $l' \in \mathscr{X}'$ and $m' \in \mathscr{W}'$, such that

$$\sigma\left(\begin{bmatrix} A & bh' \\ gl' & W_0 + gm' \end{bmatrix}\right) = \Lambda.$$

Since

$$\bigcap_{i=1}^{\nu+1} \operatorname{Ker}(CA^{i-1}) = 0$$

the conditions of Corollary 3.2 are satisfied (with the replacement of $\mathscr{T}$ by $\mathscr{W}$, τ by ν, and T by $W_0 + gm'$). It follows that

$$\begin{bmatrix} A & bh' \\ gl' & W_0 + gm' \end{bmatrix} \simeq \begin{bmatrix} A + bc' & bh' \\ KC & W \end{bmatrix} \tag{8.14}$$

for suitable c', K, W, and the proof is finished. □

From the proof it is clear that the role of Corollary 3.2, and thus of Lemma 3.8 is to provide an equivalence between the physically unrealizable composite system on the left side of (8.14), and the realizable system on the right. Here "realizability" is understood, of course, in the sense of respecting the processing constraint that only (y, w) can be directly measured.

A computational procedure for compensator design is summarized in Exercise 3.12.

3.9 Observer for a Single Linear Functional

As a second application of the definition of observability index, we show in this section how to construct a dynamic observer of order $\nu := \kappa_0 - 1$ which asymptotically evaluates a given functional $f'x$ on the state of the system

$$\dot{x} = Ax + v, \qquad y = Cx.$$

For this, assume (C, A) is observable, and introduce the observer equation

$$\dot{w} = Tw + Ry + Vv$$

and output functional $k'y + h'w$. The error with which this functional evaluates $f'x$ is then

$$e = f'x - k'y - h'w,$$

and it is enough to arrange that $e(t) \to 0$ as $t \to \infty$, with exponents in an assigned "good" subset $\mathbb{C}_g \subset \mathbb{C}^- := \{s: \Re e\ s < 0\}$.

Introduce the observer state space $\mathscr{W}$ independent of $\mathscr{X}$, with $d(\mathscr{W}) = \nu$. On $\mathscr{X} \oplus \mathscr{W}$ consider the map

$$A_e := \begin{bmatrix} A & 0 \\ RC & T \end{bmatrix}.$$

Choose $T\colon \mathscr{W} \to \mathscr{W}$ cyclic, with minimal polynomial

$$\theta(\lambda) := \lambda^{\nu} - (t_1 + t_2\lambda + \cdots + t_{\nu}\lambda^{\nu-1})$$

having all its roots in $\mathbb{C}_g$; and fix h' arbitrarily such that (h', T) is observable.

Assume temporarily that the external input $v(t) \equiv 0$. For the desired behavior of $e(\cdot)$, it clearly suffices to arrange that $\theta(d/dt)e(t) = 0$, i.e., to choose $R\colon \mathscr{Y} \to \mathscr{W}$ and $k' \in \mathscr{Y}'$, such that

$$[f' - k'C, \ -h']\theta(A_e) = 0. \tag{9.1}$$

To see that such a choice is possible, notice that

$$\theta(A_e) = \begin{bmatrix} \theta(A) & 0 \\ Q & 0 \end{bmatrix}, \tag{9.2}$$

where

$$Q := \sum_{i=1}^{\nu} \theta_i(T)RCA^{i-1} \tag{9.3}$$

and

$$\theta_i(\lambda) := \lambda^{\nu-i} - (t_{i+1} + t_{i+2}\lambda + \cdots + t_{\nu}\lambda^{\nu-1-i}). \tag{9.4}$$

By (9.4), the functionals

$$w_i' := h'\theta_i(T), \qquad i \in \mathbf{\nu}, \tag{9.5}$$

span $\mathscr{W}'$. Also, by definition of observability index, we have

$$f'\theta(A) = \sum_{i=1}^{\nu+1} e_i'CA^{i-1} \tag{9.6}$$

for suitable $e_i' \in \mathscr{Y}'$ $(i \in \mathbf{\nu + 1})$. Now (9.2)–(9.6) imply that (9.1) will follow if

$$k' = e_{\nu+1}' \tag{9.7a}$$

and

$$w_i'R = e_i' + t_i e_{\nu+1}', \qquad i \in \mathbf{\nu}. \tag{9.7b}$$

With k' and R uniquely determined by (9.7), it is now not difficult to verify that, for arbitrary $v(\cdot)$, $e(\cdot)$ will continue to satisfy the differential equation

$$\theta\left(\frac{d}{dt}\right)e(t) = 0,$$

provided $V\colon \mathscr{X} \to \mathscr{W}$ is defined by

$$h'T^{i-1}V = (f' - k'C)A^{i-1} - \sum_{j=1}^{i-1} h'T^{j-1}RCA^{i-1-j}, \qquad i \in \mathbf{\nu}.$$

While the problem of this section admits the straightforward computational solution just provided, it may be of interest to the reader to develop a geometric treatment in the style of Section 3.4. For this, one may introduce the subspace $\mathscr{V}' := \operatorname{Im} V' \subset \mathscr{X}'$, note that

$$(\mathscr{V}')^{\perp} \cap \operatorname{Ker} C \subset \operatorname{Ker} f',$$

and seek $K: \mathscr{Y} \to \mathscr{X}$, such that $(A - KC)'\mathscr{V}' \subset \mathscr{V}'$.

3.10 Preservation of Observability and Detectability

It is often useful to know that "desirable" properties like observability or detectability are preserved when the pair (C, A) is modified in various standard ways. As an obvious dual of Lemma 2.1, we have, for instance, that if (C, A) is observable (or detectable) then so is $(C, A + KC)$ for every $K: \mathscr{Y} \to \mathscr{X}$.

In the following we regard $\mathscr{X}$ as an inner product space over $\mathbb{R}$. If $M: \mathscr{X} \to \mathscr{X}$ and $M \geq 0$, $\sqrt{M}$ denotes the positive semidefinite square root. We have

Theorem 3.6

i. *If $C_1' C_1 = C_2' C_2$ and (C_1, A) is observable (resp. detectable) then (C_2, A) is observable (resp. detectable).*
ii. *If $M \geq 0$ and $(\sqrt{M}, A)$ is observable (resp. detectable), then for all $Q \geq 0$, $N > 0$ and all B, F, the pair $(\sqrt{M + Q + F'NF}, A + BF)$ is observable (resp. detectable).*

PROOF. Write

$$\mathscr{N}(C) := \bigcap_{i=1}^{n} \operatorname{Ker}(CA^{i-1}),$$

$$W(C) := \sum_{i=1}^{n} A'^{i-1} C' C A^{i-1}.$$

Clearly, $\mathscr{N}(C) = \operatorname{Ker} W(C)$ for every $C: \mathscr{X} \to \mathscr{Y}$. Then,

$$\mathscr{N}(C_1) = \operatorname{Ker} W(C_1) = \operatorname{Ker} W(C_2) = \mathscr{N}(C_2),$$

proving (i). For (ii), since

$$\begin{aligned}
\operatorname{Ker} \sqrt{M + Q + F'NF} &\subset \operatorname{Ker} \sqrt{M} \cap \operatorname{Ker} F \\
&\subset \operatorname{Ker}(K\sqrt{M}) \cap \operatorname{Ker}(BF) \\
&\subset \operatorname{Ker}(K\sqrt{M} - BF)
\end{aligned}$$

for all K, the equation

$$\tilde{A}(\tilde{K}) := (A + BF) + \tilde{K}\sqrt{M + Q + F'NF} = A + K\sqrt{M}$$

is solvable for $\tilde{K}$, given arbitrary K. It follows that $\tilde{A}(\tilde{K})$ can be assigned an arbitrary symmetric spectrum (resp. can be stabilized) by suitable choice of $\tilde{K}$, whenever $A + K\sqrt{M}$ has the same property relative to K. □

It is clear that the foregoing discussion is not changed if $\mathscr{X}^+$, $\mathscr{X}^-$ are replaced by $\mathscr{X}_b$, $\mathscr{X}_g$ as defined in Section 2.3. Thus, "the bad modes of the system (C, A) are observable" if

$$\mathscr{N}(C) \subset \mathscr{X}_g(A),$$

i.e. A is well-behaved on the unobservable subspace.

3.11 Exercises

3.1. Prove Lemma 3.1.

3.2. With $\mathscr{N}$ as in Section 3.2, let $\mathscr{X} = \mathscr{M} \oplus \mathscr{N}$, $x = x_1 + x_2$. Let $Q_1: \mathscr{X} \to \mathscr{M}$, $Q_2: \mathscr{X} \to \mathscr{N}$ be the natural projections, and write

$$A_1 = Q_1 A \,|\, \mathscr{M}, \qquad C_1 = C \,|\, \mathscr{M}$$

$$A_{21} = Q_2 A \,|\, \mathscr{M}, \qquad A_2 = Q_2 A \,|\, \mathscr{N}.$$

Show that (C_1, A_1) is observable and draw a signal flow graph for the system equations expressed in a basis adapted to $\mathscr{M} \oplus \mathscr{N}$. Indicate how a dynamic observer should be coupled to the system in order to yield an asymptotic identification of x_1.

3.3. Consider the dual maps $A': \mathscr{X}' \to \mathscr{X}'$ and $C': \mathscr{Y}' \to \mathscr{X}'$. The *observable subspace* of (C, A) is defined to be $\langle A' \,|\, \operatorname{Im} C' \rangle \subset \mathscr{X}'$. Prove that (C, A) is observable if and only if its observable subspace is all of $\mathscr{X}'$.

3.4. Consider the system

$$\dot{x} = Ax + v$$

$$y = Cx$$

$$z = Dx.$$

In the notation of Section 3.1, show that there exists a functional Ω, such that

$$\Omega[t, \omega_\alpha(t, \cdot)v(\cdot), \omega_\alpha(t, \cdot)y(\cdot)] = z(t)$$

for all $x(0) \in \mathscr{X}$ and $t \geq \alpha$, if and only if

$$\mathscr{N} \subset \operatorname{Ker} D,$$

where

$$\mathscr{N} = \bigcap_{i=1}^{n} \operatorname{Ker}(CA^{i-1}).$$

Thus, "$z(\cdot)$ is observable from $y(\cdot)$ and $v(\cdot)$ if and only if D annihilates the unobservable subspace."

3.5. Verify in detail the following synthesis of the matrices of a minimal-order observer. Assume (C, A) observable, with C: $p \times n$ and Rank $C = p$; A: $n \times n$; and $\Lambda \subset \mathbb{C}$ symmetric, with $|\Lambda| = n - p = r$. We want T: $r \times r$, V: $r \times n$, and K: $n \times p$, with the properties

$$\operatorname{Rank}\begin{bmatrix} C \\ V \end{bmatrix} = n, \qquad V(A - KC) = TV, \qquad \sigma(T) = \Lambda.$$

Step 1. Choose D: $r \times n$, such that

$$\operatorname{Rank}\begin{bmatrix} C \\ D \end{bmatrix} = n.$$

Step 2. Define

$$W = \begin{bmatrix} C \\ D \end{bmatrix}$$

and transform A and C according to

$$\tilde{A} = WAW^{-1} = \begin{bmatrix} \tilde{A}_{11}^{p \times p} & \tilde{A}_{12}^{p \times r} \\ \tilde{A}_{21}^{r \times p} & \tilde{A}_{22}^{r \times r} \end{bmatrix},$$

$$\tilde{C} = CW^{-1} = [1^{p \times p} \quad 0^{p \times r}].$$

The pair $(\tilde{A}_{12}, \tilde{A}_{22})$ is observable.

Step 3. Using a pole assignment procedure, compute $\tilde{K}_0$: $r \times p$, such that

$$\sigma(\tilde{A}_{22} - \tilde{K}_0 \tilde{A}_{12}) = \Lambda.$$

Step 4. Compute

$$\tilde{T} = \tilde{A}_{22} - \tilde{K}_0 \tilde{A}_{12},$$

$$\tilde{V} = [-K_0 \quad 1^{r \times r}],$$

$$\tilde{K} = \begin{bmatrix} \tilde{A}_{11} + \tilde{A}_{12}\tilde{K}_0 \\ \tilde{A}_{21} + \tilde{A}_{22}\tilde{K}_0 \end{bmatrix}.$$

Step 5. Compute $T = \tilde{T}$, $V = \tilde{V}W$, $VK = \tilde{V}\tilde{K}$.

Step 6. The observer is

$$\dot{z} = Tz + VKy + Vv$$

and

$$x(t) \doteqdot \begin{bmatrix} C \\ V \end{bmatrix}^{-1} \begin{bmatrix} y(t) \\ z(t) \end{bmatrix}, \qquad t \to \infty.$$

Also

$$\begin{bmatrix} C \\ V \end{bmatrix}^{-1} = W^{-1} \begin{bmatrix} 1^{p \times p} & 0 \\ \tilde{K}_0 & 1^{r \times r} \end{bmatrix}.$$

3.6. For the system triple (C, A, B) as in (1.1), let $\mathcal{N}$ be the unobservable subspace and $\mathcal{R}$ the controllable subspace. Let

$$\mathcal{X} = \mathcal{X}_1 \oplus \mathcal{X}_2 \oplus \mathcal{X}_3 \oplus \mathcal{X}_4,$$

where

$$\mathscr{X}_1 = \mathscr{N} \cap \mathscr{R}, \qquad \mathscr{X}_1 \oplus \mathscr{X}_2 = \mathscr{R}$$

$$\mathscr{X}_1 \oplus \mathscr{X}_3 = \mathscr{N}, \qquad (\mathscr{N} + \mathscr{R}) \oplus \mathscr{X}_4 = \mathscr{X}.$$

Write down the system equations using a basis for $\mathscr{X}$ adapted to the $\mathscr{X}_i$. Interpret each $\mathscr{X}_i$ in terms of controllability and observability of the corresponding subsystem: e.g. $\mathscr{X}_1$ is the "controllable but unobservable" component of the state space. Draw the signal flow graph of the composite system. Next, compute the transfer matrix $C(\lambda 1 - A)^{-1}B$. Verify that it is determined solely by the "controllable, observable" component of the system and that it is invariant under a change of basis in $\mathscr{X}$. The converse problem of obtaining a controllable, observable state description from a given transfer matrix is termed the "minimal realization problem."

3.7. For the system $\dot{x} = Ax + Bu$, $y = Cx$, assume that $u = Fw + Gy$, where $w(\cdot)$ is the state of a dynamic compensator of the form

$$\dot{w} = Tw + Hy.$$

Show that, for every choice of (F, G, H, T), the spectrum of the closed-loop system map of the composite system (with state $x \oplus w$) must include $\sigma(A_0) \cup \sigma(\bar{A})$, where $A_0 = A|\mathscr{N}$, $\bar{A}$ is the map induced by A on $\mathscr{X}/(\langle A|\mathscr{B}\rangle + \mathscr{N})$, and $\mathscr{N}$ is the unobservable subspace of (C, A). Briefly, "only the controllable, observable poles can be shifted by output feedback." From this result show that stabilization is possible by means of dynamic compensation if and only if (C, A) is detectable and (A, B) is stabilizable.

3.8. For the system $\dot{x} = Ax + Bu$, $y = Cx$, show that there exists $K: \mathscr{Y} \to \mathscr{U}$ such that $A + BKC$ is stable, if

$$\mathscr{X}^+(A) \subset \langle A|\mathscr{B}\rangle$$

and

$$\mathscr{X}^+(A) \cap \langle A|\text{Ker } C\rangle = 0.$$

Convenient necessary and sufficient conditions for the solvability of this problem are not known.

3.9. Show that the matrix pair (C, A) is observable if and only if, with $A: n \times n$,

$$\text{Rank}_{\mathbb{C}}\begin{bmatrix} A - \lambda 1 \\ C \end{bmatrix} = n$$

for all $\lambda \in \sigma(A)$. What is the corresponding criterion for detectability?

3.10. Show that (C, A) is observable if and only if, for all $\mathscr{T}$ and maps $T: \mathscr{T} \to \mathscr{T}$,

$$\text{Ker}(C \otimes 1_{\mathscr{T}'}) \cap \text{Ker}(A \otimes 1_{\mathscr{T}'} - 1_{\mathscr{X}} \otimes T') = 0.$$

What is the corresponding criterion for detectability? Hint: Interpret the given condition in terms of an A-invariant subspace contained in Ker C.

3.11. Interpret Lemma 3.8 as a statement about the existence of a "subsystem" within the composite system with signal flow graph shown in Fig. 3.2.

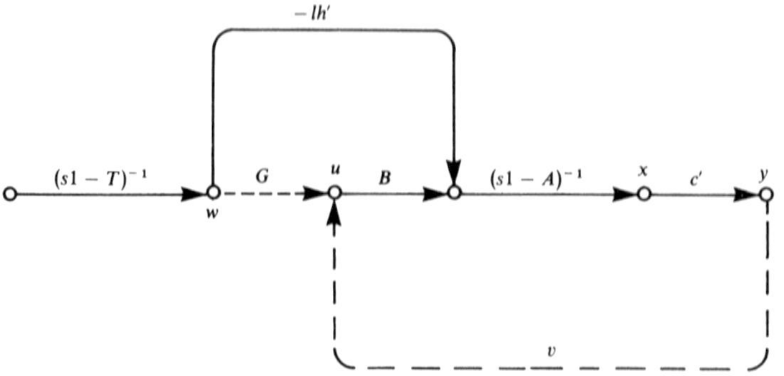

Figure 3.2 Composite System: Lemma 3.8.

3.12. Develop a computational procedure for synthesizing the compensator of Section 3.8. Hint:

Step 1. Pick H_0 "at random" to make $A_{\text{new}} := A + BH_0C$ cyclic, and set $A = A_{\text{new}}$.

Step 2. Pick $b = Bv$ "at random" to make (A, b) controllable.

Step 3. Pick

$$W_0 = \begin{bmatrix} 0 & 1 & 0 & \cdot & \cdot & \cdot & \cdot & 0 \\ 0 & 0 & 1 & \cdot & \cdot & \cdot & \cdot & 0 \\ \cdot & \cdot & \cdot & \cdot & \cdot & \cdot & \cdot & \cdot \\ 0 & \cdot & \cdot & \cdot & \cdot & \cdot & 0 & 1 \\ 0 & \cdot & \cdot & \cdot & \cdot & \cdot & \cdot & 0 \end{bmatrix}_{\nu \times \nu},$$

$$g = \begin{bmatrix} 0 \\ \vdots \\ 0 \\ 1 \end{bmatrix}_{\nu \times 1}, \qquad h' = [1 \quad 0 \quad \cdots \quad 0]_{1 \times \nu}$$

Step 4. By pole assignment compute l': $1 \times n$, m': $1 \times \nu$, so that

$$\begin{bmatrix} A & bh' \\ 0 & W_0 \end{bmatrix} + \begin{bmatrix} 0 \\ g \end{bmatrix} [l' \quad m']$$

has the desired spectrum Λ, $|\Lambda| = n + \nu$.

Step 5. Compute $\hat{e}'$, $\hat{e}_i'$: $1 \times p$ $(i \in \mathbf{v})$, not necessarily unique, such that

$$l' - \hat{e}'CA^\nu = \hat{e}_1'C + \hat{e}_2'CA + \cdots + \hat{e}_\nu'CA^{\nu-1}.$$

Step 6. Set $\tilde{A} := A + b\hat{e}'C$ and compute e', e_i': $1 \times p$ $(i \in \mathbf{v})$, such that

$$l - e'C\tilde{A}^\nu = e_1'C + e_2'C\tilde{A} + \cdots + e_\nu'C\tilde{A}^{\nu-1}.$$

Step 7. Set $T := W_0 + gm'$, $k_i' := h'T^{i-1}K$ $(i \in \mathbf{v})$ and compute k_i' from

$$k_{\nu-i+1}' = e_i' + \theta_i e' + \theta_{i+1}k_1' + \theta_{i+2}k_2' + \cdots + \theta_\nu k_{\nu-i}',$$

$(i = \nu, \nu - 1, \ldots, 1)$. [These equations result by eliminating $\mathbf{R}$ from the appropriately dualized version of (8.12)].

Step 8. Set $r_i' := h'T^{i-1}R$ $(i \in \mathbf{v})$ and compute r_i' from

$$r_1' = \hat{e}'C, \qquad r_{i+1}' = r_i'\tilde{A} + k_i'C, \qquad i \in \mathbf{v-1}.$$

Step 9. Compute

$$[K, R] = \begin{bmatrix} h' \\ h'T \\ \vdots \\ h'T^{\nu-1} \end{bmatrix}^{-1} \begin{bmatrix} k_1' & r_1' \\ \vdots & \vdots \\ k_\nu' & r_\nu' \end{bmatrix},$$

$W = T - Rbh'$, $G = vh'$, $H = v\hat{e}'$.

Step 10. As a numerical example, let

$$A = \begin{bmatrix} 0 & 0 & 0 & 0 & 0 \\ 1 & 0 & 0 & 0 & 0 \\ 0 & 1 & 0 & 0 & 0 \\ 0 & 0 & 0 & 0 & 0 \\ 0 & 0 & 0 & 1 & 0 \end{bmatrix}, \quad B = \begin{bmatrix} 1 & 0 \\ 0 & 0 \\ 0 & 0 \\ 0 & 1 \\ 0 & 0 \end{bmatrix}, \quad C = \begin{bmatrix} 0 & 1 & 0 & 0 & 0 \\ 0 & 0 & 1 & 0 & 0 \\ 0 & 0 & 0 & 0 & 1 \end{bmatrix}$$

Then, $\nu = 1$ and one can take

$$H_0 = \begin{bmatrix} 0 & 0 & 0 \\ 0 & 1 & 0 \end{bmatrix}, \quad b = \begin{bmatrix} 1 \\ 0 \\ 0 \\ 0 \\ 0 \end{bmatrix}, \quad W_0 = 0, h' = 1, g = 1.$$

Setting double poles at $s = -1, -1 \pm i$, results finally in

$$K = [74 \quad 28 \quad 92], \qquad W = -6,$$

$$G = \begin{bmatrix} 1 \\ 0 \end{bmatrix}, \qquad H = \begin{bmatrix} -17 & 0 & -16 \\ 0 & 1 & 0 \end{bmatrix}$$

3.12 Notes and References

The definition and fundamental properties of an observable system are due to Kalman [1]; see also Kalman, Ho and Narendra [1], where the decomposition given in Exercise 3.6 was introduced. The minimal realization problem is discussed by Kalman, Falb and Arbib [1]. Minimal order dynamic observers were first described by Luenberger [1], [2]; the treatment here and Lemma 3.5 follow Wonham [5]. The synthesis procedure of Exercise 3.5 mimics the algebraic development; see also Newmann [1]. The definition of detectability and the results of Sections 3.6 and 3.10 are due to Wonham [2].

The main result in Section 3.8 (Theorem 3.5) is due to Brasch and Pearson [1], while the neat proof of Lemma 3.6 (and thus Lemma 3.7) was suggested by Jan C. Willems [private communication]. Further details on compensator design (to supplement Exercise 3.12) together with numerical examples are given in a thesis by Van Den Kieboom [1]. See Shaw [1] for a caveat on the sensitivity of pole-shifting compensators with respect to small parameter changes. A "geometric" treatment of the problem of Section 3.9 is given by Wonham and Morse [2]; for an alternative, computational approach, see e.g. Murdoch [1], and for further discussion of low-order observers, Fortmann and Williamson [1], and Kimura [1].

The problem of stabilization by static output feedback addressed in Exercise 3.8 is evidently rather difficult: the first condition given is clearly necessary, but the second (sufficient) condition is quite strong. For a more refined discussion, see e.g. Denham [1], Kimura [2], [3] and, for an approach via algebraic geometry and decision theory, Anderson and Scott [1]. The observability criterion of Exercise 3.9 is due to Hautus [1].

4 Disturbance Decoupling and Output Stabilization

In this chapter we first discuss a simple feedback synthesis problem, concerned with decoupling from the system output the effect of disturbances acting at the input. Examination of this problem leads naturally to the fundamental geometric concept of (A, B)-invariant subspace, which underlies many of our constructions and results in later chapters. As an immediate application, we show how state feedback may be utilized to stabilize outputs or, more generally, to realize a given set of characteristic exponents in the time response of the output.

4.1 Disturbance Decoupling Problem (DDP)

Consider the system

$$\dot{x}(t) = Ax(t) + Bu(t) + Sq(t), \qquad t \geq 0, \tag{1.1}$$

$$z(t) = Dx(t), \qquad t \geq 0. \tag{1.2}$$

The new term $q(t)$ in (1.1) represents a disturbance which is assumed not to be directly measurable by the controller. Our problem is to find (if possible) state feedback F, such that $q(\cdot)$ has no influence on the controlled output $z(\cdot)$. Let us assume that $q(\cdot)$ belongs to a fairly rich function class $\mathbf{Q}$, the choice reflecting in some measure our ignorance of the specific features of the disturbances to be encountered. This choice is not crucial: we adopt the continuous $\mathbb{R}^\nu$-valued functions on $[0, \infty)$, set $\mathcal{Q} := \mathbb{R}^\nu$, and assume that $S: \mathcal{Q} \to \mathcal{X}$ is a time-invariant map.

Suppose next that state feedback F has been incorporated so that, in (1.1), $u(t) = Fx(t)$. We say that the system (1.1), (1.2) is *disturbance decoupled* relative to the pair $q(\cdot)$, $z(\cdot)$ if, for each initial state $x(0) \in \mathscr{X}$, the output $z(t)$, $t \geq 0$, is the same for every $q(\cdot) \in \mathbf{Q}$. Thus, disturbance decoupling simply means that the forced response

$$z(t) = D \int_0^t e^{(t-s)(A+BF)} Sq(s)\, ds = 0 \tag{1.3}$$

for all $q(\cdot) \in \mathbf{Q}$ and $t \geq 0$.

Write $\mathscr{K} := \operatorname{Ker} D$ and $\mathscr{S} := \operatorname{Im} S$. The following result is easy to see from (1.3).

Lemma 4.1. *The system* (1.1), (1.2) *is disturbance decoupled if and only if* $\langle A + BF \,|\, \mathscr{S} \rangle \subset \mathscr{K}$.

Thus, in algebraic terms the realization of disturbance decoupling by state feedback amounts to the following.

Disturbance Decoupling Problem (DDP). *Given* $A: \mathscr{X} \to \mathscr{X}$, $B: \mathscr{U} \to \mathscr{X}$, $\mathscr{S} \subset \mathscr{X}$, *and* $\mathscr{K} \subset \mathscr{X}$, *find* (*if possible*) $F: \mathscr{X} \to \mathscr{U}$, *such that*

$$\langle A + BF \,|\, \mathscr{S} \rangle \subset \mathscr{K}. \tag{1.4}$$

Observe that the subspace on the left in (1.4) is $(A + BF)$-invariant and, if (1.4) is true, belongs to $\mathscr{K}$. Intuitively, DDP will be solvable if and only if the "largest" subspace having these properties contains $\mathscr{S}$. To make this remark precise we must introduce a new concept.

4.2 (A, B)-Invariant Subspaces

Let $A: \mathscr{X} \to \mathscr{X}$ and $B: \mathscr{U} \to \mathscr{X}$. We say that a subspace $\mathscr{V} \subset \mathscr{X}$ is (A, B)-*invariant* if there exists a map $F: \mathscr{X} \to \mathscr{U}$ such that

$$(A + BF)\mathscr{V} \subset \mathscr{V}. \tag{2.1}$$

We denote the class of (A, B)-invariant subspaces of $\mathscr{X}$ by $\mathfrak{I}(A, B; \mathscr{X})$, or simply $\mathfrak{I}(\mathscr{X})$ when A and B are fixed in the discussion.

Observe that any A-invariant subspace is automatically (A, B)-invariant: just put $F = 0$. Now suppose $\mathscr{V}$ is (A, B)-invariant and that state feedback F is chosen to satisfy (2.1). The (disturbance-free) closed-loop system

$$\dot{x} = Ax + Bu, \qquad u = Fx,$$

then has the property that if $x(0) = x_0 \in \mathscr{V}$,

$$x(t) = e^{t(A+BF)} x_0 \in \mathscr{V}$$

for all t: namely $\mathscr{V}$ is invariant under the motion. Thus $\mathscr{V}$ has the property that if $x(0) \in \mathscr{V}$ then there exists a control $u(t)$ $(t \geq 0)$ such that $x(t) \in \mathscr{V}$ for all $t \geq 0$. In other words, the state $x(\cdot)$ can be held in $\mathscr{V}$ by suitable choice of $u(\cdot)$.

That the property just stated actually characterizes (A, B)-invariance is a consequence of the following result, which provides an explicit test for determining whether a given subspace is (A, B)-invariant or not.

Lemma 4.2. *Let $\mathscr{V} \subset \mathscr{X}$ and write $\mathscr{B} = \operatorname{Im} B$. Then $\mathscr{V} \in \mathscr{I}(A, B; \mathscr{X})$ if and only if*

$$A\mathscr{V} \subset \mathscr{V} + \mathscr{B}. \tag{2.2}$$

PROOF. Suppose $\mathscr{V} \in \mathscr{I}(A, B; \mathscr{X})$ and let $v \in \mathscr{V}$. By (2.1), $(A + BF)v = w$ for some $w \in \mathscr{V}$, or

$$Av = w - BFv \in \mathscr{V} + \mathscr{B}.$$

Conversely, suppose (2.2) is true and let $\{v_1, \ldots, v_\mu\}$ be a basis for $\mathscr{V}$. By (2.2) there exist $w_i \in \mathscr{V}$ and $u_i \in \mathscr{U}$ $(i \in \mathbf{\mu})$ such that

$$Av_i = w_i - Bu_i, \qquad i \in \mathbf{\mu}.$$

Define $F_0 : \mathscr{V} \to \mathscr{U}$ by

$$F_0 v_i = u_i, \qquad i \in \mathbf{\mu},$$

and let F be any extension of F_0 to $\mathscr{X}$. Then $(A + BF)v_i = w_i \in \mathscr{V}$, i.e. $(A + BF)\mathscr{V} \subset \mathscr{V}$, so that $\mathscr{V} \in \mathscr{I}(A, B; \mathscr{X})$. ☐

If $\mathscr{V} \in \mathfrak{I}(A, B; \mathscr{X})$ we write $\mathbf{F}(A, B; \mathscr{V})$, or simply $\mathbf{F}(\mathscr{V})$, for the class of maps $F: \mathscr{X} \to \mathscr{U}$, such that $(A + BF)\mathscr{V} \subset \mathscr{V}$. The notation $F \in \mathbf{F}(\mathscr{V})$ is read, "F is a friend of $\mathscr{V}$." From the proof of Lemma 4.2 we see that if $F \in \mathbf{F}(\mathscr{V})$ then $\tilde{F} \in \mathbf{F}(\mathscr{V})$ if and only if $(\tilde{F} - F)\mathscr{V} \subset B^{-1}\mathscr{V}$; in particular, $(\tilde{F} - F)|\mathscr{V} = 0$ if B is monic and $\mathscr{B} \cap \mathscr{V} = 0$.

The following observation is not required for the solution of DDP, but will find application later.

Proposition 4.1. *Let (A, B) be controllable and let $\mathscr{V} \in \mathfrak{I}(A, B; \mathscr{X})$, with $d(\mathscr{V}) = v$. If $F_0 \in \mathbf{F}(A, B; \mathscr{V})$ and $\bar{\Lambda}$ is a symmetric set of $n - v$ complex numbers, there exists $F: \mathscr{X} \to \mathscr{U}$, such that*

$$F|\mathscr{V} = F_0|\mathscr{V} \tag{2.3}$$

and

$$\sigma(A + BF) = \sigma[(A + BF)|\mathscr{V}] \cup \bar{\Lambda}.$$

PROOF. Let $P: \mathscr{X} \to \mathscr{X}/\mathscr{V}$ be the canonical projection, write

$$\bar{A}_0 = \overline{A + BF_0}$$

for the map induced by $A + BF_0$ in $\mathscr{X}/\mathscr{V}$, and let $\bar{B} := PB$. By Proposition 1.2 the pair $(\bar{A}_0, \bar{B})$ is controllable, hence by Theorem 2.1 there exists

$\bar{F}_1 : \mathcal{X}/\mathcal{V} \to \mathcal{U}$, such that

$$\sigma(\bar{A}_0 + \bar{B}\bar{F}_1) = \bar{\Lambda}.$$

Define

$$F := F_0 + \bar{F}_1 P.$$

Clearly, (2.3) holds, so that $F \in \mathbf{F}(\mathcal{V})$. If $\bar{A}_F$ is the map induced in $\mathcal{X}/\mathcal{V}$ by $A + BF$, we have that $\bar{A}_F$ is defined uniquely by the relation $\bar{A}_F P = P(A + BF)$. But

$$(\bar{A}_0 + \bar{B}\bar{F}_1)P = P(A + BF_0) + PB\bar{F}_1 P = P(A + BF)$$

and therefore, $\bar{A}_F = \bar{A}_0 + \bar{B}\bar{F}_1$. There follows

$$\sigma(A + BF) = \sigma[(A + BF)\,|\,\mathcal{V}] \cup \sigma(\bar{A}_F)$$
$$= \sigma[(A + BF)\,|\,\mathcal{V}] \cup \bar{\Lambda}$$

as claimed.   □

The following closure property will be crucial.

Lemma 4.3. *The class of subspaces* $\mathfrak{I}(A, B; \mathcal{X})$ *is closed under the operation of subspace addition.*

PROOF. From (2.2) it is clear that if $\mathcal{V}_1, \mathcal{V}_2 \in \mathfrak{I}(\mathcal{X})$, then

$$A(\mathcal{V}_1 + \mathcal{V}_2) = A\mathcal{V}_1 + A\mathcal{V}_2$$
$$\subset \mathcal{V}_1 + \mathcal{V}_2 + \mathcal{B},$$

hence, $\mathcal{V}_1 + \mathcal{V}_2 \in \mathfrak{I}(\mathcal{X})$.   □

Lemma 4.3 can be phrased more technically by saying that $\mathfrak{I}(A, B; \mathcal{X})$ is an upper semilattice relative to subspace inclusion and addition. But it is not true, in general, that the property of (A, B)-invariance is preserved by subspace intersection, and therefore, $\mathfrak{I}(A, B; \mathcal{X})$ is not a sublattice of the lattice of all subspaces of $\mathcal{X}$.

If $\mathfrak{B}$ is a family of subspaces of $\mathcal{X}$, we define the *largest* or *supremal* element $\mathcal{V}^*$ of $\mathfrak{B}$ to be that member of $\mathfrak{B}$ (when it exists) which contains every member of $\mathfrak{B}$. Thus, $\mathcal{V}^* \in \mathfrak{B}$, and if $\mathcal{V} \in \mathfrak{B}$ then $\mathcal{V} \subset \mathcal{V}^*$. It is clear that $\mathcal{V}^*$ is unique. We write

$$\mathcal{V}^* = \sup\{\mathcal{V} : \mathcal{V} \in \mathfrak{B}\},$$

or simply

$$\mathcal{V}^* = \sup \mathfrak{B}.$$

Lemma 4.4. *Let* $\mathfrak{B}$ *be a nonempty class of subspaces of* $\mathcal{X}$, *closed under addition. Then* $\mathfrak{B}$ *contains a supremal element* $\mathcal{V}^*$.

PROOF. Since $\mathscr{X}$ is finite-dimensional there is an element $\mathscr{V}^* \in \mathfrak{V}$ of greatest dimension. If $\mathscr{V} \in \mathfrak{V}$ we have that $\mathscr{V} + \mathscr{V}^* \in \mathfrak{V}$ and so $d(\mathscr{V}^*) \geq d(\mathscr{V} + \mathscr{V}^*) \geq d(\mathscr{V}^*)$; that is, $\mathscr{V}^* = \mathscr{V} + \mathscr{V}^*$, hence, $\mathscr{V}^* \supset \mathscr{V}$ and so $\mathscr{V}^*$ is supremal. □

Now let $\mathscr{K} \subset \mathscr{X}$ be arbitrary, and let $\mathfrak{I}(A, B; \mathscr{K})$ denote the subclass of (A, B)-invariant subspaces contained in $\mathscr{K}$:

$$\mathfrak{I}(A, B; \mathscr{K}) := \{\mathscr{V} : \mathscr{V} \in \mathfrak{I}(A, B; \mathscr{X}) \ \& \ \mathscr{V} \subset \mathscr{K}\}.$$

With A and B fixed, we write simply $\mathfrak{I}(\mathscr{K}) := \mathfrak{I}(A, B; \mathscr{K})$. Now trivially, $0 \in \mathfrak{I}(\mathscr{K})$, so $\mathfrak{I}(\mathscr{K}) \neq \emptyset$. Since $\mathscr{K}$ is a subspace, Lemma 4.3 implies that $\mathfrak{I}(\mathscr{K})$ is closed under addition. Then, Lemma 4.4 guarantees the existence of the supremal element

$$\mathscr{V}^* := \sup \mathfrak{I}(\mathscr{K}).$$

This simple but fundamental result is important enough to state formally.

Theorem 4.1. *Let $A: \mathscr{X} \to \mathscr{X}$ and $B: \mathscr{U} \to \mathscr{X}$. Every subspace $\mathscr{K} \subset \mathscr{X}$ contains a unique supremal (A, B)-invariant subspace [written* $\sup \mathfrak{I}(A, B; \mathscr{K})$, *or simply* $\sup \mathfrak{I}(\mathscr{K})$ *when A, B are understood from context].*

In the case where $\mathscr{K} = \operatorname{Ker} D$ and $z = Dx$, a choice of feedback control $F \in \mathbf{F}(\mathscr{V}^*)$, where $\mathscr{V}^* = \sup \mathfrak{I}(\operatorname{Ker} D)$, amounts to rendering the system maximally unobservable from z.

In order to give Theorem 4.1 a systemic application we now return to the problem of disturbance decoupling introduced in Section 4.1.

4.3 Solution of DDP

From the preceding considerations there follows immediately

Theorem 4.2. *DDP is solvable if and only if*

$$\mathscr{V}^* \supset \mathscr{S}, \tag{3.1}$$

where

$$\mathscr{V}^* := \sup \mathfrak{I}(A, B; \mathscr{K}).$$

PROOF. (If) Choose, by Lemma 4.2, $F \in \mathbf{F}(\mathscr{V}^*)$, i.e. $(A + BF)\mathscr{V}^* \subset \mathscr{V}^*$. Using (3.1), we have

$$\langle A + BF \,|\, \mathscr{S} \rangle \subset \langle A + BF \,|\, \mathscr{V}^* \rangle = \mathscr{V}^* \subset \mathscr{K}.$$

(Only if) If F solves DDP, the subspace

$$\mathscr{V} := \langle A + BF \,|\, \mathscr{S} \rangle$$

clearly belongs to $\mathfrak{I}(\mathscr{K})$, and therefore,

$$\mathscr{V}^* \supset \mathscr{V} \supset \mathscr{S}. \qquad \square$$

So far our approach to DDP has been somewhat abstract. To conclude this section we give an algorithm by which $\mathscr{V}^*$ can be computed efficiently in a finite number of steps. With $\mathscr{V}^*$ so determined, checking the condition (3.1) of Theorem 4.2 becomes trivial. If that condition is satisfied, any $F \in \mathbf{F}(\mathscr{V}^*)$ provides a solution to DDP, and such F is easy to construct, as in the proof of Lemma 4.2.

For the computation of $\mathscr{V}^*$ we have the following.

Theorem 4.3. *Let* $A: \mathscr{X} \to \mathscr{X}$, $B: \mathscr{U} \to \mathscr{X}$, *and* $\mathscr{K} \subset \mathscr{X}$. *Define the sequence* $\mathscr{V}^\mu$ *according to*

$$\mathscr{V}^0 = \mathscr{K}$$

$$\mathscr{V}^\mu = \mathscr{K} \cap A^{-1}(\mathscr{B} + \mathscr{V}^{\mu-1}); \qquad \mu = 1, 2, \ldots.$$

Then $\mathscr{V}^\mu \subset \mathscr{V}^{\mu-1}$, *and for some* $k \le d(\mathscr{K})$,

$$\mathscr{V}^k = \sup \mathfrak{I}(A, B; \mathscr{K}).$$

Proof. Recall the properties of the function A^{-1} (Section 0.4). We first observe that $\mathscr{V}^\mu\!\downarrow$, i.e. the sequence $\mathscr{V}^\mu$ is nonincreasing: clearly $\mathscr{V}^1 \subset \mathscr{V}^0$, and if $\mathscr{V}^\mu \subset \mathscr{V}^{\mu-1}$, then

$$\begin{aligned} \mathscr{V}^{\mu+1} &= \mathscr{K} \cap A^{-1}(\mathscr{B} + \mathscr{V}^\mu) \\ &\subset \mathscr{K} \cap A^{-1}(\mathscr{B} + \mathscr{V}^{\mu-1}) \\ &= \mathscr{V}^\mu. \end{aligned}$$

Thus, for some $k \le d(\mathscr{K})$, $\mathscr{V}^\mu = \mathscr{V}^k$ $(\mu \ge k)$. Now, $\mathscr{V} \in \mathfrak{I}(\mathscr{K})$ if and only if

$$\mathscr{V} \subset \mathscr{K}, \qquad \mathscr{V} \subset A^{-1}(\mathscr{V} + \mathscr{B}). \tag{3.2}$$

From (3.2), $\mathscr{V} \subset \mathscr{V}^0$, and if $\mathscr{V} \subset \mathscr{V}^{\mu-1}$,

$$\begin{aligned} \mathscr{V} &\subset \mathscr{K} \cap A^{-1}(\mathscr{V} + \mathscr{B}) \\ &\subset \mathscr{K} \cap A^{-1}(\mathscr{V}^{\mu-1} + \mathscr{B}) \\ &= \mathscr{V}^\mu. \end{aligned}$$

Therefore, $\mathscr{V} \subset \mathscr{V}^k \in \mathfrak{I}(\mathscr{K})$, and as $\mathscr{V}$ was arbitrary the result follows. $\square$

Theorems 4.2 and 4.3 furnish a constructive solution to the disturbance decoupling problem. Numerical examples are provided in Exercises 4.2 and 4.8. The principle of solution can be summarized as follows. If $\mathscr{V} \in \mathfrak{I}(\mathscr{X})$ and $q(\cdot) = 0$ then by suitable choice of $u(\cdot)$ the system state $x(\cdot)$ can always be held in $\mathscr{V}$ if it starts there. If $q(\cdot) \ne 0$ but $\operatorname{Im} S \subset \mathscr{V}$ then the contribution to $\dot{x}(t)$ by the disturbance $Sq(t)$ (i.e. the first-order effect of $q(\cdot)$ on $x(\cdot)$) is also localized to $\mathscr{V}$. Under these conditions the integrated contribution to $x(\cdot)$

by $q(\cdot)$ can be controlled to remain in $\mathscr{V}$. This contribution is unobservable at z just when $\mathscr{V} \subset \operatorname{Ker} D$, and so it is enough to work with $\mathscr{V}^*$. Of course, in actual operation with a control of form $u = Fx + Gv$, where $F \in \mathbf{F}(\mathscr{V}^*)$ and $v(\cdot)$ is a new external input, the system state $x(\cdot)$ will generally not remain in $\mathscr{V}^*$; however, linearity ensures that the contribution to $x(\cdot)$ from $q(\cdot)$ is held in $\mathscr{V}^*$, which is all that is required to decouple $q(\cdot)$ from $z(\cdot)$.

This decoupling action is achieved by signal cancellation around the feedback loop. It is possible only if $\operatorname{Im} S \subset \operatorname{Ker} D$, i.e. $DS = 0$. The latter condition means that an impulsive disturbance $q(\cdot)$ is not transmitted as an instantaneous step change in $z(\cdot)$. Thus, DDP is solvable only if $z(\cdot)$ is effectively separated from $q(\cdot)$ in the signal flow by at least two stages of integration. Now, if the necessary condition $DS = 0$ is satisfied, then DDP is certainly solvable provided $\mathscr{V}^* = \operatorname{Ker} D$, namely

$$A \operatorname{Ker} D \subset \operatorname{Ker} D + \mathscr{B}. \tag{3.3}$$

It is easy to see (Exercise 4.9) that (3.3) is true generically in the space of data points (A, B), if and only if

$$d(\mathscr{B}) + d(\operatorname{Ker} D) \geq d(\mathscr{X}). \tag{3.4}$$

Let $d(\mathscr{X}) = n$, $d(\mathscr{B}) = m$, $d(\mathscr{Z}) = q$ and D be epic. Then, (3.4) means that $m \geq q$, or the number of controls is at least as great as the number of independent outputs to be controlled, a condition that is obviously reasonable on intuitive grounds alone.

Thus, DDP is almost always solvable if $DS = 0$ and $m \geq q$. However, there is no guarantee that state feedback F can be chosen to satisfy additional reasonable requirements, for example, that $A + BF$ be stable. A more realistic version of DDP that includes a stability requirement will be solved by the use of controllability subspaces in Chapter 5. Meanwhile, the reader is invited to generalize DDP to include the possibility of direct control feedthrough at the output (Exercise 4.7), or of disturbance feedforward (Exercise 4.10).

We turn now to another simple but interesting application of the concept of supremal (A, B)-invariant subspace, that will play a fundamental role in the more realistic but difficult problems of later chapters.

4.4 Output Stabilization Problem (OSP)

Consider the system

$$\dot{x} = Ax + Bu, \qquad t \geq 0, \tag{4.1a}$$

$$z = Dx, \qquad t \geq 0. \tag{4.1b}$$

We pose the problem of stabilizing the output $z(\cdot)$ by means of state feedback: precisely, in terms of the triple (D, A, B) find conditions for the existence of state feedback F, such that

$$De^{t(A+BF)} \to 0, \qquad t \to \infty. \tag{4.2}$$

More generally, we may seek F such that the characteristic exponents of the time function of (4.2) belong to a "good" subset $\mathbb{C}_g \subset \mathbb{C}$. Our problem is thus to generalize the condition of Theorem 2.2.

We begin by translating the systems problem into purely algebraic terms. For arbitrary $F: \mathscr{X} \to \mathscr{U}$, write

$$\mathscr{N}_F := \bigcap_{i=1}^{n} \operatorname{Ker}[D(A+BF)^{i-1}].$$

Since $\mathscr{N}_F$ is the unobservable subspace of the pair $(D, A+BF)$, it is almost obvious that the exponents which appear in $De^{t(A+BF)}$ are simply the eigenvalues of the map $\overline{A+BF}$ induced by $A+BF$ in $\bar{\mathscr{X}} := \mathscr{X}/\mathscr{N}_F$. To justify this remark, recall from Lemma 3.2 that $(\bar{D}, \overline{A+BF})$ is observable. Introduce the complexifications $\mathscr{X}_{\mathbb{C}}$ and $\mathscr{Z}_{\mathbb{C}}$ of $\mathscr{X}$ and $\mathscr{Z}$, and suppose $\lambda \in \sigma(\overline{A+BF})$. Then, $\overline{A+BF}\bar{x}_0 = \lambda\bar{x}_0$ for some $\bar{x}_0 \in \bar{\mathscr{X}}_{\mathbb{C}}$, $\bar{x}_0 \neq 0$. With $u = Fx$ in (4.1a), we have

$$\dot{\bar{x}}(t) = \overline{A+BF}\bar{x}(t), \qquad t \geq 0;$$

so, if $\bar{x}(0) = \bar{x}_0$, there results

$$z(t) = \bar{D}e^{\lambda t}\bar{x}_0, \qquad t \geq 0.$$

It is clear that $z(\cdot)$ is identically zero only if there is a nontrivial $\overline{A+BF}$-invariant subspace in $\operatorname{Ker} \bar{D}$, namely $\operatorname{Span}\{\bar{x}_0\}$; and observability rules this out. So, the exponents which appear in $De^{t(A+BF)}$ all belong to $\mathbb{C}_g$ if and only if

$$\sigma(\overline{A+BF}) \subset \mathbb{C}_g, \tag{4.3}$$

as claimed.

Now revert to the usual setting with field $\mathbb{R}$. To express (4.3) in a more geometric form we shall need the polynomials α_g, α_b and modal subspaces $\mathscr{X}_g(A) = \operatorname{Ker} \alpha_g(A)$ etc. introduced in Section 2.3. Then we have

Lemma 4.5. *Let $\mathscr{S} \subset \mathscr{X}$, $A\mathscr{S} \subset \mathscr{S}$ and $\bar{\mathscr{X}} = \mathscr{X}/\mathscr{S}$. Let $P: \mathscr{X} \to \bar{\mathscr{X}}$ be the canonical projection, and $\bar{A}$ the map induced by A in $\bar{\mathscr{X}}$. Then, $\sigma(\bar{A}) \subset \mathbb{C}_g$ if and only if $\mathscr{X}_b(A) \subset \mathscr{S}$.*

Proof. (If) Let $\bar{x} \in \bar{\mathscr{X}}$. Then,

$$\begin{aligned} \alpha_g(\bar{A})\bar{x} &= \alpha_g(\bar{A})Px \\ &= P\alpha_g(A)x \in P \operatorname{Ker} \alpha_b(A) \\ &= P\mathscr{X}_b(A) \subset P\mathscr{S} = 0. \end{aligned}$$

Thus, the m.p. of $\bar{A}$ divides α_g; that is, $\sigma(\bar{A}) \subset \mathbb{C}_g$.

(Only if) Let $x \in \mathscr{X}_b(A) = \operatorname{Ker} \alpha_b(A)$. For suitable polynomials μ, ν we have

$$\begin{aligned} x &= \mu(A)\alpha_g(A)x + \nu(A)\alpha_b(A)x \\ &= \mu(A)\alpha_g(A)x \end{aligned}$$

and so $\bar{x} = Px = \mu(\bar{A})\alpha_g(\bar{A})\bar{x}$. But $\sigma(\bar{A}) \subset \mathbb{C}_g$ implies that the m.p. $\bar{\alpha}$ of $\bar{A}$ is a divisor of α_g. Thus, $\alpha_g(A)\bar{x} = 0$, so that $Px = 0$, i.e. $x \in \mathscr{S}$. □

Applying Lemma 4.5 to $A + BF$ and with $\mathscr{N}_F$ in place of $\mathscr{S}$, we see that (4.3) is true if and only if

$$\mathscr{X}_b(A + BF) \subset \mathscr{N}_F, \tag{4.4}$$

namely, "the bad modes of $A + BF$ are unobservable at the output z." Notice finally that $\mathscr{X}_b(A + BF)$ is an $(A + BF)$-invariant subspace, while $\mathscr{N}_F$ is the largest $(A + BF)$-invariant subspace of Ker D. Therefore, (4.4) can hold if and only if $\mathscr{X}_b(A + BF) \subset \operatorname{Ker} D$. On this basis we can state our original, generalized problem as follows:

Given the maps $A: \mathscr{X} \to \mathscr{X}$, $B: \mathscr{U} \to \mathscr{X}$, and $D: \mathscr{X} \to \mathscr{Z}$, together with a symmetric partition $\mathbb{C} = \mathbb{C}_g \cup \mathbb{C}_b$, find $F: \mathscr{X} \to \mathscr{U}$ such that

$$\mathscr{X}_b(A + BF) \subset \operatorname{Ker} D. \tag{4.5}$$

We shall refer to the foregoing as the **Output Stabilization Problem (OSP).** Here, "stabilization" is to be understood in the general sense indicated.

Theorem 4.4. *OSP is solvable if and only if*

$$\mathscr{X}_b(A) \subset \langle A \mid \mathscr{B} \rangle + \mathscr{V}^*, \tag{4.6}$$

where

$$\mathscr{V}^* := \sup \mathfrak{J}(A, B; \operatorname{Ker} D).$$

Intuitively, (4.6) states that "each bad mode of A is either controllable (hence can be pole-shifted), or can be made unobservable at the output." In view of Theorem 4.3 the condition is entirely constructive.

For the proof we shall need

Lemma 4.6. *Let $\mathscr{V}$ be any subspace such that $A\mathscr{V} \subset \mathscr{V}$; write $\bar{\mathscr{X}} = \mathscr{X}/\mathscr{V}$; let $P: \mathscr{X} \to \mathscr{X}/\mathscr{V}$ be the canonical projection; write $\bar{A}$ for the map induced in $\bar{\mathscr{X}}$; and relative to $\bar{A}$ define $\bar{\alpha}_b$, $\bar{\mathscr{X}}_b(A)$ etc. as in Section 2.3. Then*

$$\bar{\mathscr{X}}_b(\bar{A}) = P\mathscr{X}_b(A).$$

Proof. Let $Px \in \bar{\mathscr{X}}_b(\bar{A})$. Since $\bar{\mathscr{X}}_b(\bar{A}) = \operatorname{Ker} \bar{\alpha}_b(\bar{A})$, we have

$$P\bar{\alpha}_b(A)x = \bar{\alpha}_b(\bar{A})Px = 0;$$

so $\bar{\alpha}_b(A)x \in \mathscr{V}$. As $\bar{\alpha}_b(\lambda) \mid \alpha_b(\lambda)$ there follows $\alpha_b(A)x \in \mathscr{V}$. Now

$$x = \rho(A)\alpha_g(A)x + \sigma(A)\alpha_b(A)x$$

for suitable $\rho, \sigma \in \mathbb{R}[\lambda]$. Since

$$\rho(A)\alpha_g(A)x \in \operatorname{Ker} \alpha_b(A), \qquad \sigma(A)\alpha_b(A)x \in \mathscr{V},$$

we have

$$Px \in P \operatorname{Ker} \alpha_b(A)$$

and therefore, $\bar{\mathscr{X}}_b(\bar{A}) \subset P\mathscr{X}_b(A)$.

For the reverse inclusion let $x \in \mathscr{X}_b(A)$, so that $\alpha_b(A)x = 0$, and if $\bar{x} = Px$,

$$\alpha_b(\bar{A})\bar{x} = P\alpha_b(A)x = 0. \tag{4.7}$$

If the m.p. of $\bar{x}$ relative to $\bar{A}$ is $\bar{\xi} = \bar{\xi}_b \bar{\xi}_g$ then (4.7) implies $\bar{\xi}_b \bar{\xi}_g \mid \alpha_b$, hence $\bar{\xi}_g = 1$, that is, $\bar{\xi} \mid \bar{\alpha}_b$. Therefore, $\bar{\alpha}_b(\bar{A})\bar{x} = 0$, or

$$\bar{x} = Px \in \bar{\mathscr{X}}_b(\bar{A}).$$

□

Proof (of Theorem 4.4). Clearly, $\mathscr{N}_F \in \mathfrak{I}(\operatorname{Ker} D)$ and therefore, $\mathscr{N}_F \subset \mathscr{V}^*$ for all F. Write

$$\mathscr{S} = \langle A \mid \mathscr{B} \rangle + \mathscr{V}^*.$$

Since $\langle A \mid \mathscr{B} \rangle = \langle A + BF \mid \mathscr{B} \rangle$ for all F, and $A\mathscr{V}^* \subset \mathscr{V}^* + \mathscr{B}$, we have

$$(A + BF)\mathscr{S} \subset \mathscr{S}$$

for all F. Thus, for every F the diagram (4.8) commutes. In (4.8) the vertical arrows represent canonical projections, bars denote the induced maps, and $Q: \mathscr{X} \to \mathscr{X}/\mathscr{S}$ is the canonical projection (cf. Exercise 0.8). The map $\overline{\overline{A + BF}}$ is uniquely determined by the relation

$$\overline{\overline{A + BF}}Q = Q(A + BF) = QA.$$

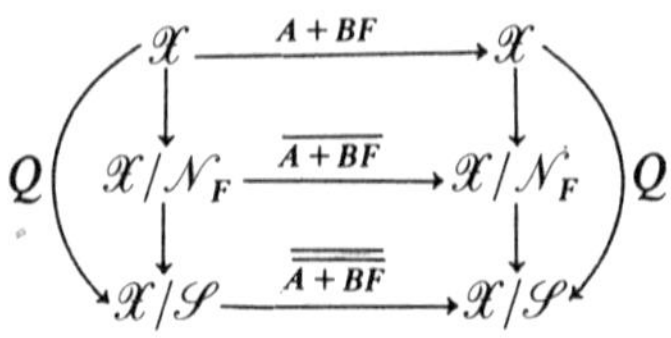

(4.8)

As $\bar{A}Q = QA$ we therefore have

$$\overline{\overline{A + BF}} = \bar{A}$$

for all F.

Now suppose OSP is solvable, so that (4.5), and hence (4.4), hold for some F. This with Lemma 4.6, yields

$$\begin{aligned} Q\mathscr{X}_b(A) &= \bar{\mathscr{X}}_b(\bar{A}) = \bar{\mathscr{X}}_b(\overline{\overline{A + BF}}) \\ &= Q\mathscr{X}_b(A + BF) \subset Q\mathscr{N}_F = 0, \end{aligned}$$

so that $\mathscr{X}_b(A) \subset \operatorname{Ker} Q = \mathscr{S}$, as claimed.

Conversely, suppose (4.6) holds, that is, $\mathscr{X}_b(A) \subset \mathscr{S}$. Choose F_0 arbitrarily such that $(A + BF_0)\mathscr{V}^* \subset \mathscr{V}^*$, and consider the diagram (4.8) with F_0 in place of F. Application of Lemma 4.6 as before yields

$$Q\mathscr{X}_b(A + BF_0) = \bar{\bar{\mathscr{X}}}_b(\overline{\overline{A + BF_0}}) = \bar{\bar{\mathscr{X}}}_b(\bar{\bar{A}})$$
$$= Q\mathscr{X}_b(A) \subset Q\mathscr{S} = 0,$$

hence $\mathscr{X}_b(A + BF_0) \subset \operatorname{Ker} Q = \mathscr{S}$. Let $P: \mathscr{X} \to \mathscr{X}/\mathscr{V}^*$ be the canonical projection and note that

$$\mathscr{V}^* = \bigcap_{i=1}^{n} (A + BF_0)^{-i+1} \operatorname{Ker} D = \mathscr{N}_{F_0}.$$

Thus, we have

$$\begin{aligned}\bar{\mathscr{X}}_b(\overline{A + BF_0}) &= P\mathscr{X}_b(A + BF_0) \subset P(\langle A \mid \mathscr{B}\rangle + \mathscr{V}^*)\\ &= P\langle A \mid \mathscr{B}\rangle = P\langle A + BF_0 \mid \mathscr{B}\rangle \\ &= \langle \overline{A + BF_0} \mid \bar{\mathscr{B}}\rangle,\end{aligned} \tag{4.9}$$

where $\bar{\mathscr{B}} := \operatorname{Im} \bar{B} = \operatorname{Im}(PB)$. Theorem 2.2 with (4.9) now implies the existence of $\bar{F}_1: \mathscr{X}/\mathscr{V}^* \to \mathscr{U}$, such that

$$\sigma(\overline{A + BF_0} + \bar{B}\bar{F}_1) \subset \mathbb{C}_g$$

or

$$\bar{\mathscr{X}}_b(\overline{A + BF_0} + \bar{B}\bar{F}_1) = 0. \tag{4.10}$$

Let $F_1 = \bar{F}_1 P$ and $F = F_0 + F_1$. Then,

$$(\overline{A + BF_0} + \bar{B}\bar{F}_1)P = P(A + BF),$$

and by uniqueness of the induced map there follows

$$\overline{A + BF_0} + \bar{B}\bar{F}_1 = \overline{A + BF}. \tag{4.11}$$

By (4.10), (4.11) and Lemma 4.6 there results

$$P\mathscr{X}_b(A + BF) = \bar{\mathscr{X}}_b(\overline{A + BF}) = 0,$$

and so

$$\mathscr{X}_b(A + BF) \subset \operatorname{Ker} P = \mathscr{V}^* \subset \operatorname{Ker} D$$

as required. □

Remark 1. In Theorem 2.2 we had, in effect, $D = 1$ and $\operatorname{Ker} D = 0$; thus, Theorem 4.4 is the generalization promised at the beginning of this section. A version of our result taking direct control feedthrough into account can be found in Exercise 4.12.

Remark 2. Whereas (4.6) is a weaker condition than that of Theorem 2.2 it guarantees only that the *output* $z(\cdot)$ is well-behaved: nothing is said about

$(A + BF)|\mathcal{N}_F$, the system map on the unobservable subspace, and this map could, for instance, be unstable. In Chapter 5 we shall see how further stability requirements can be accommodated.

Remark 3. If the assumption of full state feedback is weakened to allow only the processing of a measured vector $y = Cx$, the problem becomes significantly more complicated. Under the title "restricted (or extended) regulator problem" it will be fully treated in Chapter 6.

4.5 Exercises

4.1. Give an example to show that if $\mathcal{V}_1$, $\mathcal{V}_2$ are (A, B)-invariant, $\mathcal{V}_1 \cap \mathcal{V}_2$ need not be. Hint: Take $d(\mathcal{X}) = 3$, $d(\mathcal{B}) = 1$, $d(\mathcal{V}_i) = 2$ $(i \in \mathbf{2})$. Thus, A, B have matrices of size 3×3, 3×1; and $\mathcal{V}_i = \operatorname{Im} V_i$, with V_i of size 3×2. A random assignment of values to the 24 entries of these matrices will almost surely satisfy the problem conditions. Why?

4.2. Develop a procedure for the numerical computation of $\mathcal{V}^* = \sup \mathfrak{I}(A, B; \operatorname{Ker} D)$. Hint: Write in matrix format the algorithm of Theorem 4.3; the ingredients are obtained from Exercise 0.6. The following terminology will be useful: if M, X, Y are matrices, with M given, a *maximal solution* of the equation $MX = 0$ (resp. $YM = 0$) means a solution X (resp. Y) of maximal rank, having linearly independent columns (resp. rows) when it is not the zero column (resp. row). With reference to Theorem 4.3, setting $\mathcal{K} = \operatorname{Ker} D$, let $\mathcal{V}^\mu = \operatorname{Im} V_\mu$, with V_0 a max. sol. of $DV_0 = 0$. Let W_μ be a max. sol. of

$$W_\mu[B, V_{\mu-1}] = 0, \qquad \mu = 1, 2, \ldots;$$

and obtain V_μ as a max. sol. of

$$\begin{bmatrix} D \\ W_\mu A \end{bmatrix} V_\mu = 0, \qquad \mu = 1, 2, \ldots.$$

At each stage one has $\mathcal{V}^\mu \subset \mathcal{V}^{\mu-1}$, i.e. (as a check),

$$\operatorname{Rank}[V_{\mu-1}, V_\mu] = \operatorname{Rank} V_{\mu-1};$$

and the stopping rule is $\mathcal{V}_\mu = \mathcal{V}_{\mu-1}$, i.e.

$$\operatorname{Rank} V_\mu = \operatorname{Rank} V_{\mu-1}.$$

As an illustration, let

$$A = \begin{bmatrix} 0 & 1 & 0 & 0 & 0 \\ 0 & 0 & 1 & 0 & 0 \\ 0 & 0 & 0 & 0 & 0 \\ 0 & 0 & 0 & 0 & 1 \\ 0 & 0 & 0 & 0 & 0 \end{bmatrix}, \qquad B = \begin{bmatrix} 0 & 0 \\ 0 & 0 \\ 1 & 0 \\ 0 & 1 \\ 0 & 0 \end{bmatrix}, \qquad D = \begin{bmatrix} 1 & 0 & 0 & 0 & 0 \\ 0 & 0 & 0 & 1 & 0 \end{bmatrix}.$$

This gives

$$V_0 = \begin{bmatrix} 0 & 0 & 0 \\ 1 & 0 & 0 \\ 0 & 1 & 0 \\ 0 & 0 & 0 \\ 0 & 0 & 1 \end{bmatrix}, \qquad W_1 = [1 \quad 0 \quad 0 \quad 0 \quad 0],$$

$$V_1 = \begin{bmatrix} 0 & 0 \\ 0 & 0 \\ 1 & 0 \\ 0 & 0 \\ 0 & 1 \end{bmatrix}, \qquad W_2 = \begin{bmatrix} 1 & 0 & 0 & 0 & 0 \\ 0 & 1 & 0 & 0 & 0 \end{bmatrix},$$

$$V_2 = \begin{bmatrix} 0 \\ 0 \\ 0 \\ 0 \\ 1 \end{bmatrix}, \qquad W_3 = \begin{bmatrix} 1 & 0 & 0 & 0 & 0 \\ 0 & 1 & 0 & 0 & 0 \end{bmatrix},$$

and $\operatorname{Im} V_3 = \operatorname{Im} V_2$; i.e. $\mathscr{V}^* = \mathscr{V}^2 = \operatorname{Im} V_2$.

4.3. Show that $(D, A + BF)$ is observable for all F if and only if $\sup \mathfrak{I}(A, B; \operatorname{Ker} D) = 0$.

4.4. *System invertibility.* Let $H(\lambda)$ be the transfer matrix (Section 0.18) of the triple (C, A, B), where $C\colon p \times n$ and $B\colon n \times m$, with $p \geq m$. Show that, as a matrix over the field $\mathbb{R}(\lambda)$, $H(\lambda)$ has a left inverse, if and only if B is monic and the subspace

$$\mathscr{B} \cap \sup \mathfrak{I}(A, B; \operatorname{Ker} C) = 0.$$

What are the dual statement and conditions, in case $m \geq p$? Hint: Consider

$$y(t) = \int_0^t Ce^{(t-\tau)A} Bu(\tau)\, d\tau$$

with $u(\cdot)$ analytic. Under what conditions does the vanishing of the derivatives $y'(0), y''(0), \ldots,$ imply that of $u(0), u'(0), \ldots$?

4.5. *$\mathscr{V}^*$ and the single-input, single-output system.* Let $\dot{x} = Ax + bu$, $z = d'x$. The transfer function from u to z is

$$h(\lambda) = d'(\lambda 1 - A)^{-1} b = \frac{\beta(\lambda)}{\alpha(\lambda)},$$

where $\alpha(\lambda)$ is the ch.p. of A. Check that if (d', A, b) is complete then α and β are coprime. Assuming completeness, show that $d(\mathscr{V}^*) = \deg \beta(\lambda)$, and that if $(A + bf')\mathscr{V}^* \subset \mathscr{V}^*$ then $\beta(\lambda)$ is just the ch.p. of $(A + bf') | \mathscr{V}^*$. Thus, synthesis of $\mathscr{V}^*$ amounts to using feedback to cancel the zeros of the closed-loop transfer function

$$d'(\lambda 1 - A - bf')^{-1} b.$$

Show that the reduced-order system thus obtained has state-space $\bar{\mathscr{X}} \simeq \mathscr{X}/\mathscr{V}^*$ and, by means of an appropriate commutative diagram, explain how its state description is calculated.

4.6. *(A, B)-invariance and direct control feedthrough.* Consider the system with direct control feedthrough:

$$\dot{x} = Ax + Bu, \qquad z = Dx + Eu.$$

Introduce the family of subspaces

$$\mathfrak{I}(A, B; D, E) := \{\mathscr{V} : \mathscr{V} \subset \mathscr{X} \ \&\ \exists F : \mathscr{X} \to \mathscr{U},$$
$$(A + BF)\mathscr{V} \subset \mathscr{V} \subset \operatorname{Ker}(D + EF)\}.$$

Show that $\mathfrak{I}$ is closed under addition, and compute its supremal element $\mathscr{V}^{\Delta} := \sup \mathfrak{I}$. Hint: Take the external direct sum $\mathscr{X} \oplus \mathscr{Z}$ and define the maps

$$A_e := \begin{bmatrix} A & 0 \\ D & 0 \end{bmatrix} : \mathscr{X} \oplus \mathscr{Z} \to \mathscr{X} \oplus \mathscr{Z}, \qquad B_e := \begin{bmatrix} B \\ E \end{bmatrix} : \mathscr{U} \to \mathscr{X} \oplus \mathscr{Z}.$$

Prove that $\mathscr{V}^{\Delta} = \sup \mathfrak{I}(A_e, B_e; \mathscr{X})$, the "ordinary" supremal (A_e, B_e)-invariant subspace contained in $\mathscr{X}$. Note that systemically our approach amounts to working with the integral of $z(\cdot)$ as a new state variable and output, rather than with $z(\cdot)$ itself.

4.7. *Disturbance decoupling with direct control feedthrough.* Use the results of Exercise 4.6 to solve the disturbance decoupling problem for the system

$$\dot{x} = Ax + Bu + Sq, \qquad z = Dx + Eu.$$

Hint: Show that there exists $F : \mathscr{X} \to \mathscr{U}$ such that

$$\langle A + BF \,|\, \operatorname{Im} S \rangle \subset \operatorname{Ker}(D + EF)$$

if and only if $\operatorname{Im} S \subset \mathscr{V}^{\Delta}$.

4.8. Construct a numerical example (say with $n = 5$, $m = 2$, $p = 3$) to illustrate the application of Theorems 4.2 and 4.3. Draw the signal flow graph and indicate the feedback branches. Hint: Let

$$A = \begin{bmatrix} 0 & 1 & 0 & 0 & 0 \\ 0 & 0 & 1 & 0 & 0 \\ 0 & 0 & 0 & 0 & 0 \\ 0 & 0 & 0 & 1 & 0 \\ 0 & 0 & 0 & 0 & 0 \end{bmatrix}, \qquad B = \begin{bmatrix} 0 & 0 \\ 0 & 0 \\ 1 & 0 \\ 0 & 0 \\ 0 & 1 \end{bmatrix},$$

$$D = \begin{bmatrix} 1 & 0 & 0 & -1 & 0 \\ 1 & -1 & 0 & 0 & 0 \\ 0 & 0 & 0 & 1 & -1 \end{bmatrix}, \qquad S = \begin{bmatrix} 1 \\ 1 \\ 1 \\ 1 \\ 1 \end{bmatrix}.$$

Verify that $\mathscr{V}^* = \operatorname{Im} S$, and that

$$F = \begin{bmatrix} 0 & 0 & 1 & 0 & 0 \\ 0 & 0 & 0 & 0 & 1 \end{bmatrix} \in \mathbf{F}(\mathscr{V}^*)$$

is a solution. Note that since $\mathscr{B} \cap \mathscr{V}^* = 0$ and B is monic, all solutions F coincide on $\mathscr{V}^*$; furthermore, $\sigma[(A + BF) | \mathscr{V}^*] = \{1\}$, i.e. disturbance decoupling is only obtained at the price of instability.

4.9. Verify that DDP is not generically solvable (in fact is generically unsolvable!) in the space of data points (A, B, D, S). Hint: Note that DDP is solvable only if $DS = 0$. But if $DS = 0$, verify the remarks in Section 4.3 on generic solvability in the space of data points (A, B).

4.10. *Disturbance decoupling with feedforward.* Consider the system

$$\dot{x} = Ax + Bu + S_1 q_1 + S_2 q_2$$

$$z = Dx,$$

where the $q_i(\cdot)$ represent independent disturbances as in Section 4.1, and we assume that q_1 can be measured by the controller but q_2 cannot. A control is to be chosen of form

$$u = Fx + Gq_1$$

in order to decouple the output z from (q_1, q_2). Show that (F, G) must satisfy the condition

$$\langle A + BF \,|\, \mathrm{Im}(BG + S_1) + \mathrm{Im}\, S_2 \rangle \subset \mathrm{Ker}\, D,$$

and that such (F, G) exist if and only if

$$\mathscr{S}_1 \subset \mathscr{V}^* + \mathscr{B}, \qquad \mathscr{S}_2 \subset \mathscr{V}^*,$$

where $\mathscr{S}_i = \mathrm{Im}\, S_i$ and $\mathscr{V}^* = \sup \Im(A, B; \mathrm{Ker}\, D)$. From these results obtain conditions for generic solvability in the relevant spaces of data points.

4.11. Construct an example to illustrate the application of Theorem 4.4 and also that $\sigma[(A + BF)\,|\,\mathscr{N}_F]$ might unavoidably be bad. Hint: The example in Exercise 4.2 will serve.

4.12. *Output stabilization with direct control feedthrough.* Use the results of Exercise 4.6 to solve the output stabilization problem for the system

$$\dot{x} = Ax + Bu, \qquad z = Dx + Eu.$$

Hint: Show that there exists $F: \mathscr{X} \to \mathscr{U}$ such that $\mathscr{X}_b(A + BF) \subset \mathrm{Ker}(D + EF)$, if and only if

$$\mathscr{X}_b(A) \subset \langle A \,|\, \mathscr{B} \rangle + \mathscr{V}^\Delta.$$

4.13. *Problem of perfect tracking.* Prove that

$$\mathscr{T}^* := \mathscr{B} \cap A^{-1}\mathscr{B}$$

is the largest subspace of $\mathscr{X}$ such that

$$A\mathscr{T} + \mathscr{T} \subset \mathscr{B}.$$

From this show that if $r(\cdot)$ is continuously differentiable, $r(t) \in \mathscr{T}^*$ for all $t \geq 0$, and $x(0) = r(0)$, there exists a continuous control $u(t)$, $t \geq 0$, such that $x(t) = r(t)$, $t \geq 0$, where

$$\dot{x}(t) = Ax(t) + Bu(t).$$

Furthermore, $\mathscr{T}^*$ is the largest subspace of $\mathscr{X}$ having this property.

4.14. Let $A\mathscr{N} \subset \mathscr{N}$. Show that there exists $F: \mathscr{X} \to \mathscr{U}$ such that $\mathscr{X}^+(A + BF) \subset \mathscr{N} \cap \operatorname{Ker} F$, if and only if

$$\mathscr{X}^+(A) \subset \langle A \mid \mathscr{B} \rangle + \mathscr{N}.$$

With the help of this result solve the following: Given $D: \mathscr{X} \to \mathscr{Z}$ and the system $\dot{x} = Ax + Bu$, $z = Dx$, find a necessary and sufficient condition for the existence of $F: \mathscr{X} \to \mathscr{U}$ such that, if $u = Fx$, then $z(t) \to 0$ and $u(t) \to 0$ $(t \to \infty)$ for every initial state $x(0)$.

4.6 Notes and References

The idea of (A, B)-invariant subspace and results equivalent to Theorems 4.2 and 4.3 were discovered independently by Basile and Marro [2], [3] and by Wonham and Morse [1]. An interesting application of DDP to control of a distillation column is presented by Takamatsu, Hashimoto and Nakai [1]. The treatment of output stabilization is adapted from Bhattacharyya, Pearson and Wonham [1]. The computational method of Exercise 4.2 is straightforward and effective for use with most systems of order up to around 10, but numerical instabilities may be encountered with large systems. See Moore and Laub [1] for a stable computational approach by more refined techniques. The geometric significance of system invertibility (Exercise 4.4) has been pointed out by Silverman and Payne [1]. For further information related to Exercise 4.6 see Morse [1] and Anderson [1]. A version of Exercise 4.10 is treated by Bhattacharyya [2]. The tracking problem of Exercise 4.13 is taken from Basile and Marro [1], and the results of Exercise 4.14 are due to Bhattacharyya [1].

5 Controllability Subspaces

Given a system pair (A, B) we consider all pairs $(A + BF, BG)$ which can be formed by means of state feedback F and the connection of a "gain" matrix G at the system input (Fig. 5.1). The controllable subspace of $(A + BF, BG)$ is called a controllability subspace (c.s.) of the original pair (A, B). The family of c.s. of a fixed pair (A, B) is a subfamily, in general proper, of the (A, B)-invariant subspaces: the importance of c.s. derives from the fact that the restriction of $A + BF$ to an $(A + BF)$-invariant c.s. can be assigned an arbitrary spectrum by suitable choice of F.

For the single-input system corresponding to a pair (A, b) the family of c.s. obviously comprises simply 0 and $\langle A | \mathscr{b} \rangle$. However, in the multi-input situation, where $d(\mathscr{B}) \geq 2$, the family of c.s. is in general nontrivial. This fact, together with the spectral assignability already mentioned, indicates that c.s.

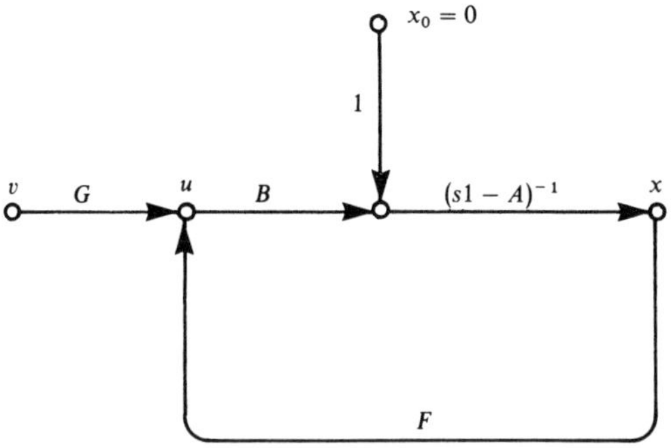

Figure 5.1 Controllability Subspace: $\mathscr{R} = \langle A + BF | \operatorname{Im}(BG) \rangle$ is the subspace of states $x(1)$ generated by allowing $v(\cdot)$ to vary over (say) all continuous inputs defined for $0 \leq t \leq 1$, with $x(0) = 0$.

is a central geometric concept in the state space theory of linear multivariable control.

This chapter is devoted to the basic properties of c.s. The main applications, to tracking, regulation and noninteraction, are treated in the chapters to follow.

5.1 Controllability Subspaces

Let $A: \mathscr{X} \to \mathscr{X}$ and $B: \mathscr{U} \to \mathscr{X}$. A subspace $\mathscr{R} \subset \mathscr{X}$ is a *controllability subspace* (c.s.) of the pair (A, B) if there exist maps $F: \mathscr{X} \to \mathscr{U}$ and $G: \mathscr{U} \to \mathscr{U}$, such that

$$\mathscr{R} = \langle A + BF \,|\, \operatorname{Im}(BG) \rangle. \tag{1.1}$$

Thus, $\mathscr{R}$ is precisely the controllable subspace of the pair $(A + BF, BG)$. We adopt the notation $\mathfrak{C}(A, B; \mathscr{X})$, or simply $\mathfrak{C}(\mathscr{X})$, for the class of c.s. of (A, B).

In terms of the system $\dot{x} = Ax + Bu$, a c.s. $\mathscr{R} \subset \mathscr{X}$ is characterized by the following property: For every $x \in \mathscr{R}$ there exists a continuous control $u(t)$ $(0 \le t \le 1)$ such that, if $x(0) = 0$ then $x(t) \in \mathscr{R}$ $(0 \le t \le 1)$ and $x(1) = x$. That is, every state $x \in \mathscr{R}$ is reachable from the origin along a controlled trajectory that does not leave $\mathscr{R}$. Indeed, if $\mathscr{R}$ is a c.s. then by (1.1) the pair

$$((A + BF)\,|\,\mathscr{R}, BG)$$

is controllable, so $\mathscr{R}$ has the reachability property just as shown in Section 1.1. The converse statement is less obvious and its proof is deferred to Exercise 5.7.

The appearance of G in (1.1) is eliminated by use of the following observation.

Proposition 5.1. *If $\hat{\mathscr{B}} \subset \mathscr{B}$ and $\langle A \,|\, \hat{\mathscr{B}} \rangle = \mathscr{R}$ then $\langle A \,|\, \mathscr{B} \cap \mathscr{R} \rangle = \mathscr{R}$. Conversely, if $\langle A \,|\, \mathscr{B} \cap \mathscr{R} \rangle = \mathscr{R}$, there exists $G: \mathscr{U} \to \mathscr{U}$, such that*

$$\langle A \,|\, \operatorname{Im}(BG) \rangle = \mathscr{R}.$$

PROOF. If $\langle A \,|\, \hat{\mathscr{B}} \rangle \subset \mathscr{R}$ then $\hat{\mathscr{B}} \subset \mathscr{R}$, i.e. $\hat{\mathscr{B}} \subset \mathscr{B} \cap \mathscr{R}$, so

$$\mathscr{R} = \langle A \,|\, \hat{\mathscr{B}} \rangle \subset \langle A \,|\, \mathscr{B} \cap \mathscr{R} \rangle.$$

Clearly, $A\mathscr{R} \subset \mathscr{R}$; hence, $\langle A \,|\, \mathscr{B} \cap \mathscr{R} \rangle \subset \mathscr{R}$; and so

$$\langle A \,|\, \mathscr{B} \cap \mathscr{R} \rangle = \mathscr{R}.$$

For the converse let $\{b_1, \ldots, b_r\}$ be a basis for $\mathscr{B} \cap \mathscr{R}$. Then, $b_i = Bu_i$ $(u_i \in \mathscr{U})$ where the u_i $(i \in \mathbf{r})$ are independent. Let $\{u_1, \ldots, u_m\}$ be a basis for $\mathscr{U}$, and define

$$Gu_i = u_i, \qquad i \in \mathbf{r}$$

$$Gu_i = 0, \qquad i = r + 1, \ldots, m.$$

Then, $\operatorname{Im}(BG) = \mathscr{B} \cap \mathscr{R}$. □

As an immediate consequence, we have

Proposition 5.2. *A subspace $\mathscr{R}$ belongs to $\mathfrak{C}(A, B; \mathscr{X})$ if and only if there exists a map $F: \mathscr{X} \to \mathscr{U}$, such that*

$$\mathscr{R} = \langle A + BF \,|\, \mathscr{B} \cap \mathscr{R} \rangle.$$

Recall the notation $\mathbf{F}(A, B; \mathscr{S})$, or $\mathbf{F}(\mathscr{S})$, for the class of maps $F: \mathscr{X} \to \mathscr{U}$, such that $(A + BF)\mathscr{S} \subset \mathscr{S}$. Thus, $\mathbf{F}(\mathscr{S}) \neq \emptyset$ if and only if $\mathscr{S}$ is (A, B)-invariant. If $\mathscr{R} \in \mathfrak{C}(A, B; \mathscr{X})$ clearly $\mathbf{F}(\mathscr{R}) \neq \emptyset$, and we have

Proposition 5.3. *If $\mathscr{R} \in \mathfrak{C}(A, B; \mathscr{X})$, then*

$$\mathscr{R} = \langle A + BF \,|\, \mathscr{B} \cap \mathscr{R} \rangle$$

for every map $F \in \mathbf{F}(\mathscr{R})$.

Proof. By Proposition 5.2 there is a map $F_0: \mathscr{X} \to \mathscr{U}$, such that

$$\mathscr{R} = \langle A + BF_0 \,|\, \mathscr{B} \cap \mathscr{R} \rangle.$$

Clearly, $F_0 \in \mathbf{F}(\mathscr{R})$. Let $F_1 \in \mathbf{F}(\mathscr{R})$ and write

$$\mathscr{R}_1 := \langle A + BF_1 \,|\, \mathscr{B} \cap \mathscr{R} \rangle. \tag{1.2}$$

Then, $\mathscr{R}_1 \subset \mathscr{R}$. For the reverse inclusion suppose

$$(A + BF_0)^{i-1}(\mathscr{B} \cap \mathscr{R}) \subset \mathscr{R}_1, \qquad i \in \mathbf{k}, \tag{1.3}$$

for some $k \in \mathbf{n}$. With (1.3) as induction hypothesis, we have

$$\begin{aligned}
&\sum_{i=1}^{k+1} (A + BF_0)^{i-1}(\mathscr{B} \cap \mathscr{R}) \\
&= \mathscr{B} \cap \mathscr{R} + (A + BF_0) \sum_{i=1}^{k} (A + BF_0)^{i-1}(\mathscr{B} \cap \mathscr{R}) \\
&\subset \mathscr{B} \cap \mathscr{R} + (A + BF_0)\mathscr{R}_1 = \mathscr{B} \cap \mathscr{R} + [A + BF_1 + B(F_0 - F_1)]\mathscr{R}_1 \\
&\subset \mathscr{B} \cap \mathscr{R} + (A + BF_1)\mathscr{R}_1 + B(F_0 - F_1)\mathscr{R}_1.
\end{aligned} \tag{1.4}$$

Let $x \in \mathscr{R}_1$. Then, $B(F_0 - F_1)x \in \mathscr{B}$ and, since $\mathscr{R}_1 \subset \mathscr{R}$,

$$B(F_0 - F_1)x = (A + BF_0)x - (A + BF_1)x \in \mathscr{R}.$$

Hence, the subspace on the right in (1.4) is contained in

$$\mathscr{B} \cap \mathscr{R} + (A + BF_1)\mathscr{R}_1 \subset \mathscr{R}_1. \tag{1.5}$$

By (1.4) and (1.5)

$$(A + BF_0)^k(\mathscr{B} \cap \mathscr{R}) \subset \mathscr{R}_1.$$

Since (1.2) implies that (1.3) is true for $k = 1$, we have that (1.3) is true for $k \in \mathbf{n}$, hence $\mathscr{R} \subset \mathscr{R}_1$. □

The foregoing result provides a way of checking whether a given subspace $\mathscr{R} \subset \mathscr{X}$ is a c.s.: Verify first that $\mathscr{R} \in \mathfrak{I}(A, B; \mathscr{X})$, i.e. $A\mathscr{R} \subset \mathscr{R} + \mathscr{B}$. If this is so, construct any F such that $(A + BF)\mathscr{R} \subset \mathscr{R}$, and then check that

$$\langle A + BF \,|\, \mathscr{B} \cap \mathscr{R} \rangle = \mathscr{R}.$$

5.2 Spectral Assignability

It will be shown that the family of controllability subspaces can be characterized in terms of the spectral assignability property of controllable pairs. A simple consequence of Theorem 2.1, these results, especially the first, are basic to the applications.

Theorem 5.1. *Let $\mathscr{R} \in \mathfrak{C}(A, B; \mathscr{X})$ with $d(\mathscr{R}) = \rho \geq 1$. Let $0 \neq b \in \mathscr{B} \cap \mathscr{R}$. For every symmetric set Λ of ρ complex numbers there exists a map $F: \mathscr{X} \to \mathscr{U}$, such that*

$$\mathscr{R} = \langle A + BF \,|\, \mathscr{b} \rangle$$

and

$$\sigma[(A + BF) \,|\, \mathscr{R}] = \Lambda.$$

Proof. Suppose

$$\mathscr{R} = \langle A + BF_0 \,|\, \mathscr{B} \cap \mathscr{R} \rangle \tag{2.1}$$

and choose $G: \mathscr{U} \to \mathscr{U}$, such that

$$\operatorname{Im}(BG) = \mathscr{B} \cap \mathscr{R}. \tag{2.2}$$

Define $A_0: \mathscr{R} \to \mathscr{R}$ and $B_0: \mathscr{U} \to \mathscr{R}$ according to

$$A_0 := (A + BF_0) \,|\, \mathscr{R}, \qquad B_0 := BG.$$

By (2.1) and (2.2), we have

$$\langle A_0 \,|\, \mathscr{B}_0 \rangle = \mathscr{R}.$$

Then, application of Theorem 2.1 to the pair (A_0, B_0) yields the existence of $F_1: \mathscr{R} \to \mathscr{U}$, such that

$$\mathscr{R} = \langle A_0 + B_0 F_1 \,|\, \mathscr{b} \rangle$$

and

$$\sigma(A_0 + B_0 F_1) = \Lambda.$$

Let $F_2: \mathscr{X} \to \mathscr{U}$ be any extension of F_1 from $\mathscr{R}$ to $\mathscr{X}$. Then,

$$F := F_0 + GF_2$$

is a map with the properties required. □

As a converse to Theorem 5.1 we prove the following criterion for a given subspace to be a c.s.

Theorem 5.2. *Let $\mathscr{R} \subset \mathscr{X}$ be a subspace with $d(\mathscr{R}) = \rho \geq 1$. Suppose that for every symmetric set Λ of ρ complex numbers there exists a map $F: \mathscr{X} \to \mathscr{U}$, such that*

$$(A + BF)\mathscr{R} \subset \mathscr{R}, \qquad \sigma[(A + BF)|\mathscr{R}] = \Lambda. \tag{2.3}$$

Then, $\mathscr{R} \in \mathfrak{C}(A, B; \mathscr{X})$.

Proof. Fix $F_0 \in \mathbf{F}(\mathscr{R})$ and write $A_0 := (A + BF_0)|\mathscr{R}$. We have $F \in \mathbf{F}(\mathscr{R})$ if and only if $B(F - F_0)\mathscr{R} \subset \mathscr{B} \cap \mathscr{R}$. Let $B_0: \mathscr{U} \to \mathscr{R}$ be an arbitrary map with $\operatorname{Im} B_0 = \mathscr{B} \cap \mathscr{R}$. Then, if $F \in \mathbf{F}(\mathscr{R})$, there exists $F_1: \mathscr{R} \to \mathscr{U}$, such that

$$B_0 F_1 = B(F - F_0)|\mathscr{R}.$$

Thus, (2.3) implies that for every Λ there exists F_1, such that

$$\sigma(A_0 + B_0 F_1) = \Lambda.$$

By Theorem 2.1, the pair (A_0, B_0) is controllable. Hence,

$$\mathscr{R} = \langle A_0 | \mathscr{B}_0 \rangle = \langle A + BF_0 | \mathscr{B} \cap \mathscr{R} \rangle$$

and therefore $\mathscr{R}$ is a c.s. □

5.3 Controllability Subspace Algorithm

In this section we characterize controllability subspaces by means of an algorithm which computes $\mathscr{R}$ without explicitly constructing $F \in \mathbf{F}(\mathscr{R})$. This result will be useful in obtaining general properties of the family $\mathfrak{C}(A, B; \mathscr{X})$.

For an arbitrary, fixed subspace $\mathscr{R} \subset \mathscr{X}$ define a family $\mathfrak{S}$ of subspaces $\mathscr{S} \subset \mathscr{X}$ according to

$$\mathfrak{S} := \{\mathscr{S}: \mathscr{S} = \mathscr{R} \cap (A\mathscr{S} + \mathscr{B})\}. \tag{3.1}$$

It will be shown that $\mathfrak{S}$ has a unique least member.

Lemma 5.1. *There is a unique element $\mathscr{S}_* \in \mathfrak{S}$ such that $\mathscr{S}_* \subset \mathscr{S}$ for every $\mathscr{S} \in \mathfrak{S}$.*

Proof. Define a sequence $\mathscr{S}^\mu \subset \mathscr{X}$ according to

$$\mathscr{S}^0 = 0; \qquad \mathscr{S}^\mu = \mathscr{R} \cap (A\mathscr{S}^{\mu-1} + \mathscr{B}), \qquad \mu \in \mathbf{n}. \tag{3.2}$$

The $\mathscr{S}^\mu$ sequence is nondecreasing: clearly $\mathscr{S}^1 \supset \mathscr{S}^0$, and if $\mathscr{S}^\mu \supset \mathscr{S}^{\mu-1}$, then

$$\mathscr{S}^{\mu+1} = \mathscr{R} \cap (A\mathscr{S}^\mu + \mathscr{B}) \supset \mathscr{R} \cap (A\mathscr{S}^{\mu-1} + \mathscr{B}) = \mathscr{S}^\mu.$$

Thus, there exists $k \in \mathbf{n}$, such that

$$\mathscr{S}^\mu = \mathscr{S}^k, \qquad \mu \geq k,$$

and we set $\mathscr{S}_* := \mathscr{S}^k$. Clearly, $\mathscr{S}_* \in \mathfrak{S}$. To show that $\mathscr{S}_*$ is infimal, let $\mathscr{S} \in \mathfrak{S}$. Then, $\mathscr{S} \supset \mathscr{S}^0$, and if $\mathscr{S} \supset \mathscr{S}^\mu$, we have

$$\mathscr{S} = \mathscr{R} \cap (A\mathscr{S} + \mathscr{B}) \supset \mathscr{R} \cap (A\mathscr{S}^\mu + \mathscr{B}) = \mathscr{S}^{\mu+1}$$

So $\mathscr{S} \supset \mathscr{S}^\mu$ for all μ, hence $\mathscr{S} \supset \mathscr{S}_*$. □

The algorithm (3.2) in the proof of Lemma 5.1 will be used often in the sequel: we call it the *controllability subspace algorithm* (CSA). Thus, we have

Lemma 5.2. *The least element $\mathscr{S}_*$ of $\mathfrak{S}$ is given by*

$$\mathscr{S}_* = \lim \mathscr{S}^\mu = \mathscr{S}^n, \tag{3.3}$$

where $\mathscr{S}^\mu$ is computed by CSA:

$$\mathscr{S}^0 = 0; \qquad \mathscr{S}^\mu = \mathscr{R} \cap (A\mathscr{S}^{\mu-1} + \mathscr{B}), \qquad \mu \in \mathbf{n}. \tag{3.2 bis}$$

The next two lemmas will link the $\mathscr{S}^\mu$ to the definition of c.s.

Lemma 5.3. *Let $\mathscr{R} \in \mathfrak{I}(A, B; \mathscr{X})$. If $F \in \mathbf{F}(\mathscr{R})$ and $\hat{\mathscr{R}} \subset \mathscr{R}$, then*

$$\mathscr{B} \cap \mathscr{R} + (A + BF)\hat{\mathscr{R}} = \mathscr{R} \cap (A\hat{\mathscr{R}} + \mathscr{B}).$$

Proof. $F \in \mathbf{F}(\mathscr{R})$ implies $(A + BF)\hat{\mathscr{R}} \subset \mathscr{R}$; also

$$A\hat{\mathscr{R}} + \mathscr{B} = (A + BF)\hat{\mathscr{R}} + \mathscr{B}.$$

By the modular distributive rule (0.3.1),

$$\begin{aligned}\mathscr{R} \cap (A\hat{\mathscr{R}} + \mathscr{B}) &= \mathscr{R} \cap [(A + BF)\hat{\mathscr{R}} + \mathscr{B}] \\ &= (A + BF)\hat{\mathscr{R}} + \mathscr{B} \cap \mathscr{R}.\end{aligned}$$ □

Lemma 5.4. *Let $\mathscr{R} \in \mathfrak{I}(A, B; \mathscr{X})$, let $F \in \mathbf{F}(\mathscr{R})$, and define $\mathscr{S}^\mu$ by CSA. Then,*

$$\mathscr{S}^\mu = \sum_{j=1}^{\mu} (A + BF)^{j-1}(\mathscr{B} \cap \mathscr{R}), \qquad \mu \in \mathbf{n}. \tag{3.4}$$

Proof. Clearly, (3.4) is true for $\mu = 1$. If it is true for $\mu = \nu$, then

$$\begin{aligned}\sum_{j=1}^{\nu+1} (A + BF)^{j-1}(\mathscr{B} \cap \mathscr{R}) &= \mathscr{B} \cap \mathscr{R} + (A + BF)\mathscr{S}^\nu \\ &= \mathscr{R} \cap (A\mathscr{S}^\nu + \mathscr{B}) \quad \text{(by Lemma 5.3)} \\ &= \mathscr{S}^{\nu+1}.\end{aligned}$$ □

We can now give the promised characterization of c.s.

Theorem 5.3. *Let $\mathscr{R} \subset \mathscr{X}$ and define the family $\mathfrak{S}$ as in (3.1). Then, $\mathscr{R} \in \mathfrak{C}(A, B; \mathscr{X})$ if and only if*

$$\mathscr{R} \in \mathfrak{I}(A, B; \mathscr{X}) \quad [\text{i.e. } A\mathscr{R} \subset \mathscr{R} + \mathscr{B}], \tag{3.5}$$

and

$$\mathscr{R} = \mathscr{S}_*. \tag{3.6}$$

Here, $\mathscr{S}_*$ is the least member of $\mathfrak{S}$ and is computable by CSA, as in (3.3).

Proof. If (3.5) is true then $\mathbf{F}(\mathscr{R}) \neq \varnothing$. Taking $F \in \mathbf{F}(\mathscr{R})$, we have from (3.6), (3.3) and (3.4), in that order,

$$\mathscr{R} = \mathscr{S}_* = \mathscr{S}^n = \langle A + BF \,|\, \mathscr{B} \cap \mathscr{R} \rangle,$$

so that $\mathscr{R}$ is a c.s. Conversely, if $\mathscr{R}$ is a c.s. then $\mathbf{F}(\mathscr{R}) \neq \varnothing$, so that (3.5) is true; and if $F \in \mathbf{F}(\mathscr{R})$

$$\mathscr{R} = \langle A + BF \,|\, \mathscr{B} \cap \mathscr{R} \rangle = \mathscr{S}^n = \mathscr{S}_*$$

by (3.4) and (3.3). □

5.4 Supremal Controllability Subspace

In this section we show that the family of c.s. of a fixed pair (A, B) is a semilattice with respect to inclusion and subspace addition, and hence that the family of c.s. which belong to a given subspace contains a supremal element. This property of c.s. is crucial in the applications.

Lemma 5.5. *The class of subspaces $\mathfrak{C}(A, B; \mathscr{X})$ is closed under the operation of subspace addition.*

Proof. The proof is based on the characterization of c.s. in Theorem 5.3. Clearly,

$$A(\mathscr{R}_1 + \mathscr{R}_2) \subset \mathscr{R}_1 + \mathscr{R}_2 + \mathscr{B}.$$

Also, $\mathscr{R}_i = \mathscr{S}_i^n$ ($i \in \mathbf{2}$), where

$$\mathscr{S}_i^0 = 0; \qquad \mathscr{S}_i^\mu = \mathscr{R}_i \cap (A\mathscr{S}_i^{\mu-1} + \mathscr{B}), \qquad \mu \in \mathbf{n}.$$

Define $\mathscr{S}^\mu$ according to

$$\mathscr{S}^0 = 0; \qquad \mathscr{S}^\mu = (\mathscr{R}_1 + \mathscr{R}_2) \cap (A\mathscr{S}^{\mu-1} + \mathscr{B}), \qquad \mu \in \mathbf{n}.$$

We have $\mathscr{S}^0 = 0 = \mathscr{S}_i^0$ ($i \in \mathbf{2}$), and if $\mathscr{S}^\mu \supset \mathscr{S}_i^\mu$ then

$$\mathscr{S}^{\mu+1} \supset \mathscr{R}_i \cap (A\mathscr{S}_i^\mu + \mathscr{B}) = \mathscr{S}_i^{\mu+1}, \qquad i \in \mathbf{2},$$

and so $\mathscr{S}^{\mu+1} \supset \mathscr{S}_1^{\mu+1} + \mathscr{S}_2^{\mu+1}$. Therefore,

$$\mathscr{R}_1 + \mathscr{R}_2 = \mathscr{S}_1^n + \mathscr{S}_2^n \subset \mathscr{S}^n \subset \mathscr{R}_1 + \mathscr{R}_2,$$

hence $\mathscr{R}_1 + \mathscr{R}_2 = \mathscr{S}^n$, and the result follows by Theorem 5.3. □

Now let $\mathscr{K} \subset \mathscr{X}$ be an arbitrary subspace, and write $\mathfrak{C}(A, B; \mathscr{K})$, or simply $\mathfrak{C}(\mathscr{K})$, for the family of c.s. in $\mathscr{K}$, i.e.

$$\mathfrak{C}(A, B; \mathscr{K}) := \{\mathscr{R}: \mathscr{R} \in \mathfrak{C}(A, B; \mathscr{X}) \ \& \ \mathscr{R} \subset \mathscr{K}\}.$$

Theorem 5.4. *Let $A: \mathscr{X} \to \mathscr{X}$ and $B: \mathscr{U} \to \mathscr{X}$. Every subspace $\mathscr{K} \subset \mathscr{X}$ contains a unique supremal controllability subspace [written* $\sup \mathfrak{C}(A, B; \mathscr{K})$, *or simply* $\sup \mathfrak{C}(\mathscr{K})$].

PROOF. The family $\mathfrak{C}(A, B; \mathscr{K})$ possesses at least one member, namely 0; and by Lemma 5.5, it is closed under addition. The result now follows by Lemma 4.4. □

The supremal element $\sup \mathfrak{C}(\mathscr{K})$ will often be denoted by $\mathscr{R}^*$. We now describe two ways of computing $\mathscr{R}^*$. Both of these require a prior computation of $\mathscr{V}^* = \sup \mathfrak{I}(A, B; \mathscr{K})$. An algorithm which computes $\mathscr{V}^*$ was presented in Theorem 4.3.

Theorem 5.5. *Let* $\mathscr{V}^* := \sup \mathfrak{I}(A, B; \mathscr{K})$, $\mathscr{R}^* := \sup \mathfrak{C}(A, B; \mathscr{K})$. *If* $F \in \mathbf{F}(A, B; \mathscr{V}^*)$ *then*

$$\mathscr{R}^* = \langle A + BF \mid \mathscr{B} \cap \mathscr{V}^* \rangle. \tag{4.1}$$

For the proof we need two preliminary results. The first of these is a generalization of Lemma 2.1.

Lemma 5.6. *Let $\mathscr{V} \in \mathfrak{I}(A, B; \mathscr{X})$, $\mathscr{B}_0 \subset \mathscr{B} \cap \mathscr{V}$, $F_0 \in \mathbf{F}(\mathscr{V})$, and define*

$$\mathscr{R} := \langle A + BF_0 \mid \mathscr{B}_0 \rangle.$$

If $F \in \mathbf{F}(\mathscr{V})$ and $B(F - F_0)\mathscr{V} \subset \mathscr{B}_0$ then

$$\mathscr{R} = \langle A + BF \mid \mathscr{B}_0 \rangle.$$

PROOF. Write $\mathscr{R}_1 = \langle A + BF \mid \mathscr{B}_0 \rangle$ and

$$\mathscr{V}^i = \sum_{j=1}^{i} (A + BF_0)^{j-1}\mathscr{B}_0, \qquad i \in \mathbf{n}.$$

Then, $\mathscr{V}^1 = \mathscr{B}_0 \subset \mathscr{R}_1$. Suppose $\mathscr{V}^i \subset \mathscr{R}_1$. We have

$$\begin{aligned}\mathscr{V}^{i+1} &= \mathscr{B}_0 + (A + BF_0)\mathscr{V}^i \\ &\subset \mathscr{B}_0 + (A + BF)\mathscr{V}^i + B(F - F_0)\mathscr{V}^i.\end{aligned}$$

Since $F \in \mathbf{F}(\mathscr{R}_1)$,

$$(A + BF)\mathscr{V}^i \subset \mathscr{R}_1;$$

and because $F \in \mathbf{F}(\mathscr{V})$ and $\mathscr{B}_0 \subset \mathscr{V}$, we have that $\mathscr{R}_1 \subset \mathscr{V}$, hence

$$B(F - F_0)\mathscr{V}^i \subset B(F - F_0)\mathscr{R}_1 \subset \mathscr{B}_0 \subset \mathscr{R}_1.$$

Therefore, $\mathscr{V}^{i+1} \subset \mathscr{R}_1$, so that $\mathscr{V}^i \subset \mathscr{R}_1$ $(i \in \mathbf{n})$ and

$$\mathscr{R} = \mathscr{V}^n \subset \mathscr{R}_1.$$

By interchanging the roles of F and F_0, we infer that $\mathscr{R}_1 \subset \mathscr{R}$, and the result follows. □

Lemma 5.7. *Let both* $\mathscr{R}$ *and* $\mathscr{V} \in \mathfrak{I}(A, B; \mathscr{X})$, *and suppose* $\mathscr{R} \subset \mathscr{V}$. *If* $F_0 \in \mathbf{F}(\mathscr{R})$ *there exists* $F \in \mathbf{F}(\mathscr{V}) \cap \mathbf{F}(\mathscr{R})$ *such that*

$$F \mid \mathscr{R} = F_0 \mid \mathscr{R}.$$

PROOF. Let $\mathscr{R} \oplus \mathscr{S} = \mathscr{V}$ and let $\{s_1, \ldots, s_q\}$ be a basis for $\mathscr{S}$. Then,

$$As_i = v_i + Bu_i, \qquad i \in \mathbf{q},$$

for some $v_i \in \mathscr{V}$ and $u_i \in \mathscr{U}$. Let $F\colon \mathscr{X} \to \mathscr{U}$ be any map such that $Fx = F_0 x$ $(x \in \mathscr{R})$ and $Fs_i = -u_i$ $(i \in \mathbf{q})$. Then, F has the required properties. □

PROOF (of Theorem 5.5). With $F \in \mathbf{F}(\mathscr{V}^*)$, write

$$\mathscr{R} = \langle A + BF \mid \mathscr{B} \cap \mathscr{V}^* \rangle.$$

Since $\mathscr{B} \cap \mathscr{V}^* = \operatorname{Im}(BG)$ for some $G\colon \mathscr{U} \to \mathscr{U}$, and since

$$(A + BF)^{j-1}(\mathscr{B} \cap \mathscr{V}^*) \subset \mathscr{V}^* \subset \mathscr{K}, \qquad j \in \mathbf{n},$$

it is clear that $\mathscr{R} \in \mathfrak{C}(\mathscr{K})$. Let $\mathscr{R}_0 \in \mathfrak{C}(\mathscr{K})$ be arbitrary. Then,

$$\mathscr{R}_0 = \langle A + BF_0 \mid \mathscr{B} \cap \mathscr{R}_0 \rangle$$

for some F_0. Since $(A + BF_0)\mathscr{R}_0 \subset \mathscr{R}_0$, clearly

$$\mathscr{R}_0 \subset \sup \mathfrak{I}(\mathscr{K}) = \mathscr{V}^*. \tag{4.2}$$

Choose, by Lemma 5.7, $F_1 \in \mathbf{F}(\mathscr{R}_0) \cap \mathbf{F}(\mathscr{V}^*)$, such that $F_1 \mid \mathscr{R}_0 = F_0 \mid \mathscr{R}_0$. If $x \in \mathscr{V}^*$

$$B(F - F_1)x = (A + BF)x - (A + BF_1)x \in \mathscr{V}^*,$$

so that $B(F - F_1)\mathscr{V}^* \subset \mathscr{B} \cap \mathscr{V}^*$. Then,

$$\begin{aligned} \mathscr{R}_0 &= \langle A + BF_1 \mid \mathscr{B} \cap \mathscr{R}_0 \rangle \\ &\subset \langle A + BF_1 \mid \mathscr{B} \cap \mathscr{V}^* \rangle && \text{by (4.2)} \\ &= \langle A + BF \mid \mathscr{B} \cap \mathscr{V}^* \rangle && \text{by Lemma 5.6} \\ &= \mathscr{R}. \end{aligned}$$

Therefore, $\mathscr{R} \in \mathfrak{C}(\mathscr{K})$ is supremal and so $\mathscr{R} = \mathscr{R}^*$. □

Corollary 5.1. *If* $\mathscr{V}^* = \sup \mathfrak{I}(\mathscr{K})$ *and* $\mathscr{R}^* = \sup \mathfrak{C}(\mathscr{K})$, *then*

$$\mathbf{F}(\mathscr{V}^*) \subset \mathbf{F}(\mathscr{R}^*). \tag{4.3}$$

In other words, any friend of $\mathscr{V}^*$ is a friend of $\mathscr{R}^*$.

PROOF. The assertion is immediate from (4.1). □

We turn now to a second method of computing $\mathscr{R}^*$, which does not require the prior computation of an $F \in \mathbf{F}(\mathscr{V}^*)$. This is a generalization of CSA.

Theorem 5.6. *Define the sequence $\mathscr{S}^\mu$ according to*

$$\mathscr{S}^0 = 0; \qquad \mathscr{S}^\mu = \mathscr{V}^* \cap (A\mathscr{S}^{\mu-1} + \mathscr{B}), \qquad \mu \in \mathbf{n}. \tag{4.4}$$

Then, $\mathscr{S}^\mu = \mathscr{R}^*$ for $\mu \geq d(\mathscr{V}^*)$.

Proof. Induction shows that $\mathscr{S}^\mu\uparrow$ and so $\mathscr{S}^\mu = \mathscr{S}^k$ for $\mu \geq k := d(\mathscr{V}^*)$. Write $\mathscr{S}_* := \mathscr{S}^k$. Since $\mathscr{S}^\mu \subset \mathscr{V}^*$ and $A\mathscr{V}^* \subset \mathscr{V}^* + \mathscr{B}$, we have

$$\begin{aligned} A\mathscr{S}^\mu &\subset (\mathscr{V}^* + \mathscr{B}) \cap (A\mathscr{S}^\mu + \mathscr{B}) \\ &= \mathscr{V}^* \cap (A\mathscr{S}^\mu + \mathscr{B}) + \mathscr{B} \\ &= \mathscr{S}^{\mu+1} + \mathscr{B}, \end{aligned}$$

so that $A\mathscr{S}_* \subset \mathscr{S}_* + \mathscr{B}$. Since

$$\mathscr{S}^\mu \subset \mathscr{S}_* \subset \mathscr{V}^*, \qquad \mu \in \mathbf{n},$$

(4.4) implies

$$\mathscr{S}^\mu = \mathscr{S}_* \cap (A\mathscr{S}^{\mu-1} + \mathscr{B}), \qquad \mu \in \mathbf{n}.$$

We conclude from Theorem 5.3 that $\mathscr{S}_* \in \mathfrak{C}(\mathscr{K})$, and so $\mathscr{S}_* \subset \mathscr{R}^*$. On the other hand, $\mathscr{R}^* = \mathscr{R}^n$, where

$$\mathscr{R}^0 = 0; \qquad \mathscr{R}^\mu = \mathscr{R}^* \cap (A\mathscr{R}^{\mu-1} + \mathscr{B}), \qquad \mu \in \mathbf{n}.$$

Since $\mathscr{R}^* \subset \mathscr{V}^*$ it follows easily by induction on μ that $\mathscr{R}^\mu \subset \mathscr{S}^\mu$ $(\mu \in \mathbf{n})$ and therefore, $\mathscr{R}^* \subset \mathscr{S}_*$. □

To conclude this section we give a result on feedback maps and spectral assignability which generalizes the discussion in Section 2.3.

Theorem 5.7. *Let $\mathscr{V} \in \mathfrak{I}(A, B; \mathscr{X})$ and let $\mathscr{R}^* := \sup \mathfrak{C}(A, B; \mathscr{V})$. For $F \in \mathbf{F}(\mathscr{V})$ write $A_F := A + BF$ and $\bar{A}_F$ for the map induced in $\mathscr{V}/\mathscr{R}^*$ by A_F. Then $\bar{A}_F$ is independent of $F \in \mathbf{F}(\mathscr{V})$.*

Proof. By Corollary 5.1 with $\mathscr{K} = \mathscr{V}$, we have $A_F\mathscr{R}^* \subset \mathscr{R}^*$, so $\bar{A}_F$ is well defined. Let $P: \mathscr{V} \to \mathscr{V}/\mathscr{R}^*$ be the canonical projection. If $F_i \in \mathbf{F}(\mathscr{V})$ $(i \in \mathbf{2})$ and $\bar{x} = Px \in \mathscr{V}/\mathscr{R}^*$ then

$$\begin{aligned} \bar{A}_{F_1}\bar{x} - \bar{A}_{F_2}\bar{x} &= PA_{F_1}x - PA_{F_2}x = P(A_{F_1} - A_{F_2})x \\ &= PB(F_1 - F_2)x \in P(\mathscr{B} \cap \mathscr{V}) \\ &\qquad [\text{since } x \in \mathscr{V} \text{ and } F_1, F_2 \in \mathbf{F}(\mathscr{V})] \\ &\subset P\mathscr{R}^* = 0. \end{aligned}$$

□

Corollary 5.2. *Under the conditions of Theorem 5.7, if $F \in \mathbf{F}(\mathscr{V})$ then*

$$\sigma[(A + BF)\,|\,\mathscr{V}] = \sigma_F \cup \sigma^*,$$

where

$$\sigma_F := \sigma[(A + BF) | \mathscr{R}^*]$$

is freely assignable by suitable choice of $F \in \mathbf{F}(\mathscr{V})$, *and*

$$\sigma^* := \sigma(\overline{A + BF})$$

is fixed for all $F \in \mathbf{F}(\mathscr{V})$.

5.5 Transmission Zeros

The situation described by Theorem 5.7 and Corollary 5.2 is depicted in the lattice diagram, Fig. 5.2. Here, we have taken a triple (D, A, B) and set $\mathscr{V}^* = \sup \mathfrak{I}(\text{Ker } D)$, $\mathscr{R}^* = \sup \mathfrak{C}(\text{Ker } D)$. Our results may be paraphrased by saying that if state feedback control is to be chosen to make $\mathscr{V}^*$ invariant then we enjoy complete freedom of spectrum assignment on $\mathscr{R}^*$, but have no residual freedom to modify in any way the (induced) dynamic action on $\mathscr{V}^*/\mathscr{R}^*$. A numerical illustration is presented in Exercise 5.6.

Now suppose (D, A, B) is complete. In analogy to the single-input, single-

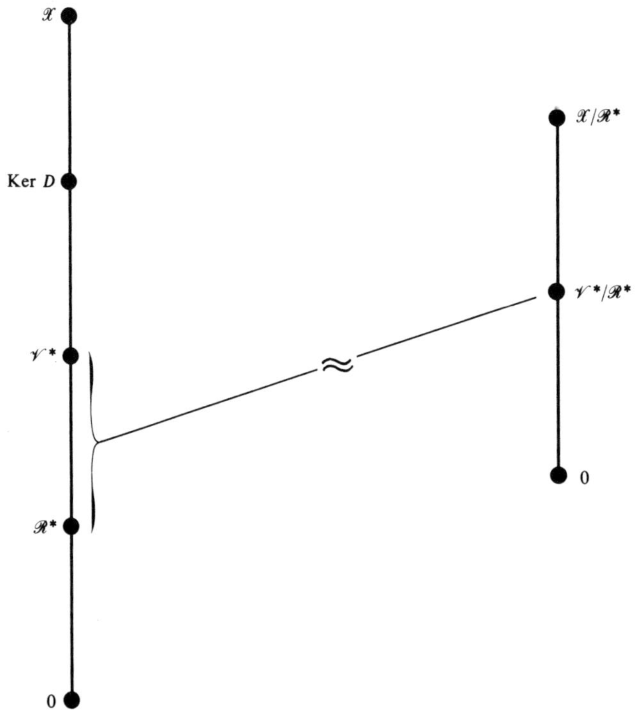

Figure 5.2 Lattice Diagram: Origin of Transmission Zeros. Transmission zeros constitute fixed spectrum of map $\bar{A}_F$ induced on $\mathscr{V}^*/\mathscr{R}^*$.

output case (where $\mathscr{R}^*$ is necessarily 0; see Exercise 4.5), the fixed spectrum $\sigma^* = \sigma^*(A, B; \operatorname{Ker} D)$ of Corollary 5.2 is termed the set of *transmission zeros* of (D, A, B). Consider the transfer matrix $H(\lambda) = D(\lambda 1 - A)^{-1}B$. It is known that, as a matrix of rational functions in λ, $H(\lambda)$ can be written in the form

$$H(\lambda) = P(\lambda)S(\lambda)Q(\lambda).$$

Here, $P(\lambda)$, $Q(\lambda)$ are square matrices having constant nonzero determinant, and $S(\lambda)$ is the *Smith-McMillan* form of $H(\lambda)$. If D: $p \times n$ and B: $n \times m$ then $S(\lambda)$ is a $p \times m$ diagonal matrix of the form

$$S(\lambda) = \operatorname{diag}\left[\frac{\epsilon_1(\lambda)}{\psi_1(\lambda)}, \ldots, \frac{\epsilon_k(\lambda)}{\psi_k(\lambda)}, \overbrace{0, \ldots, 0}^{r}\right].$$

Here $k + r = \min(p, m)$ and the ϵ_i and ψ_i are monic polynomials such that each pair (ϵ_i, ψ_i) is coprime, and the following divisibility relations are satisfied:

$$\epsilon_1 | \epsilon_2, \ldots, \epsilon_{k-1} | \epsilon_k;$$
$$\psi_2 | \psi_1, \ldots, \psi_k | \psi_{k-1}.$$

It can then be shown that k is the cyclic index of A, and ψ_i is the ith invariant factor of A. Finally, if $\epsilon_i = 1$ $(i \in \mathbf{l})$ and $\deg \epsilon_i \geq 1$ $(l + 1 \leq i \leq k)$, then $\epsilon_{l+1}, \ldots, \epsilon_k$ are precisely the invariant factors (listed in reverse order) of the map $\overline{A + BF}$ induced in $\mathscr{V}^*/\mathscr{R}^*$ for any $F \in \mathbf{F}(\mathscr{V}^*)$. Thus, the set $\sigma^*(A, B; \operatorname{Ker} D)$ of transmission zeros of the complete triple (D, A, B) is just the list of zeros of the nontrivial numerator polynomials in the Smith-McMillan form of the corresponding transfer matrix $H(\lambda)$. This statement provides the multivariable generalization of Exercise 4.5. While the proof is relatively straightforward it is somewhat outside our scope, and the reader is referred to the literature (Section 5.9).

5.6 Disturbance Decoupling with Stability

Most of the preceding ideas are illustrated by the problem of disturbance decoupling introduced in Section 4.1. But this time we impose the additional, realistic constraint that the closed loop system map $A + BF$ be stable or, more generally, have its spectrum in the "good" part of the complex plane. Thus, let $A: \mathscr{X} \to \mathscr{X}$, $B: \mathscr{U} \to \mathscr{X}$, $\mathscr{S} \subset \mathscr{X}$, and $\mathscr{K} \subset \mathscr{X}$, and let $\mathbb{C} = \mathbb{C}_g \cup \mathbb{C}_b$ be a symmetric partition of the complex plane. We pose the problem of

Disturbance Decoupling with Stability (DDPS). *Find (if possible) $F: \mathscr{X} \to \mathscr{U}$ such that*

$$\langle A + BF | \mathscr{S} \rangle \subset \mathscr{K} \quad \textit{and} \quad \sigma(A + BF) \subset \mathbb{C}_g. \tag{6.1}$$

To solve DDPS we start with the same heuristic approach as in Chapter 4. Noting that the subspace $\langle A + BF | \mathscr{S} \rangle$ is (A, B)-invariant and ought to be contained in $\mathscr{K}$, we introduce the family of subspaces

$$\mathfrak{V} := \{\mathscr{V} : \mathscr{V} \in \mathfrak{I}(\mathscr{K}), \ \& \ \exists F \in \mathbf{F}(\mathscr{V}), \sigma[(A + BF)|\mathscr{V}] \subset \mathbb{C}_g\}. \tag{6.2}$$

It is natural to conjecture that $\mathfrak{V}$ has a largest member $\mathscr{V}_g^*$. If $\mathscr{V}_g^*$ can be found, it is clearly enough to check that $\mathscr{V}_g^* \supset \mathscr{S}$, and then that the spectrum of the induced map in $\mathscr{X}/\mathscr{V}_g^*$ can also be assigned to $\mathbb{C}_g$.

We carry out this program as follows. Let

$$\mathscr{V}^* := \sup \mathfrak{I}(A, B; \mathscr{K}),$$

$$\mathscr{R}^* := \sup \mathfrak{C}(A, B; \mathscr{K}).$$

Choose $F_0 \in \mathbf{F}(\mathscr{V}^*)$, write $A_0 = A + BF_0$, let $P: \mathscr{X} \to \mathscr{X}/\mathscr{R}^*$ be the canonical projection, and let $\bar{A}_0$ be the map induced in $\mathscr{X}/\mathscr{R}^*$ by A_0: since by (4.3), $F_0 \in \mathbf{F}(\mathscr{R}^*)$, $\bar{A}_0$ is well-defined. We have that $\mathscr{V}^*/\mathscr{R}^*$ is $\bar{A}_0$-invariant, for

$$\bar{A}_0 \left(\frac{\mathscr{V}^*}{\mathscr{R}^*} \right) = \bar{A}_0 P\mathscr{V}^* = PA_0 \mathscr{V}^* \subset P\mathscr{V}^* = \frac{\mathscr{V}^*}{\mathscr{R}^*}.$$

Also, by Theorem 5.7 the restriction of $\bar{A}_0$ to $\mathscr{V}^*/\mathscr{R}^*$ is independent of the choice of $F_0 \in \mathbf{F}(\mathscr{V}^*)$. Let $\beta(\lambda)$ be the m.p. of $\bar{A}_0 | (\mathscr{V}^*/\mathscr{R}^*)$. Factor $\beta(\lambda) = \beta_g(\lambda)\beta_b(\lambda)$, where the zeros of β_g (resp. β_b) in $\mathbb{C}$ belong to $\mathbb{C}_g$ (resp. $\mathbb{C}_b$); and write

$$\bar{\mathscr{X}}_g^* := \frac{\mathscr{V}^*}{\mathscr{R}^*} \cap \operatorname{Ker} \beta_g(\bar{A}_0), \qquad \bar{\mathscr{X}}_b^* := \frac{\mathscr{V}^*}{\mathscr{R}^*} \cap \operatorname{Ker} \beta_b(\bar{A}_0).$$

Let us now take stock. In $\mathscr{R}^*$, we have that $\sigma(A_0 | \mathscr{R}^*)$ can be assigned arbitrarily, hence to $\mathbb{C}_g$, by suitable choice of $F_0 \in \mathbf{F}(\mathscr{V}^*)$; whereas nothing can be done with such F_0 to change $\sigma[\bar{A}_0 | \mathscr{V}^*/\mathscr{R}^*]$. Now, β_g and β_b coprime implies

$$\frac{\mathscr{V}^*}{\mathscr{R}^*} = \bar{\mathscr{X}}_g^* \oplus \bar{\mathscr{X}}_b^*. \tag{6.3}$$

Thus, the subspace of "good" modes of $A_0 | \mathscr{V}^*$ can be made to be just that subspace $\mathscr{V} \subset \mathscr{V}^*$ (whose existence will shortly be established) for which

$$\sigma(A_0 | \mathscr{V}) = \sigma(A_0 | \mathscr{R}^*) \cup \sigma(\bar{A}_0 | \bar{\mathscr{X}}_g^*). \tag{6.4}$$

With (6.4) as objective, define

$$\mathscr{V}_g^* := P^{-1}\bar{\mathscr{X}}_g^*. \tag{6.5}$$

Now we can prove

Lemma 5.8. *The subspace* $\mathscr{V}_g^*$ *defined by* (6.5) *is the largest member of the family* $\mathfrak{V}$ *defined by* (6.2).

Proof. We first check that $\mathscr{V}_g^* \in \mathfrak{V}$. Now,

$$P\mathscr{V}_g^* = P(P^{-1}\bar{\mathscr{X}}_g^*) = \bar{\mathscr{X}}_g^* \subset P\mathscr{V}^*, \tag{6.6}$$

and therefore, $\mathscr{V}_g^* \subset \mathscr{V}^* + \mathscr{R}^* = \mathscr{V}^* \subset \mathscr{X}$. Also,

$$PA_0\mathscr{V}_g^* = PA_0P^{-1}\bar{\mathscr{X}}_g^* = \bar{A}_0P(P^{-1}\bar{\mathscr{X}}_g^*) \subset \bar{A}_0\bar{\mathscr{X}}_g^* \subset \bar{\mathscr{X}}_g^*,$$

hence, $A_0\mathscr{V}_g^* \subset P^{-1}\bar{\mathscr{X}}_g^* = \mathscr{V}_g^*$, and so $A\mathscr{V}_g^* \subset \mathscr{V}_g^* + \mathscr{B}$. Since $F_0 \in \mathbf{F}(\mathscr{V}^*)$ was arbitrary, we have also shown that $\mathbf{F}(\mathscr{V}^*) \subset \mathbf{F}(\mathscr{V}_g^*)$. Next, as $\mathscr{V}_g^* \supset \text{Ker } P = \mathscr{R}^*$, we can indeed choose

$$F_0 \in \mathbf{F}(\mathscr{V}^*) \subset \mathbf{F}(\mathscr{R}^*) \cap \mathbf{F}(\mathscr{V}_g^*),$$

such that $\sigma(A_0 \,|\, \mathscr{R}^*) \subset \mathbb{C}_g$. Then (6.3), (6.5) and (6.6) imply that (6.4) holds with $\mathscr{V} = \mathscr{V}_g^*$, and so finally, $\mathscr{V}_g^* \in \mathfrak{V}$.

To show that $\mathscr{V}_g^*$ is supremal, choose an arbitrary $\mathscr{V} \in \mathfrak{V}$. By Lemma 5.7 we choose $F \in \mathbf{F}(\mathscr{V}) \cap \mathbf{F}(\mathscr{V}^*)$, and write $A_F := A + BF$. Since $F \in \mathbf{F}(\mathscr{R}^*)$, the induced map $\bar{A}_F$ on $\mathscr{X}/\mathscr{R}^*$ exists, and by Theorem 5.7 coincides on $\mathscr{V}^*/\mathscr{R}^*$ with $\bar{A}_0$. Then, as $\bar{A}_F P\mathscr{V} = PA_F\mathscr{V} \subset P\mathscr{V}$, $P\mathscr{V}$ is also $\bar{A}_0$-invariant. From (6.3) and the fact that β_g and β_b are coprime, we can therefore write

$$P\mathscr{V} = (P\mathscr{V}) \cap \bar{\mathscr{X}}_g^* \oplus (P\mathscr{V}) \cap \bar{\mathscr{X}}_b^*. \tag{6.7}$$

From (6.7) and the assumption $\mathscr{V} \in \mathfrak{V}$ there follows $(P\mathscr{V}) \cap \bar{\mathscr{X}}_b^* = 0$, hence $P\mathscr{V} \subset \bar{\mathscr{X}}_g^*$. Then,

$$\mathscr{V} \subset P^{-1}\bar{\mathscr{X}}_g^* = \mathscr{V}_g^*,$$

hence $\mathscr{V}_g^*$ is supremal, as claimed. □

The inclusion relations involved in the foregoing discussion are summarized in the lattice diagrams, Fig. 5.3.

It is now easy to prove the main result.

Theorem 5.8. *Suppose (A, B) is controllable. Then DDPS is solvable if and only if*

$$\mathscr{V}_g^* \supset \mathscr{S}. \tag{6.8}$$

Since $\mathscr{V}^* \supset \mathscr{V}_g^*$, the condition (6.8) strengthens the condition $\mathscr{V}^* \supset \mathscr{S}$ given by Theorem 4.2 for solvability of DDP.

PROOF. Suppose (6.8) holds. We have already shown that there exists $F_0 \in \mathbf{F}(\mathscr{V}_g^*)$, such that

$$\sigma[(A + BF_0) \,|\, \mathscr{V}_g^*] \subset \mathbb{C}_g.$$

Since $(A + BF_0, B)$ is controllable, by Proposition 1.2 the corresponding pair $(\tilde{A}_0, \tilde{B})$ induced in $\mathscr{X}/\mathscr{V}_g^*$ is also controllable, hence there exists $\tilde{F}: \mathscr{X}/\mathscr{V}_g^* \to \mathscr{U}$, such that

$$\sigma(\tilde{A}_0 + \tilde{B}\tilde{F}) \subset \mathbb{C}_g. \tag{6.9}$$

Let $Q: \mathscr{X} \to \mathscr{X}/\mathscr{V}_g^*$ be the canonical projection and set $F = F_0 + \tilde{F}Q$. Then it follows immediately that F has the required properties (6.1).

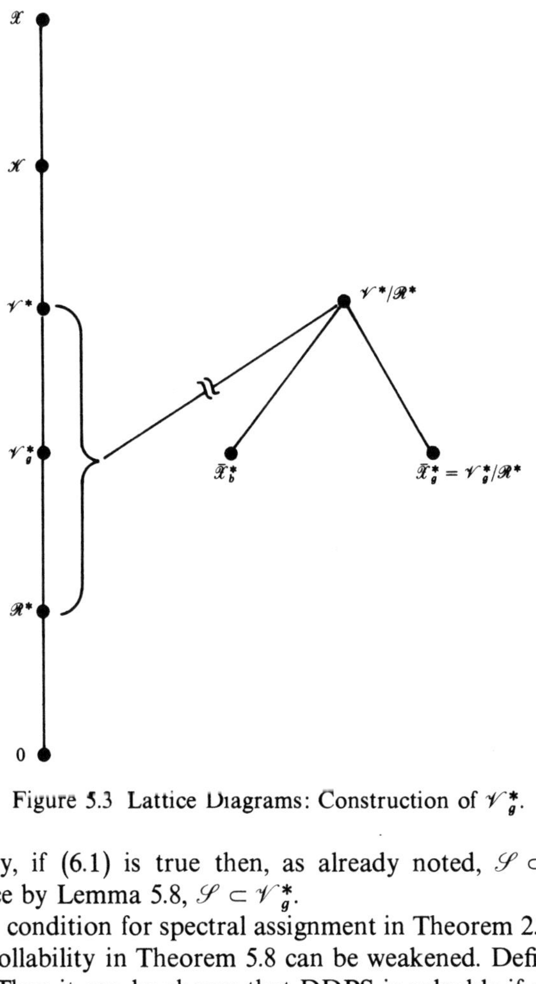

Figure 5.3 Lattice Diagrams: Construction of $\mathscr{V}_g^*$.

Conversely, if (6.1) is true then, as already noted, $\mathscr{S} \subset \mathscr{V}$ for some $\mathscr{V} \in \mathfrak{V}$, hence by Lemma 5.8, $\mathscr{S} \subset \mathscr{V}_g^*$. □

Using the condition for spectral assignment in Theorem 2.2, the hypothesis of controllability in Theorem 5.8 can be weakened. Define $\mathscr{X}_b(A)$ as in Section 2.3. Then it can be shown that DDPS is solvable if and only if

$$\mathscr{V}_g^* \supset \mathscr{S} \quad \text{and} \quad \mathscr{X}_b(A) \subset \langle A \mid \mathscr{B} \rangle. \tag{6.10}$$

A numerical illustration of DDPS is provided in Exercise 5.10, and a modification of the result for the case of direct control feedthrough is stated in Exercise 5.13.

5.7 Controllability Indices

In this chapter we have studied several constructions involving the transformation of a given controllable pair (A, B) into a new pair $(A + BF, BG)$. It is of some interest to ask what properties, if any, of (A, B) remain invariant under *all* such transformations. We shall investigate this question under the restriction that the admissible maps $G: \mathscr{U} \to \mathscr{U}$ are precisely the isomorphisms of $\mathscr{U}$; it is then clear that, as far as BG is concerned, the sole invariant is the "geometric" invariant $\mathscr{B} = \operatorname{Im} B \subset \mathscr{X}$. As for $A + BF$, we already know that $\sigma(A + BF)$ can be assigned arbitrarily by suitable choice of F. This suggests that most of the linear algebraic structure of A is erased if we let F vary through the set of all possible maps from $\mathscr{X}$ to $\mathscr{U}$. What is left will depend on how $\mathscr{B}$ is intertwined with the geometric structure of A.

To make the problem precise, we fix n and m $(1 \le m \le n)$, and consider the subset of all matrix pairs

$$(A, B) \in \mathbb{R}^{n \times n} \times \mathbb{R}^{n \times m}$$

for which B is monic and (A, B) is controllable, namely

$$\operatorname{Rank} B = m \tag{7.1a}$$

and

$$\operatorname{Rank}[B, AB, \ldots, A^{n-1}B] = n. \tag{7.1b}$$

Listing the elements of A and B in some arbitrary order, we regard (A, B) as a point $\mathbf{p} \in \mathbb{R}^N$, $N = n^2 + nm$. Denote by $\mathbf{S}$ the subset of $\mathbb{R}^N$ determined by (7.1): clearly, $\mathbf{S}$ is the complement of a proper algebraic variety in $\mathbb{R}^N$. Next, introduce the class $\mathfrak{G}$ of transformations $\mathfrak{g}$ on $\mathbb{R}^N$ defined by

$$\mathfrak{g}\mathbf{p} := \mathfrak{g}(A, B) := (T^{-1}(A + BF)T,\ T^{-1}BG). \tag{7.2}$$

That is, each $\mathfrak{g} \in \mathfrak{G}$ is represented by a distinct triple of matrices

$$(T, F, G) \in \mathbb{R}^{n \times n} \times \mathbb{R}^{m \times n} \times \mathbb{R}^{m \times m},$$

where T, and also G, are required to be nonsingular. It is easy to see that the action (7.2) assigns to $\mathfrak{G}$ the structure of a *transformation group*, with identity

$$1 = (1_n, 0, 1_m),$$

composition rule

$$\mathfrak{g}_2 \circ \mathfrak{g}_1 = (T_1 T_2,\ F_1 + G_1 F_2 T_1^{-1},\ G_1 G_2)$$

and inverse

$$\mathfrak{g}^{-1} = (T^{-1},\ -G^{-1}FT,\ G^{-1}).$$

If (A, B) is controllable and B is monic, clearly the pair $\mathfrak{g}(A, B)$ given by (7.2) also has these properties for each $\mathfrak{g} \in \mathfrak{G}$; namely, $\mathbf{S}$ is invariant under the action of $\mathfrak{G}$:

$$\mathfrak{G}\mathbf{S} := \bigcup_{\mathfrak{g} \in \mathfrak{G}} \mathfrak{g}\mathbf{S} = \mathbf{S}.$$

Fix $\mathbf{p} \in \mathbf{S}$. The subset

$$\mathfrak{G}\mathbf{p} := \{\mathfrak{g}\mathbf{p} : \mathfrak{g} \in \mathfrak{G}\} \subset \mathbf{S}$$

is the *orbit* of $\mathbf{p}$ under $\mathfrak{G}$. It is easy to see that the orbits partition $\mathbf{S}$ into disjoint subsets. In other words, $\mathfrak{G}$ induces on $\mathbf{S}$ an equivalence relation by the definition: $\mathbf{p}_1, \mathbf{p}_2 \in \mathbf{S}$ are *equivalent* if and only if $\mathbf{p}_1, \mathbf{p}_2$ belong to the same orbit, namely, $\mathbf{p}_2 = \mathfrak{g}\mathbf{p}_1$ for some $\mathfrak{g} \in \mathfrak{G}$. Denote the set of orbits by $\boldsymbol{\Omega} := \mathbf{S}/\mathfrak{G}$. Translated into the present setting, our original problem is simply to parametrize $\boldsymbol{\Omega}$ in some effective way, and to identify with each $\boldsymbol{\omega} \in \boldsymbol{\Omega}$ a unique element $\mathbf{p}^* = \mathbf{p}^*(\boldsymbol{\omega}) \in \boldsymbol{\omega}$ which exhibits the invariant structure of all $\mathbf{p} = (A, B)$ on the orbit $\boldsymbol{\omega}$. Without posing this problem formally and abstractly, we shall pass immediately to the main results, which are self-explanatory.

Theorem 5.9. *There is a bijection from $\boldsymbol{\Omega}$ to the set of all lists of integers*

$$\kappa = (\kappa_1, \kappa_2, \ldots, \kappa_m)$$

with the properties

$$\kappa_1 \geq \kappa_2 \geq \cdots \geq \kappa_m \geq 1$$

and

$$\kappa_1 + \kappa_2 + \cdots + \kappa_m = n.$$

Thus, if $n = 8$ and $m = 3$, there are exactly five distinct orbits, corresponding to the lists

$$(3, 3, 2), (4, 2, 2), (4, 3, 1), (5, 2, 1), (6, 1, 1).$$

From now on we write $\kappa^*(\boldsymbol{\omega})$ for the list κ which labels $\boldsymbol{\omega} \in \boldsymbol{\Omega}$. Our second result assigns to $\kappa^*(\boldsymbol{\omega})$ its geometric meaning relative to A and B.

Theorem 5.10. *Let $\mathbf{p} = (A, B)$ in $\boldsymbol{\omega}$ and let $\kappa^*(\boldsymbol{\omega}) = (\kappa_1, \ldots, \kappa_m)$. There are (A, B)-controllability subspaces $\mathscr{R}_1, \ldots, \mathscr{R}_m$ with $d(\mathscr{R}_i) = \kappa_i$, such that*

$$\bigoplus_{i=1}^{m} \mathscr{R}_i = \mathscr{X}. \tag{7.3}$$

Furthermore, for any decomposition of $\mathscr{X}$ as the direct sum of m nonzero independent c.s. $\mathscr{R}_i$, it is necessarily true, possibly after reordering, that

$$d(\mathscr{R}_i) = \kappa_i, \qquad i \in \mathbf{m}.$$

Corollary 5.3. *Under the conditions of Theorem 5.10, there are bases for $\mathscr{X}$ and $\mathscr{U}$, and a map $F: \mathscr{X} \to \mathscr{U}$, such that*

$$\text{Mat}(A + BF) = \text{Mat } A^* := \text{diag}[A_1, \ldots, A_m]$$
$$\text{Mat } B = \text{Mat } B^* := \text{diag}[b_1, \ldots, b_m].$$

Here, A_i and b_i are given by

$$A_i = \begin{bmatrix} 0 & 1 & 0 & \cdot & \cdot & \cdot & 0 \\ 0 & 0 & 1 & \cdot & \cdot & \cdot & 0 \\ \cdot & \cdot & \cdot & \cdot & \cdot & \cdot & \cdot \\ 0 & \cdot & \cdot & \cdot & \cdot & 0 & 1 \\ 0 & \cdot & \cdot & \cdot & \cdot & \cdot & 0 \end{bmatrix}_{\kappa_i \times \kappa_i}, \qquad b_i = \begin{bmatrix} 0 \\ \vdots \\ 0 \\ 1 \end{bmatrix}_{\kappa_i \times 1}$$

Thus, any matrix pair (A, B) in $\mathbf{S}$ can be transformed into exactly one pair (A^*, B^*) by suitable choice of (T, F, G). Conversely, to any list κ there corresponds exactly one orbit in $\mathbf{S}$, namely the set of all transforms under $\mathfrak{G}$ of the corresponding matrix pair (A^*, B^*). If $\mathbf{p} = (A, B) \in \boldsymbol{\omega}$ we define $\mathbf{p}^*(\boldsymbol{\omega}) = (A^*, B^*)$, and call this pair the *canonical form* of (A, B) under $\mathfrak{G}$. The canonical form exhibits the original "system" $\dot{x} = Ax + Bu$, after transformation, as a parallel array of m decoupled "subsystems" of dynamic order $\kappa_1, \ldots, \kappa_m$.[1]

If $\mathbf{p} = (A, B) \in \boldsymbol{\omega}$ and $\kappa^*(\boldsymbol{\omega}) = (\kappa_1, \ldots, \kappa_m)$, the integers κ_i $(i \in \mathbf{m})$ are the *controllability indices* of (A, B). With some abuse of terminology κ_1, the largest controllability index, is called "*the*" *controllability index* of (A, B); as will be clear from (7.8), below,

$$\kappa_1 = \min\{j: 1 \le j \le n, \mathscr{B} + A\mathscr{B} + \cdots + A^{j-1}\mathscr{B} = \mathscr{X}\}.$$

For the proof of Theorems 5.9 and 5.10, we need

Lemma 5.9. *If $b \in \mathscr{B}$ has the properties*

$$A^k b \in \mathscr{B} + A\mathscr{B} + \cdots + A^{k-1}\mathscr{B},$$

and

$$A^{k-1} b \notin \mathscr{B} + A\mathscr{B} + \cdots + A^{k-2}\mathscr{B}$$

for some k $(2 \le k \le n)$, then there exists $F: \mathscr{X} \to \mathscr{U}$, such that

$$(A + BF)^k b = 0.$$

Further, the vectors

$$b, (A + BF)b, \ldots, (A + BF)^{k-1} b$$

are linearly independent.

PROOF. By assumption, there exist vectors $b_1, \ldots, b_k \in \mathscr{B}$, such that

$$A(\cdots A((Ab - b_1) - b_2) \cdots - b_{k-1}) - b_k = 0;$$

and the vectors $x_1, \ldots, x_k$, given by

$$b, Ab - b_1, \ldots, A^{k-1}b - A^{k-2}b_1 - \cdots - b_{k-1},$$

are linearly independent. Define $F: \mathscr{X} \to \mathscr{U}$, such that $BFx_i = -b_i$ $(i \in \mathbf{k})$; then F has the required properties. □

We define the controllability indices κ_i of (A, B) as follows. Write

$$\mathscr{S}_j := \mathscr{B} + A\mathscr{B} + \cdots + A^j\mathscr{B} \tag{7.4}$$

for $j = 0, 1, \ldots, n - 1$; and let

$$\rho_0 := m; \qquad \rho_j := d\left(\frac{\mathscr{S}_j}{\mathscr{S}_{j-1}}\right), \qquad j \in \mathbf{n - 1}. \tag{7.5}$$

[1] In applications, however, these subsystems need not have any particular physical identity. Moreover, the $\mathscr{R}_i$ themselves, unlike κ, are in general not unique.

It is easy to check (Exercise 5.15) that

$$\rho_0 \geq \rho_1 \geq \cdots \geq \rho_{n-1} \geq 0 \tag{7.6}$$

and

$$\rho_0 + \rho_1 + \cdots + \rho_{n-1} = n. \tag{7.7}$$

Now define, for $i \in \mathbf{m}$,

$$\kappa_i := \text{number of integers in the set } \{\rho_0, \rho_1, \ldots, \rho_{n-1}\} \text{ which are } \geq i. \tag{7.8}$$

Thus,

$$\kappa_1 \geq \kappa_2 \geq \cdots \geq \kappa_m$$

and by (7.7),

$$\kappa_1 + \kappa_2 + \cdots + \kappa_m = n.$$

Since for any $F: \mathscr{X} \to \mathscr{U}$,

$$\mathscr{S}_j = \mathscr{B} + (A + BF)\mathscr{B} + \cdots + (A + BF)^j\mathscr{B} \tag{7.9}$$

it is clear that the ρ_j and κ_j are invariant under a transformation $A \mapsto A + BF$.

PROOF (of Theorems 5.9 and 5.10). Let $b_1, \ldots, b_m$ be a basis for $\mathscr{B}$; write down the list

$$b_1, \ldots, b_m;\ Ab_1, \ldots, Ab_m;\ \ldots;\ A^{n-1}b_1, \ldots, A^{n-1}b_m;$$

and working from left to right, delete each vector which is linearly dependent on its predecessors. By a relabeling of the b_i, if necessary, we may arrange that terms in $b_m, \ldots, b_1$, respectively, disappear from the list in that order. The reduced list will then look like

$$\begin{array}{l}
b_1, \ldots, b_m;\ \ldots;\ A^{\kappa_m-1}b_1, \ldots, A^{\kappa_m-1}b_m; \\
A^{\kappa_m}b_1, \ldots, A^{\kappa_m}b_{m-1};\ \ldots;\ A^{\kappa_{m-1}-1}b_1, \ldots, A^{\kappa_{m-1}-1}b_{m-1}; \\
\cdot \quad \cdot \quad \cdot \quad \cdot \quad \cdot \quad \cdot \quad \cdot \quad \cdot \\
A^{\kappa_3}b_1, A^{\kappa_3}b_2;\ \ldots;\ A^{\kappa_2-1}b_1, A^{\kappa_2-1}b_2; \\
A^{\kappa_2}b_1;\ \ldots;\ A^{\kappa_1-1}b_1.
\end{array}$$

Here, the κ_i ($i \in \mathbf{m}$) are the controllability indices of (A, B): that the list takes the form shown is an immediate consequence of their definition.[1] Thus, for $j \in \mathbf{m}$,

$$A^{\kappa_j}b_j \in \mathscr{B} + A\mathscr{B} + \cdots + A^{\kappa_j-1}\mathscr{B} + A^{\kappa_j}\operatorname{Span}\{b_1, \ldots, b_{j-1}\},$$

but

$$A^{\kappa_j-1}b_j \notin \mathscr{B} + A\mathscr{B} + \cdots + A^{\kappa_j-2}\mathscr{B} + A^{\kappa_j-1}\operatorname{Span}\{b_1, \ldots, b_{j-1}\}.$$

From these relations it can be seen that if each b_j is replaced by a suitable vector

$$(b_j)_{\text{new}} = b_j - \tilde{b}_j,\ \tilde{b}_j \in \operatorname{Span}\{b_1, \ldots, b_{j-1}\},$$

then we obtain

$$A^{\kappa_j}(b_j)_{\text{new}} \in \mathscr{B} + A\mathscr{B} + \cdots + A^{\kappa_j-1}\mathscr{B} \tag{7.10}$$

[1] If $\kappa_{j-1} = \kappa_j$ for some j, the array should be contracted by suitable row deletions.

with

$$A^{\kappa_j-1}(b_j)_{\text{new}} \notin \mathscr{B} + A\mathscr{B} + \cdots + A^{\kappa_j-2}\mathscr{B}. \tag{7.11}$$

From now on write simply b_j for $(b_j)_{\text{new}}$. Applying Lemma 5.9 to each of the vectors $b_1, \ldots, b_m$, we obtain maps $F_i: \mathscr{X} \to \mathscr{U}$ and subspaces

$$\mathscr{R}_i = \langle A + BF_i | b_i \rangle, \qquad i \in \mathbf{m}, \tag{7.12}$$

such that $d(\mathscr{R}_i) = \kappa_i$ and

$$(A + BF_i)^{\kappa_i}\mathscr{R}_i = 0, \qquad i \in \mathbf{m}. \tag{7.13}$$

Now the vectors in the above array remain linearly independent if each element, say $A^j b_i$, is replaced by a vector of form

$$A^j b_i + \sum_{r=1}^{j} A^{r-1}\hat{b}_r$$

for arbitrary $\hat{b}_r \in \mathscr{B}$. Using this fact, we see that the $\mathscr{R}_i$ are linearly independent. Defining $F: \mathscr{X} \to \mathscr{U}$ by setting $F | \mathscr{R}_i = F_i | \mathscr{R}_i$ $(i \in \mathbf{m})$, we have that (7.12) and (7.13) are true with F_i replaced by F.

By the remark following (7.9) it is obvious that $\kappa = (\kappa_1, \ldots, \kappa_m)$ is invariant on the $\mathfrak{G}$-orbit of the matrix pair (A, B). Furthermore, distinct orbits $\boldsymbol{\omega}_1$, $\boldsymbol{\omega}_2$ yield distinct evaluations of $\kappa^*(\boldsymbol{\omega}_1)$, $\kappa^*(\boldsymbol{\omega}_2)$: for if $\mathbf{p}_i \in \boldsymbol{\omega}_i$ and $\kappa^*(\boldsymbol{\omega}_i) = \kappa$ $(i \in \mathbf{2})$, there exist $\mathfrak{g}_i \in \mathfrak{G}$ such that $\mathfrak{g}_i \mathbf{p}_i = \mathbf{p}^*$, where $\mathbf{p}^*$ is the canonical pair (A^*, B^*) with indices κ; therefore, $\mathbf{p}_2 = \mathfrak{g}_2^{-1}\mathfrak{g}_1\mathbf{p}_1$ and so $\boldsymbol{\omega}_2 = \boldsymbol{\omega}_1$.

It remains to check the uniqueness assertion of Theorem 5.10. Suppose $\mathscr{X}$ admits a second decomposition of type (7.3) into c.s. $\hat{\mathscr{R}}_i$, with $d(\hat{\mathscr{R}}_i) = \mu_i$ $(i \in \mathbf{m})$ arranged in descending order. Since the κ_i are feedback invariants we can replace A by $A + BF$, where $(A + BF)^{\mu_i}\hat{\mathscr{R}}_i = 0$ $(i \in \mathbf{m})$. Defining $\rho_j = d(\mathscr{S}_j/\mathscr{S}_{j-1})$ as before, we find by an easy direct computation that

$$\rho_j = \begin{cases} m, & j = 0, 1, \ldots, \mu_m - 1 \\ m - 1, & j = \mu_m, \ldots, \mu_{m-1} - 1 \\ \vdots & \\ 1, & j = \mu_2, \ldots, \mu_1 - 1 \\ 0, & j = \mu_1, \ldots, n - 1. \end{cases}$$

It follows immediately that $\kappa_i = \mu_i$ $(i \in \mathbf{m})$ and the proof is complete. □

From symmetry considerations one would expect that generically, the dimensions κ_i of the $\mathscr{R}_i$ in (7.3) would all be about equal. This is the content of

Corollary 5.4. *Let $1 \le m \le n$ and suppose $n = km + v$ for some v, $0 \le v < m$. Then in $\mathbb{R}^N$ $(N = n^2 + nm)$, generically,*

$$\kappa_i = \begin{cases} k + 1, & 1 \le i \le v \\ k, & v + 1 \le i \le m. \end{cases}$$

PROOF. The result is immediate by (7.4)–(7.8), if we note that, generically,

$$d(\mathscr{B} + A\mathscr{B} + \cdots + A^{j-1}\mathscr{B}) = \min(n, jm). \qquad \square$$

To conclude this section we cite the following amusing application.

Theorem 5.11. *Let* (A, B) *be controllable, with controllability indices* $(\kappa_1, \ldots, \kappa_m)$. *Then the possible dimensions of the nonzero c.s. of* (A, B) *are given by the list:*

$$\begin{aligned}
&\kappa_m;\\
&\kappa_{m-1}, \kappa_{m-1} + 1, \ldots, \kappa_{m-1} + \kappa_m;\\
&\kappa_{m-2}, \kappa_{m-2} + 1, \ldots, \kappa_{m-2} + \kappa_{m-1} + \kappa_m;\\
&\cdot \quad \cdot \quad \cdot \quad \cdot \quad \cdot \quad \cdot\\
&\kappa_1, \kappa_1 + 1, \ldots, \kappa_1 + \kappa_2 + \cdots + \kappa_m.
\end{aligned} \tag{7.14}$$

There is exactly one c.s. of dimension $r \neq 0$ *if* (i) $r = n$, *or* (ii) *for some* $j \in \mathbf{m-1}$,

$$\kappa_j > r = \kappa_{j+1} + \kappa_{j+2} + \cdots + \kappa_m. \tag{7.15}$$

If $r \neq n$ *and* (7.15) *fails, but* r *is in the list* (7.14), *there are nondenumerably many distinct c.s. of dimension* r.

For example, if $\kappa = (5, 2, 1)$, there are nonzero c.s. of dimension r if and only if $1 \le r \le 8$ and $r \neq 4$. These c.s. are unique if $r = 1$, 3, or 8 but are nondenumerably many if $r = 2$, 5, 6, or 7.

The proof is left as Exercise 5.16.

5.8 Exercises

5.1. A necessary condition that $\mathscr{R}$ be a c.s. is that $\mathscr{R} + \mathscr{B} = A\mathscr{R} + \mathscr{B}$. Is this condition sufficient?

5.2. Show by an example that the intersection of two c.s. need not be a c.s., or even be (A, B)-invariant. Show, however, that the family of c.s. of (A, B) is a lattice relative to the operations $+$ and $\wedge$, where

$$\mathscr{R}_1 \wedge \mathscr{R}_2 := \sup \mathfrak{C}(A, B; \mathscr{R}_1 \cap \mathscr{R}_2).$$

Is this lattice modular?

5.3. Let $\mathscr{S} \subset \mathscr{X}$ be a fixed subspace and consider the family $\mathscr{F}$ of c.s. $\mathscr{R}$, such that $\mathscr{R} \supset \mathscr{S}$. Show by example that, in general, $\mathscr{F}$ does not possess a (unique) infimal element. Show, however, that if $\mathscr{F}$ is nonempty it possesses at least one *minimal* element $\mathscr{R}_0$ such that $\mathscr{R}_0 \supset \mathscr{R} \in \mathscr{F}$ implies $\mathscr{R} = \mathscr{R}_0$.

5.4. Suppose (A, B) is controllable and $A\mathscr{R} \subset \mathscr{R}$. Write $\mathscr{X} = \mathscr{R} \oplus \mathscr{S}$, and in a compatible basis let

$$A = \begin{bmatrix} A_1 & A_3 \\ 0 & A_2 \end{bmatrix}, \qquad B = \begin{bmatrix} B_1 \\ B_2 \end{bmatrix}.$$

Show that (A_1, B_1) is controllable if $\mathscr{R}$ is a c.s. In that case there exists G, such that $\hat{B} := BG$ has the matrix

$$\hat{B} = \begin{bmatrix} \hat{B}_{11} & \hat{B}_{12} \\ 0 & \hat{B}_{22} \end{bmatrix},$$

where $(A_1, \hat{B}_{11})$ is controllable.

5.5. Develop a procedure for the numerical computation of $\mathscr{R}^* := \sup \mathfrak{C}(A, B; \operatorname{Ker} D)$. Hint: First compute $\mathscr{V}^* := \sup \mathfrak{I}(A, B; \operatorname{Ker} D)$ by the procedure of Exercise 4.2. Then implement the algorithm of Theorem 5.6, as follows. Let $\mathscr{V}^* = \operatorname{Im} V^*$ and let W^* be a maximal solution of $W^*V^* = 0$. Define $\mathscr{S}_\mu = \operatorname{Im} S_\mu$ and let T_μ be a max. sol. of $T_\mu[AS_{\mu-1}, B] = 0$, with $S_0 = 0$. Then, S_μ is obtained as a max. sol. of

$$\begin{bmatrix} W^* \\ T_\mu \end{bmatrix} S_\mu = 0, \qquad \mu = 1, 2, \ldots.$$

Since $\mathscr{S}_{\mu+1} \supset \mathscr{S}_\mu$ one has as a check,

$$\operatorname{Rank}[S_{\mu+1}, S_\mu] = \operatorname{Rank} S_{\mu+1};$$

and the stopping test is $\operatorname{Rank} S_{k+1} = \operatorname{Rank} S_k$, i.e. $\mathscr{R}^* = \operatorname{Im} S_k$.

As an illustration suppose

$$A = \begin{bmatrix} 1 & 0 & 0 & 0 & 1 & 0 \\ 0 & 0 & 1 & 0 & 0 & 0 \\ 0 & -1 & 0 & 0 & 0 & 0 \\ 0 & 0 & 0 & 0 & 1 & 0 \\ 0 & 0 & 0 & 1 & 0 & 0 \\ 0 & 0 & 0 & 0 & 0 & 1 \end{bmatrix}, \qquad B = \begin{bmatrix} 0 & 1 & 0 \\ 0 & 0 & 0 \\ 1 & 0 & 0 \\ 0 & 0 & 1 \\ 0 & 0 & 0 \\ 0 & 0 & 0 \end{bmatrix},$$

$$V^* = \begin{bmatrix} 1 & 0 & 0 & 1 & 2 \\ 0 & 1 & 0 & 0 & 0 \\ 0 & 0 & 1 & 0 & 0 \\ 0 & 0 & 0 & 0 & 0 \\ 1 & 0 & 0 & 0 & 0 \\ 0 & 0 & 0 & 0 & 1 \end{bmatrix}.$$

Then $W^* = [0 \quad 0 \quad 0 \quad 1 \quad 0 \quad 0]$ and

$$T_1 = \begin{bmatrix} 0 & 1 & 0 & 0 & 0 & 0 \\ 0 & 0 & 0 & 0 & 1 & 0 \\ 0 & 0 & 0 & 0 & 0 & 1 \end{bmatrix}, \qquad S_1 = \begin{bmatrix} 1 & 0 \\ 0 & 0 \\ 0 & 1 \\ 0 & 0 \\ 0 & 0 \\ 0 & 0 \end{bmatrix},$$

$$T_2 = \begin{bmatrix} 0 & 0 & 0 & 0 & 1 & 0 \\ 0 & 0 & 0 & 0 & 0 & 1 \end{bmatrix}, \qquad S_2 = \begin{bmatrix} 1 & 0 & 0 \\ 0 & 1 & 0 \\ 0 & 0 & 1 \\ 0 & 0 & 0 \\ 0 & 0 & 0 \\ 0 & 0 & 0 \end{bmatrix}.$$

This leads to $\operatorname{Rank} S_3 = \operatorname{Rank} S_2$, i.e. $\mathscr{R}^* = \operatorname{Im} S_2$.

5.6. Construct an example to illustrate Theorem 5.7. Hint: Continuing from Exercise 5.5, first compute $\mathbf{F}(\mathscr{V}^*)$, namely those $F: \mathscr{X} \to \mathscr{U}$, such that $(A + BF)\mathscr{V}^* \subset \mathscr{V}^*$. In matrices, these F are the solutions of

$$W^*(A + BF)V^* = 0,$$

or $W^*BFV^* = -W^*AV^*$. In this case, we get

$$[0 \quad 0 \quad 1]F^{3\times 6}\begin{bmatrix} 1 & 0 & 0 & 1 & 2 \\ 0 & 1 & 0 & 0 & 0 \\ 0 & 0 & 1 & 0 & 0 \\ 0 & 0 & 0 & 0 & 0 \\ 1 & 0 & 0 & 0 & 0 \\ 0 & 0 & 0 & 0 & 1 \end{bmatrix} = -[1 \quad 0 \quad 0 \quad 0 \quad 0].$$

Writing $F = [f_{ij}]$ ($i \in \mathbf{3}, j \in \mathbf{6}$) and solving,

$$F = \begin{bmatrix} f_{11} & f_{12} & f_{13} & f_{14} & f_{15} & f_{16} \\ f_{21} & f_{22} & f_{23} & f_{24} & f_{25} & f_{26} \\ 0 & 0 & 0 & f_{34} & -1 & 0 \end{bmatrix},$$

where the elements written f_{ij} are unrestricted. The general structure of $A + BF$, $F \in \mathbf{F}(\mathscr{V}^*)$ is now, by inspection,

$$A + BF = \begin{bmatrix} 1+f_{21} & f_{22} & f_{23} & f_{24} & 1+f_{25} & f_{26} \\ 0 & 0 & 1 & 0 & 0 & 0 \\ f_{11} & -1+f_{12} & f_{13} & f_{14} & f_{15} & f_{16} \\ 0 & 0 & 0 & f_{34} & 0 & 0 \\ 0 & 0 & 0 & 1 & 0 & 0 \\ 0 & 0 & 0 & 0 & 0 & 1 \end{bmatrix}.$$

Next, compute the matrix of $(A + BF)|\mathscr{V}^*$ and exhibit its action on $\mathscr{R}^*$: write $\mathscr{V}^* = \mathscr{R}^* \oplus \mathscr{S}$, where $\mathscr{S}$ is an arbitrary complement of $\mathscr{R}^*$ in $\mathscr{V}^*$, select a basis of $\mathscr{V}^*$ adapted to this decomposition, and transform the matrix to this basis. For example, with $\mathscr{R}^* = \text{Im } S_2$ above,

$$\mathscr{V}^* = \text{Im}\begin{bmatrix} 1 & 0 & 0 \\ 0 & 1 & 0 \\ 0 & 0 & 1 \\ 0 & 0 & 0 \\ 0 & 0 & 0 \\ 0 & 0 & 0 \end{bmatrix} \oplus \text{Im}\begin{bmatrix} 1 & 2 \\ 0 & 0 \\ 0 & 0 \\ 0 & 0 \\ 1 & 0 \\ 0 & 1 \end{bmatrix}.$$

Computing the action of $A + BF$ on these basis vectors yields the matrix

$$(A + BF)|\mathscr{V}^* = \left[\begin{array}{ccc|cc} 1+f_{21} & f_{22} & f_{23} & 2+f_{21}+f_{25} & 2f_{21}+f_{26} \\ 0 & 0 & 1 & 0 & 0 \\ f_{11} & -1+f_{12} & f_{13} & f_{11}+f_{15} & 2f_{11}+f_{16} \\ \hline 0 & 0 & 0 & 0 & 0 \\ 0 & 0 & 0 & 0 & 1 \end{array}\right].$$

From this the matrix of the fixed induced map $\overline{A + BF}$ on $\mathscr{V}^*/\mathscr{R}^*$ is read off as $\left[\begin{smallmatrix} 0 & 0 \\ 0 & 1 \end{smallmatrix}\right]$. Finally, it can be verified that the spectrum of $(A + BF)|\mathscr{R}^*$, i.e. of the upper left block, is arbitrarily assignable by setting $f_{13} = f_{23} = 0, f_{22} = 1$, and suitably choosing f_{11}, f_{12} and f_{21}.

5.7. Given $\dot{x} = Ax + Bu$ and $\mathscr{K} \subset \mathscr{X}$, show that the largest c.s. $\mathscr{R}^* \subset \mathscr{K}$ is characterized by the following property: $\mathscr{R}^*$ is the largest subspace $\mathscr{S} \subset \mathscr{K}$, such that, if $x(0) = 0$ and $x \in \mathscr{S}$, there exists a continuous control $u(t)$, $0 \leq t \leq 1$, for which $x(t) \in \mathscr{K}$, $0 \leq t \leq 1$, and $x(1) = x$. Thus, $\mathscr{R}^*$ is the largest subspace of $\mathscr{K}$ all of whose states can be reached from $x(0) = 0$ along a controlled trajectory lying entirely in $\mathscr{K}$.

5.8. Let $A\mathscr{N} \subset \mathscr{N} \subset \mathscr{K} \subset \mathscr{X}$ and $P: \mathscr{X} \to \bar{\mathscr{X}} := \mathscr{X}/\mathscr{N}$ the canonical projection. If $\mathscr{V}^*$, $\mathscr{R}^*$ (resp. $(\bar{\mathscr{V}})^*$, $(\bar{\mathscr{R}})^*$) have their usual meaning relative to $\mathscr{K}$ (resp. $\bar{\mathscr{K}}$), show that

$$\overline{\mathscr{V}^*} = (\bar{\mathscr{V}})^* \quad \text{and} \quad \overline{\mathscr{R}^*} = (\bar{\mathscr{R}})^*.$$

5.9. *Controllability subspaces and direct control feedthrough.* Extend Exercise 4.6 to include controllability subspaces. Namely, given

$$\dot{x} = Ax + Bu, \qquad z = Dx + Eu,$$

say that $\mathscr{R} \subset \mathscr{X}$ is a *generalized* c.s. (g.c.s.) relative to (A, B, D, E), and write $\mathscr{R} \in \mathfrak{C}(A, B; D, E)$, if the following is true. $\mathscr{R}$ is a c.s. and, for some $F: \mathscr{X} \to \mathscr{U}$ and $G: \mathscr{U} \to \mathscr{U}$, we have

$$\mathscr{R} = \langle A + BF \,|\, \mathrm{Im}(BG) \rangle,$$

$(D + EF)\mathscr{R} = 0$, $EG = 0$. Thus, if $u = Fx + Gv$, then

$$\dot{x} = (A + BF)x + BGv, \qquad z = (D + EF)x + EGv;$$

and our conditions mean that all states $x \in \mathscr{R}$ are reachable from $x = 0$ subject to the constraint that the output $z(\cdot)$ is maintained at $z = 0$. Just as in Exercise 4.6, bring in the extended maps

$$A_e: \mathscr{X} \oplus \mathscr{Z} \to \mathscr{X} \oplus \mathscr{Z}, \qquad B_e: \mathscr{U} \to \mathscr{X} \oplus \mathscr{Z}$$

defined by the matrices

$$A_e = \begin{bmatrix} A & 0 \\ D & 0 \end{bmatrix}, \qquad B_e = \begin{bmatrix} B \\ E \end{bmatrix}.$$

Show that $\mathscr{R}$ is a g.c.s. if and only if $\mathscr{R}$ is an (ordinary) c.s. for (A_e, B_e), such that $\mathscr{R} \subset \mathscr{X}$, namely

$$\mathscr{R} \in \mathfrak{C}(A_e, B_e; \mathscr{X}).$$

Thus, $\mathscr{R}^\Delta := \sup \mathfrak{C}(A, B; D, E)$ exists and can be computed as $\mathscr{R}^\Delta = \sup \mathfrak{C}(A_e, B_e; \mathscr{X})$.

5.10. Construct an example to illustrate DDPS and Theorem 5.8. Hint: Arrange, for instance, that $\mathscr{V}^* = \mathscr{R}^* \oplus \mathscr{S}_g \oplus \mathscr{S}_b$, $\mathscr{X} = \mathscr{V}^* \oplus \mathscr{W}$, with $d(\mathscr{R}^*) = d(\mathscr{S}_g) = d(\mathscr{S}_b) = 1$, $d(\mathscr{W}) = 2$. It can be checked that the following data satisfy these conditions.

$$A = \left[\begin{array}{c|c|c|cc} 0 & 0 & 0 & 0 & 0 \\ \hline 0 & -1 & 0 & 0 & 0 \\ \hline 0 & 0 & 0 & 1 & 0 \\ \hline 0 & 0 & 0 & 0 & 1 \\ 0 & 0 & 0 & 0 & 0 \end{array}\right], \qquad B = \begin{bmatrix} 1 & 0 \\ 0 & 0 \\ 0 & 0 \\ 0 & 0 \\ 0 & 1 \end{bmatrix}, \qquad D = [0 \;\; 0 \;\; 0 \;\; 1 \;\; 0].$$

Here, "good" means "stable." A solution is

$$F = \begin{bmatrix} -1 & 0 & 0 & 0 & 0 \\ 0 & 0 & -1 & -3 & -3 \end{bmatrix};$$

all eigenvalues are then assigned to $s = -1$. The decoupled disturbances are those for which

$$\mathscr{S} \subset \mathscr{V}_g^* = \mathscr{R}^* \oplus \mathscr{S}_g = \text{Im} \begin{bmatrix} 1 & 0 \\ 0 & 1 \\ 0 & 0 \\ 0 & 0 \\ 0 & 0 \end{bmatrix}.$$

Note that while (A, B) is not controllable, the condition (6.10) is satisfied. Note also that the example could be made to look more impressive by disguising the structure: replace A by $A + B\hat{F}$ for a random $\hat{F}$, and apply a random similarity transformation

$$(D, A + B\hat{F}, B) \mapsto (DT^{-1}, T(A + B\hat{F})T^{-1}, TB).$$

Now solve the disguised version by computing, in this order: $\mathscr{V}^*$, $\mathscr{R}^*$, $F_0 \in \mathbf{F}(\mathscr{V}^*)$, $\mathscr{V}_g^*$, $Q: \mathscr{X} \to \mathscr{X}/\mathscr{V}_g^*$, $\tilde{F}: \mathscr{X}/\mathscr{V}_g^* \to \mathscr{U}$, and finally $F = F_0 + \tilde{F}Q$, just as in the proof of Theorem 5.8. Here, F_0 is chosen to make $(A + BF_0)|\mathscr{R}^*$ stable.

5.11. Verify the assertion at the end of Section 5.5.

5.12. *DDPS with feedforward.* Generalize the problem of disturbance decoupling with stability (DDPS, Section 5.6) to include the possibility of disturbance feedforward as in Exercise 4.10. In the notation used there, show that the generalized problem is solvable if and only if

$$\mathscr{S}_1 \subset \mathscr{V}_g^* + \mathscr{B}, \qquad \mathscr{S}_2 \subset \mathscr{V}_g^*,$$

where $\mathscr{V}_g^*$ is defined as in Section 5.6.

5.13. *DDPS with direct control feedthrough.* Define the obvious version of DDPS for the case where $z = Dx + Eu$. Using the results of Exercises 4.6 and 5.9, show that this problem is solvable if and only if $\mathscr{V}_g^\Delta \supset \mathscr{S}$, where $\mathscr{V}_g^\Delta$ is defined in terms of $\mathscr{V}^\Delta$ and $\mathscr{R}^\Delta$ in exactly the way $\mathscr{V}_g^*$ was defined by (6.5) in terms of $\mathscr{V}^*$ and $\mathscr{R}^*$.

5.14. Construct a numerical illustration of Theorems 5.9 and 5.10 by working through the proof with $n = 8$, $m = 3$, and randomly chosen A, B.

5.15. Show that, for arbitrary subspaces $\mathscr{R} \subset \mathscr{S} \subset \mathscr{X}$ and a map $P: \mathscr{X} \to \mathscr{X}$,

$$d\left(\frac{\mathscr{S}}{\mathscr{R}}\right) = d\left(\frac{P\mathscr{S}}{P\mathscr{R}}\right) + d\left(\frac{\mathscr{S} \cap \text{Ker } P}{\mathscr{R} \cap \text{Ker } P}\right),$$

and if $\mathscr{T} \subset \mathscr{X}$,

$$d\left(\frac{\mathscr{S}}{\mathscr{R}}\right) = d\left(\frac{\mathscr{S} + \mathscr{T}}{\mathscr{R} + \mathscr{T}}\right) + d\left(\frac{\mathscr{S} \cap \mathscr{T}}{\mathscr{R} \cap \mathscr{T}}\right).$$

Applying these results to the $\mathscr{S}_j$ defined by (7.4) show that (for $j = 0, 1, \ldots, n-1$)

$$d\left(\frac{\mathscr{S}_j}{\mathscr{S}_{j-1}}\right) = d\left(\frac{\mathscr{S}_{j+1}}{\mathscr{S}_j}\right) + d\left(\frac{\mathscr{B} \cap A\mathscr{S}_j}{\mathscr{B} \cap A\mathscr{S}_{j-1}}\right) + d\left(\frac{\mathscr{S}_j \cap \operatorname{Ker} A}{\mathscr{S}_{j-1} \cap \operatorname{Ker} A}\right),$$

where $\mathscr{S}_{-1} := 0$.

5.16. Prove Theorem 5.11. Hint: First, prove the result for $m = 2$. In this case, to construct a c.s. $\mathscr{R}$ with $d(\mathscr{R}) = \kappa_1 + k$ $(1 \le k \le \kappa_2 - 1)$ take (A, B) in canonical form, and as a basis for $\mathscr{R}$ the vectors $x_1, \ldots, x_{\kappa_1 + k}$ given by

$$b_1, Ab_1, \ldots, A^{\mu-1}b_1,$$

$$A^{\mu}b_1 + b_2, A^{\mu+1}b_1 + Ab_2, \ldots, A^{\kappa_1-1}b_1 + A^{\kappa_1-\mu-1}b_2,$$

$$A^{\kappa_1-\mu}b_2, \ldots, A^{\kappa_2-1}b_2,$$

where $\mu := \kappa_1 + k - \kappa_2$; then define $F: \mathscr{X} \to \mathscr{U}$ according to

$$BFx_i = 0, \qquad i \ne \mu; \qquad BFx_\mu = b_2.$$

Show next that there is no nonzero c.s. with dimension k': $k' < \kappa_2$ or $\kappa_2 < k' < \kappa_1$; and finally, check uniqueness. The generalization to arbitrary m is now straightforward.

5.17. *Dualization of* (A, B)*-invariance.* For the triple (D, A, B) consider the family of subspaces

$$\mathfrak{I}'(D, A; \operatorname{Im} B) := \{\mathscr{W} : \mathscr{B} + A(\mathscr{W} \cap \operatorname{Ker} D) \subset \mathscr{W} \subset \mathscr{X}\}.$$

i. By dualizing the defining relation, show that $\mathscr{W} \in \mathfrak{I}'(D, A; \operatorname{Im} B)$ if and only if $\mathscr{W}^\perp \in \mathfrak{I}(A', \operatorname{Im} D'; \mathscr{B}^\perp)$.
ii. Check that $\mathscr{W}_* := \inf \mathfrak{I}'(D, A; \operatorname{Im} B)$ exists and can be computed by the algorithm

$$\mathscr{W}_{\mu+1} = \mathscr{B} + A(\mathscr{W}_\mu \cap \operatorname{Ker} D), \qquad \mathscr{W}_0 = 0.$$

iii. With $\mathscr{V}^*$ and $\mathscr{R}^*$ defined as usual with respect to (D, A, B), prove that

$$\mathscr{R}^* = \mathscr{V}^* \cap \mathscr{W}_*.$$

Hint: For $\mathscr{R}^* \subset \mathscr{W}_*$ use the algorithms for $\mathscr{R}^\mu$, $\mathscr{W}_\mu$ and prove by induction that $\mathscr{R}^\mu \subset \mathscr{W}_\mu$. For $\mathscr{V}^* \cap \mathscr{W}_* \subset \mathscr{R}^*$ use the algorithm for $\mathscr{W}_*$ plus the fact that $\mathscr{R}^* = \mathscr{V}^* \cap (A\mathscr{R}^* + \mathscr{B})$.
iv. Show that $\mathscr{B} \cap \mathscr{V}^* = 0$ if and only if $\mathscr{W}_* \cap \mathscr{V}^* = 0$, and that $\mathscr{W}_* + \operatorname{Ker} D = \mathscr{X}$ if and only if $\mathscr{W}_* + \mathscr{V}^* = \mathscr{X}$. Hint: For the first statement consider $\mathscr{W}_\mu \cap \mathscr{V}^*$; for the second, dualize.
v. To dualize the definition of $\mathscr{R}^*$, consider

$$\mathscr{S}_*^\perp := \sup \mathfrak{C}(A', \operatorname{Im} D'; \mathscr{B}^\perp) \subset \mathscr{X}'.$$

Show that $\mathscr{S}_* = \lim \mathscr{S}_\mu$, where

$$\mathscr{S}_{\mu+1} = \mathscr{W}_* + [(A^{-1}\mathscr{S}_\mu) \cap \operatorname{Ker} D], \qquad \mathscr{S}_0 = \mathscr{X}.$$

Then, verify that $\mathscr{S}_* = \mathscr{W}_* + \mathscr{V}^*$.

vi. Let D be epic and B monic. With reference to Exercise 4.4, show that $H(\lambda) := D(\lambda 1 - A)^{-1}B$ is left-invertible if and only if $\mathscr{W}_* \cap \mathscr{V}^* = 0$, and is right-invertible if and only if $\mathscr{W}_* + \mathscr{V}^* = \mathscr{X}$.

vii. Show that there exist $F: \mathscr{X} \to \mathscr{U}$ and $K: \mathscr{Y} \to \mathscr{X}$, such that $\mathscr{V}^*$ and $\mathscr{W}^*$ are each $(A + BF + KD)$-invariant. Compute the matrices of the new triple $(D, A + BF + KD, B)$ in a basis adapted to a decomposition of the form

$$\mathscr{X} = \hat{\mathscr{V}}^* \oplus \mathscr{R}^* \oplus \hat{\mathscr{W}}_* \oplus \mathscr{T},$$

where $\hat{\mathscr{V}}^* \oplus \mathscr{R}^* = \mathscr{V}^*$ and $\mathscr{R}^* \oplus \hat{\mathscr{W}}_* = \mathscr{W}_*$. In terms of these matrices compute the transfer matrix of the triple.

5.9 Notes and References

The concept of controllability subspace was introduced by Wonham and Morse [1]. Most of the results of this chapter were given in the reference cited, and by Morse and Wonham [1]. The identification of "transfer matrix zeros" with the zeros of its Smith-McMillan form is due to Rosenbrock [1]. Several alternative and inequivalent definitions of the transmission zeros of a (not necessarily complete) triple have been proposed in the literature; for a survey of the connections see Francis and Wonham [2]. A clear exposition of the relationship of σ^* to the Smith-McMillan form is given by Hosoe [1] and by Anderson [2]. For the numerical computation of transmission zeros see Laub and Moore [1]. The results of Section 5.5 were reported informally by Wonham [6].

Section 5.7 is adapted from Brunovsky [1] and Wonham and Morse [2], except for Theorem 5.11, which is due to Warren and Eckberg [1]. For alternative approaches to the subject of controllability indices, via Kronecker or "minimal" indices, see Rosenbrock [1], Kalman [4] and Warren and Eckberg [1]. The canonical structure can be exploited to obtain a more detailed version of the pole assignment theorem, as in Rosenbrock [1] and Dickinson [1]. Some topological features of the feedback group are described by Brockett [1]. Additional information on the structure of c.s. is presented by Heymann [3] in terms of the concepts of input chain and controllability chain; by such means the problem is solved of determining when a given linear system can be simulated by another.

The "naive" computational procedure of Exercise 5.5 is adequate for low-order systems, but a more sophisticated technique is described by Moore and Laub [1]. The result of Exercise 5.7 is due to Morse and Wonham [1]. The approach to generalized c.s. in Exercise 5.9 was suggested by Anderson's discussion [1] of "output-nulling" c.s. The duality relations of Exercise 5.17 were explored in detail by Morse [2]; this paper contains further information on canonical structure, under a generalization of the feedback group.

6 Tracking and Regulation I: Output Regulation

A typical multivariable control problem requires the design of dynamic compensation to guarantee the following desirable behavior of the closed loop system.

1. Each of an assigned set of output variables converges to, or *tracks*, a corresponding observed reference input from a specified function class. This is the *servo* problem.
2. Each of an assigned set of output variables converges to zero from arbitrary initial values, when the system is perturbed by possibly nonmeasurable disturbances in a specified function class. This is the *regulator* problem; the convergent variables are said to be *regulated*; and zero represents the desired or *set-point* value.

Very often both features are present in the same problem: the system is required to track in the presence of disturbances. Indeed, the formal distinction between the servo and regulator problems disappears if we regard each tracking error (the difference between a reference input and the corresponding output) as a variable to be regulated. For this reason, we may safely restrict attention to the regulator problem alone.

We shall also assume that the disturbance variables (including reference inputs) satisfy known, time-invariant, linear differential equations of finite order, which we lump with the equations of the plant. The combined system equations then take the standard form

$$\dot{x}(t) = Ax(t) + Bu(t), \qquad t \geq 0, \tag{0.1a}$$

$$y(t) = Cx(t), \qquad t \geq 0, \tag{0.1b}$$

and

$$z(t) = Dx(t), \qquad t \geq 0. \tag{0.1c}$$

Here, $y(\cdot)$ is the vector of directly measured outputs and $z(\cdot)$ is the variable to be regulated. Thus, it is required that for every initial state $x(0+)$, we have

$$z(t) \to 0, \qquad t \to \infty. \tag{0.2}$$

We refer to this as the property of *output regulation.*

As an illustration, suppose that the problem requires that a scalar plant output track a ramp reference signal while rejecting a step disturbance. Then, (0.1) could be written in more detail as

$$\frac{d}{dt}\begin{bmatrix} x_1 \\ x_2 \\ x_3 \\ x_4 \end{bmatrix} = \begin{bmatrix} A_{11} & 0 & 0 & a_{14} \\ 0 & 0 & 1 & 0 \\ 0 & 0 & 0 & 0 \\ 0 & 0 & 0 & 0 \end{bmatrix}\begin{bmatrix} x_1 \\ x_2 \\ x_3 \\ x_4 \end{bmatrix} + \begin{bmatrix} B_1 \\ 0 \\ 0 \\ 0 \end{bmatrix} u$$

$$y = Cx, \qquad z = [d_1' \quad -1 \quad 0 \quad 0]x.$$

Here, x_1 is the state vector of the plant, (x_2, x_3) are the state variables of the reference (ramp) generator and x_4 is that of the disturbance (step) generator.

In this example, the dynamics associated with (x_2, x_3, x_4) serve to model the "outside world" as seen by the controlled plant; such a model we shall call the *exosystem.* The exosystem is typically a convenient fiction by which the designer may specify the class of tracking and disturbance rejection tasks which the controlled system is to accomplish with zero asymptotic error. The exosystem is coupled to the controlled system both at points of disturbance injection and, implicitly, via the definition of the regulated variable.

The setup is visualized to work as follows. Each tracking or disturbance rejection "task" is initiated by placing a suitable initial condition on the exosystem dynamics; the system is to be designed so that then $z(t) \to 0$ as $t \to \infty$, or more realistically $z(t) \doteqdot 0$ for t larger than some "settling time" T determined by the exponents assigned to the response $z(\cdot)$. In practice, these exponents would be selected so that T is small compared to the estimated time interval between successive "tasks."

In our general description (0.1) the triple (C, A, B) need not be controllable and observable, or even stabilizable and detectable. If it were, the problem of achieving output regulation could be solved by an immediate application of the results of Sections 3.5–3.7. As in our illustration, stabilizability may fail because the exosystem is typically unstable and cannot be controlled; and detectability may fail if, for instance, certain disturbance variables happen to be decoupled from the measured outputs. In this chapter we shall discuss in detail how these issues affect solvability of the problem.

Before going further let us note a design requirement that is usually imposed in addition to the property of output regulation: namely, *internal stability*, or stability of controllable and observable modes. Typically, this amounts to the requirement that the combined system comprising plant and controller (but not exosystem) be stable, so that the corresponding state subvector tends to zero when no tracking or disturbance rejection task is present. The problem of output regulation combined with internal stability (RPIS) is investigated in Chapters 7 and 8. In this chapter we confine attention to the less practical but logically simpler problem of output regulation alone. It is suggested that the reader interested mainly in RPIS proceed directly to Chapter 7 after completing Section 6.1 on the use of observers.

6.1 Restricted Regulator Problem (RRP)

Let $\mathscr{N}$ denote the unobservable subspace of (C, A):

$$\mathscr{N} := \bigcap_{i=1}^{n} \operatorname{Ker}(CA^{i-1}).$$

Write $\bar{\mathscr{X}} := \mathscr{X}/\mathscr{N}$ and let $P: \mathscr{X} \to \bar{\mathscr{X}}$ be the canonical projection. As noted in Section 3.2, the "observable subsystem" (strictly, factor system) may be identified with the triple $(\bar{C}, \bar{A}, \bar{B})$ in the commutative diagram below:

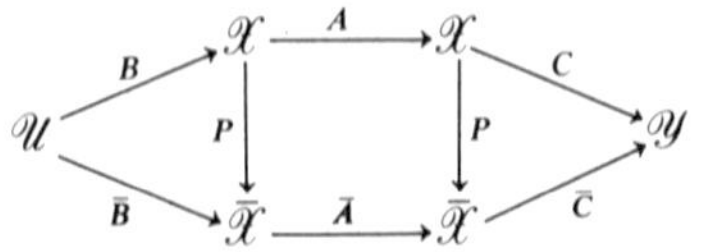

Since $A\mathscr{N} \subset \mathscr{N}$, the induced map $\bar{A}: \bar{\mathscr{X}} \to \bar{\mathscr{X}}$ is well defined, and as $\operatorname{Ker} C \supset \mathscr{N}$, there exists $\bar{C}: \bar{\mathscr{X}} \to \mathscr{Y}$ with $\bar{C}P = C$; finally, we set $PB = \bar{B}$.

Since the pair $(\bar{C}, \bar{A})$ is observable we know from Section 3.3 that maps $\bar{J}: \bar{\mathscr{X}} \to \bar{\mathscr{X}}$ and $\bar{K}: \mathscr{Y} \to \bar{\mathscr{X}}$ exist, such that the auxiliary system

$$\dot{\bar{w}}(t) = \bar{J}\bar{w}(t) + \bar{K}y(t) + \bar{B}u(t), \qquad t \geq 0,$$

is a dynamic observer for the system (0.1a,b) reduced mod $\mathscr{N}$. Namely, we select $\bar{K}$ such that

$$\sigma(\bar{A} - \bar{K}\bar{C}) \subset \mathbb{C}^-$$

and set $\bar{J} := \bar{A} - \bar{K}\bar{C}$, $\bar{e}(t) := \bar{x}(t) - \bar{w}(t)$. Since

$$\dot{\bar{x}}(t) = \bar{A}\bar{x}(t) + \bar{B}u(t), \qquad t \geq 0,$$

there follows $\dot{\bar{e}}(t) = \bar{J}\bar{e}(t)$ $(t \geq 0)$, so that $\bar{e}(t) \to 0$ $(t \to \infty)$. Since $\sigma(\bar{J})$ can be assigned arbitrarily by suitable choice of $\bar{K}$, convergence can in principle be made arbitrarily exponentially fast.[1] Now suppose $u(t) = \bar{F}\bar{w}(t)$ for some $\bar{F}: \bar{\mathscr{X}} \to \mathscr{U}$. Setting $F = \bar{F}P$ and noting that

$$u = \bar{F}(\bar{x} - \bar{e}) = \bar{F}(Px - \bar{e}) = Fx - \bar{F}\bar{e},$$

[1] In practice, the rate of convergence is limited by the bandwidth permitted by noise.

we obtain from (0.1a),

$$\dot{x} = (A + BF)x - B\bar{F}\bar{e}, \qquad t \geq 0. \tag{1.1}$$

From (0.1c), (0.2) and (1.1) we conclude that regulation is achieved with the control law $u = \bar{F}\bar{w}$ if and only if

$$De^{t(A+BF)}\left[x(0+) - \int_0^t e^{-\tau(A+BF)}B\bar{F}\bar{e}(\tau)\,d\tau\right] \to 0$$

as $t \to \infty$, for all $x(0+)$, $\bar{e}(0+)$. Equivalently,

$$De^{t(A+BF)} \to 0, \qquad t \to \infty, \tag{1.2}$$

and

$$\int_0^t De^{(t-\tau)(A+BF)}B\bar{F}e^{\tau\bar{J}}\,d\tau \to 0, \qquad t \to \infty. \tag{1.3}$$

Suppose (1.2) is true. Since $\bar{J}$ is stable, the integral in (1.3) is the convolution of functions having Laplace transforms which are analytic in $\mathbb{C}^+$, and (1.3) follows. On this basis we can (and shall) ignore the term containing $\bar{e}$ in (1.1), assume that the state $\bar{x}$ of the observable reduced system is directly observable at the start, and admit a priori all controls of the form $u = \bar{F}\bar{x}$. A control $u = Fx$ can be written in this form if and only if $F = \bar{F}P$ for some $\bar{F}$, that is, $\operatorname{Ker} F \supset \mathcal{N}$, and this version of the observability constraint will be used in the sequel. We remark that exactly the same reasoning applies if the observer is chosen to be of minimal dynamic order given by

$$d(\operatorname{Ker} \bar{C}) = d(\operatorname{Ker} C) - d(\mathcal{N})$$

along the lines of Section 3.4.

Finally, as was shown in Section 4.4, a condition equivalent to (1.2) is $\mathcal{X}^+(A + BF) \subset \operatorname{Ker} D$.

In this way we are led to formulate the

Restricted Regulator Problem (RRP). *Given the maps $A: \mathcal{X} \to \mathcal{X}$, $B: \mathcal{U} \to \mathcal{X}$, $D: \mathcal{X} \to \mathcal{Z}$, and a subspace $\mathcal{N} \subset \mathcal{X}$ with $A\mathcal{N} \subset \mathcal{N}$, find $F: \mathcal{X} \to \mathcal{U}$, such that*

$$\operatorname{Ker} F \supset \mathcal{N} \tag{1.4}$$

and

$$\mathcal{X}^+(A + BF) \subset \operatorname{Ker} D. \tag{1.5}$$

RRP is "restricted" in the sense that no provision is made for dynamic compensation other than that tacitly introduced by the observer. Actually, we shall exploit dynamic compensation later, using a technique of state space extension which will bring our results to a satisfactory completion.

6.2 Solvability of RRP

In this section, we obtain necessary and sufficient conditions for the solvability of RRP. As they stand, these conditions are not constructive in the sense of providing an algorithmic solution of the problem when a solution exists; nevertheless they can be made so in combination with state space extension, as will be shown in Section 6.3.

Theorem 6.1. *RRP is solvable if and only if there exists a subspace* $\mathscr{V} \subset \mathscr{X}$, *such that*

$$\mathscr{V} \subset \operatorname{Ker} D \cap A^{-1}(\mathscr{V} + \mathscr{B}) \tag{2.1}$$

$$\mathscr{X}^+(A) \cap \mathscr{N} + A(\mathscr{V} \cap \mathscr{N}) \subset \mathscr{V} \tag{2.2}$$

and

$$\mathscr{X}^+(A) \subset \langle A \mid \mathscr{B} \rangle + \mathscr{V}. \tag{2.3}$$

Notice that condition (2.1) is equivalent to $\mathscr{V} \in \mathfrak{J}(A, B; \operatorname{Ker} D)$; we employ the more explicit version for greater convenience. The subspace $\mathscr{V}$ can be thought of as either $\mathscr{X}^+(A + BF)$ or some subspace containing it, and conditions (2.1)–(2.3) as simply the result of eliminating F from (1.4) and (1.5), as we shall see.

Before proving Theorem 6.1 we note various structural features of conditions (2.1)–(2.3). Introduce the family of subspaces

$$\mathfrak{V} = \{\mathscr{V} : \mathscr{V} \in \mathfrak{J}(A, B; \operatorname{Ker} D) \ \&\ A(\mathscr{V} \cap \mathscr{N}) \subset \mathscr{V}\}.$$

In general, $\mathfrak{V}$ is not closed under addition and it is not true that $\mathfrak{V}$ contains a supremal element (in the sense of Section 4.2). However, as $\mathfrak{V}$ is nonempty ($0 \in \mathfrak{V}$) it always has, possibly many, maximal elements: by definition, $\mathscr{V}^M \in \mathfrak{V}$ is *maximal* if $\mathscr{V} \in \mathfrak{V}$ and $\mathscr{V} \supset \mathscr{V}^M$ imply $\mathscr{V} = \mathscr{V}^M$. We have

Corollary 6.1. *RRP is solvable if and only if*

$$\mathscr{X}^+(A) \cap \mathscr{N} \subset \operatorname{Ker} D, \tag{2.4}$$

and for some maximal element $\mathscr{V}^M \in \mathfrak{V}$,

$$\mathscr{X}^+(A) \subset \langle A \mid \mathscr{B} \rangle + \mathscr{V}^M. \tag{2.5}$$

The difficulty in verifying (2.5) is that of effectively parametrizing the subfamily of all $\mathscr{V}^M$. Actually, in many cases which arise in practice it happens that $\mathfrak{V}$ does contain a (unique) supremal element, namely the familiar

$$\mathscr{V}^* := \sup \mathfrak{J}(A, B; \operatorname{Ker} D).$$

That is, $\mathscr{V}^*$ satisfies the second defining condition in (2.2)

$$A(\mathscr{V}^* \cap \mathscr{N}) \subset \mathscr{V}^*. \tag{2.6}$$

Then, of course, the $\mathscr{V}^M$ all coincide with $\mathscr{V}^*$ and RRP is constructively solvable by the algorithm of Theorem 4.3:

$$\begin{aligned} \mathscr{V}^0 &= \operatorname{Ker} D \\ \mathscr{V}^j &= \operatorname{Ker} D \cap A^{-1}(\mathscr{V}^{j-1} + \mathscr{B}), \qquad j = 1, 2, \ldots \\ \mathscr{V}^* &= \mathscr{V}^n. \end{aligned} \tag{2.7}$$

A sufficient condition for (2.6) to be true is included in the following.

Corollary 6.2. *Suppose*

$$A(\mathscr{N} \cap \operatorname{Ker} D) \subset \operatorname{Ker} D. \tag{2.8}$$

Then RRP is solvable if and only if

$$\mathscr{X}^+(A) \cap \mathscr{N} \subset \operatorname{Ker} D$$

and

$$\mathscr{X}^+(A) \subset \langle A \mid \mathscr{B} \rangle + \mathscr{V}^*.$$

The proof of these results depends on the following two lemmas.

Lemma 6.1. *Let* $A\colon \mathscr{X} \to \mathscr{X}$, $A_1\colon \mathscr{X} \to \mathscr{X}$ *and* $\mathscr{N} \subset \mathscr{X}$, *with* $A\mathscr{N} \subset \mathscr{N}$ *and* $A_1 \mid \mathscr{N} = A \mid \mathscr{N}$. *Then,*

$$\mathscr{X}^+(A_1) \cap \mathscr{N} = \mathscr{X}^+(A) \cap \mathscr{N}.$$

Proof. Denote by α_1^+ (resp. α^+) the unstable factor of the m.p. of A_1 (resp. A). Let $x \in \mathscr{X}^+(A_1) \cap \mathscr{N}$. Then, $x \in \operatorname{Ker} \alpha_1^+(A_1)$. Since A coincides with A_1 on $\mathscr{N}$, $A^j x = A_1^j x$ $(j = 1, 2, \ldots)$ and, therefore,

$$\alpha_1^+(A)x = \alpha_1^+(A_1)x = 0.$$

Let $\alpha_x(\lambda)$ be the m.p. of x with respect to A. Then, $\alpha_x \mid \alpha_1^+$, that is, the complex zeros of α_x belong to $\mathbb{C}^+$. Let $\alpha = \alpha^+\alpha^-$ be the m.p. of A. Then also $\alpha_x \mid \alpha$, and therefore, $\alpha_x \mid \alpha^+$. Thus, $x \in \operatorname{Ker} \alpha^+(A)$; that is, $x \in \mathscr{X}^+(A) \cap \mathscr{N}$. We have shown that

$$\mathscr{X}^+(A_1) \cap \mathscr{N} \subset \mathscr{X}^+(A) \cap \mathscr{N},$$

and the reverse inclusion follows by symmetry. □

Lemma 6.2. *For arbitrary* $F\colon \mathscr{X} \to \mathscr{U}$,

$$\mathscr{X}^+(A + BF) + \langle A \mid \mathscr{B} \rangle = \mathscr{X}^+(A) + \langle A \mid \mathscr{B} \rangle. \tag{2.9}$$

In particular, if for some $\mathscr{V} \subset \mathscr{X}$,

$$\mathscr{X}^+(A) \subset \langle A \mid \mathscr{B} \rangle + \mathscr{V} \tag{2.10}$$

then

$$\mathscr{X}^+(A + BF) \subset \langle A + BF \mid \mathscr{B} \rangle + \mathscr{V}. \tag{2.11}$$

PROOF. If $P: \mathscr{X} \to \mathscr{X}/\langle A \mid \mathscr{B} \rangle$ is the canonical projection and a bar denotes the induced map in $\mathscr{X}/\langle A \mid \mathscr{B} \rangle$, then $\overline{A + BF} = \bar{A}$ and

$$\begin{aligned} P\mathscr{X}^+(A + BF) &= \bar{\mathscr{X}}^+(\overline{A + BF}) \quad \text{(by Lemma 4.6)} \\ &= \bar{\mathscr{X}}^+(\bar{A}) = P\mathscr{X}^+(A). \end{aligned} \tag{2.12}$$

From (2.12), (2.9) follows at once. Adding $\langle A \mid \mathscr{B} \rangle$ to the left side of (2.10) and using (2.9) now yields (2.11).

PROOF (of Theorem 6.1). (If) Let $\mathscr{V}$ have the properties (2.1)–(2.3). Then,

$$A\mathscr{V} \subset \mathscr{V} + \mathscr{B}, \qquad A(\mathscr{V} \cap \mathscr{N}) \subset \mathscr{V}. \tag{2.13}$$

By (2.13) there exists $F_0: \mathscr{X} \to \mathscr{U}$, such that

$$(A + BF_0)\mathscr{V} \subset \mathscr{V}, \qquad F_0(\mathscr{V} \cap \mathscr{N}) = 0.$$

Let

$$\mathscr{V} + \mathscr{N} = \hat{\mathscr{V}} \oplus \mathscr{V} \cap \mathscr{N} \oplus \hat{\mathscr{N}},$$

where $\hat{\mathscr{V}} \subset \mathscr{V}$ and $\hat{\mathscr{N}} \subset \mathscr{N}$. Define $F_1: \mathscr{X}' \to \mathscr{U}$, such that $F_1 \mid \mathscr{V} = F_0 \mid \mathscr{V}$ and $F_1 \mid \hat{\mathscr{N}} = 0$. Then, $F_1\mathscr{N} = 0$ and $(A + BF_1)\mathscr{V} \subset \mathscr{V}$. Write $A_1 = A + BF_1$, let $P: \mathscr{X} \to \mathscr{X}/\mathscr{V}$ be the canonical projection, and let $\bar{A}_1$ be the induced map on $\bar{\mathscr{X}} = \mathscr{X}/\mathscr{V}$. Now $A_1 \mid \mathscr{N} = A \mid \mathscr{N}$, so by Lemma 6.1 and (2.2)

$$\mathscr{X}^+(A_1) \cap \mathscr{N} = \mathscr{X}^+(A) \cap \mathscr{N} \subset \mathscr{V}.$$

Thus,

$$\mathscr{N} = \mathscr{N} \cap \mathscr{X}^+(A_1) \oplus \mathscr{N} \cap \mathscr{X}^-(A_1) \subset \mathscr{X}^-(A_1) + \mathscr{V},$$

so that

$$\bar{\mathscr{N}} := P\mathscr{N} \subset \bar{\mathscr{X}}^-(\bar{A}_1). \tag{2.14}$$

Also, (2.3) with Lemma 6.2 yields

$$\mathscr{X}^+(A_1) \subset \langle A_1 \mid \mathscr{B} \rangle + \mathscr{V}$$

so

$$\bar{\mathscr{X}}^+(\bar{A}_1) \subset \langle \bar{A}_1 \mid \bar{\mathscr{B}} \rangle. \tag{2.15}$$

By (2.14) and (2.15) there exists $\bar{F}_2: \bar{\mathscr{X}} \to \mathscr{U}$, such that $\operatorname{Ker} \bar{F}_2 \supset \bar{\mathscr{N}}$ and $\bar{A}_1 + \bar{B}\bar{F}_2$ is stable. Define $F_2 = \bar{F}_2 P$. Then, $F_2\mathscr{N} = \bar{F}_2\bar{\mathscr{N}} = 0$. Let $F = F_1 + F_2$. Then $F\mathscr{N} = 0$. Also, $F_2\mathscr{V} = 0$ implies

$$(A + BF) \mid \mathscr{V} = A_1 \mid \mathscr{V},$$

so $(A + BF)\mathscr{V} \subset \mathscr{V}$. For the induced map $\overline{A + BF}$ on $\bar{\mathscr{X}}$, we have

$$\bar{\mathscr{X}}^+(\overline{A + BF}) = \bar{\mathscr{X}}^+(\bar{A}_1 + \bar{B}\bar{F}_2) = 0,$$

so that

$$\mathscr{X}^+(A + BF) \subset \mathscr{V} \subset \operatorname{Ker} D$$

as required.

(Only if) Let $\mathscr{V} = \mathscr{X}^+(A + BF)$. Then, (2.1) is clear from (1.5). Since $\operatorname{Ker} F \supset \mathscr{N}$, we have

$$(A + BF)|\mathscr{N} = A|\mathscr{N}$$

and by Lemma 6.1

$$\mathscr{V} \cap \mathscr{N} = \mathscr{X}^+(A + BF) \cap \mathscr{N} = \mathscr{X}^+(A) \cap \mathscr{N},$$

so that $A(\mathscr{V} \cap \mathscr{N}) \subset \mathscr{V} \cap \mathscr{N}$, proving (2.2). Finally,

$$\mathscr{V} + \langle A|\mathscr{B}\rangle = \mathscr{X}^+(A + BF) + \langle A|\mathscr{B}\rangle = \mathscr{X}^+(A) + \langle A|\mathscr{B}\rangle,$$

by Lemma 6.2, and this verifies (2.3). □

Proof (of Corollary 6.1). If (2.1)–(2.3) hold for some $\mathscr{V}$, then $\mathscr{V} \in \mathfrak{V}$. There is some $\mathscr{V}^M \in \mathfrak{V}$ with $\mathscr{V}^M \supset \mathscr{V}$, and (2.1)–(2.3) clearly hold for such $\mathscr{V}^M$. Furthermore,

$$\mathscr{X}^+(A) \cap \mathscr{N} \subset \mathscr{V} \subset \operatorname{Ker} D.$$

For the converse, set $\mathscr{V} = \mathscr{V}^M$. By (2.4), $\mathscr{X}^+(A) \cap \mathscr{N}$ is an A-invariant subspace of $\mathscr{N} \cap \operatorname{Ker} D$, hence certainly belongs to each $\mathscr{V}^M$: indeed, if $\mathscr{V} \in \mathfrak{V}$, then

$$[\mathscr{V} + \mathscr{X}^+(A) \cap \mathscr{N}] \cap \mathscr{N} = \mathscr{V} \cap \mathscr{N} + \mathscr{X}^+(A) \cap \mathscr{N},$$

and therefore, $\mathscr{V} + \mathscr{X}^+(A) \cap \mathscr{N} \in \mathfrak{V}$. □

Proof (of Corollary 6.2). By (2.8), $A(\mathscr{N} \cap \operatorname{Ker} D) \subset \mathscr{N} \cap \operatorname{Ker} D$, so $\mathscr{N} \cap \operatorname{Ker} D \subset \mathscr{V}^*$. Therefore,

$$\mathscr{V}^* \cap \mathscr{N} \subset \operatorname{Ker} D \cap \mathscr{N} \subset \mathscr{V}^* \cap \mathscr{N}.$$

Thus, $\mathscr{V}^* \cap \mathscr{N} = \operatorname{Ker} D \cap \mathscr{N}$ is A-invariant, so $\mathscr{V}^* \in \mathfrak{V}$. Therefore, $\mathscr{V}^M = \mathscr{V}^*$ for all $\mathscr{V}^M$, and the result follows by Corollary 6.1. □

To conclude this section, we give a description of the subspaces $\mathscr{V}^M$ that will be useful later.

Proposition 6.1. *Let*

$$\mathscr{V}_0 := \bigcap_{i=1}^{n} A^{-i+1}(\mathscr{N} \cap \operatorname{Ker} D).$$

Each subspace $\mathscr{V}^M$ is of the form

$$\mathscr{V}^M = \mathscr{V}_0 \oplus \mathscr{V}_1,$$

where

$$\mathscr{V}_1 = \sup\{\mathscr{V} : \mathscr{V} \subset \mathscr{W} \cap A^{-1}(\mathscr{B} + \mathscr{V}_0 + \mathscr{V})\}$$

and $\mathscr{W}$ *is a suitable complement of* $\mathscr{N} \cap \operatorname{Ker} D$ *in* $\operatorname{Ker} D$.

The idea is illustrated by the lattice diagram, Fig. 6.1.

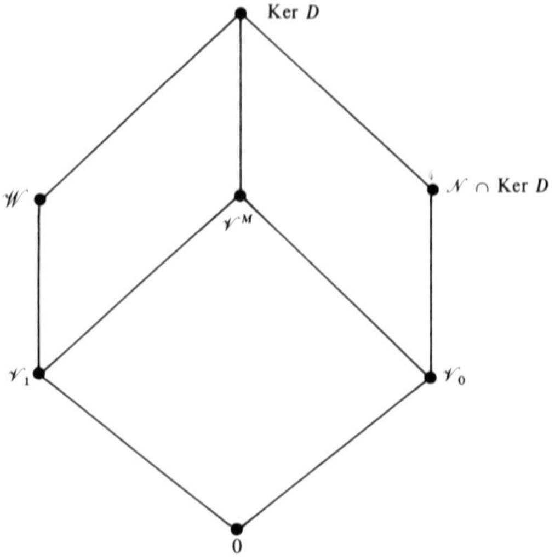

Figure 6.1 Lattice Diagram: Structure of $\mathscr{V}^M$.

PROOF. Suppose $\mathscr{V}^M$ is maximal, and write

$$\mathscr{V}^M = \mathscr{V}^M \cap \mathscr{N} \oplus \mathscr{V}_1^M,$$

$$\operatorname{Ker} D = \operatorname{Ker} D \cap \mathscr{N} \oplus \mathscr{V}_1^M \oplus \mathscr{W}_1$$

and

$$\mathscr{V}_1^M \oplus \mathscr{W}_1 = \mathscr{W}.$$

It will be shown that

$$\mathscr{V}^M \cap \mathscr{N} = \mathscr{V}_0 \tag{2.16}$$

and

$$\mathscr{V}_1^M = \mathscr{V}_1. \tag{2.17}$$

As to (2.16), observe that

$$\mathscr{V}_0 = \sup\{\mathscr{V} : \mathscr{V} \subset \mathscr{N} \cap \operatorname{Ker} D,\ A\mathscr{V} \subset \mathscr{V}\}, \tag{2.18}$$

hence, the subspace

$$(\mathscr{V}^M + \mathscr{V}_0) \cap \mathscr{N} = \mathscr{V}^M \cap \mathscr{N} + \mathscr{V}_0$$

is A-invariant, so $\mathscr{V}^M + \mathscr{V}_0 \in \mathfrak{V}$. By maximality of $\mathscr{V}^M$, $\mathscr{V}^M \supset \mathscr{V}_0$, i.e. $\mathscr{V}^M \cap \mathscr{N} \supset \mathscr{V}_0$. But as $\mathscr{V}^M \cap \mathscr{N}$ is A-invariant and belongs to $\mathscr{N} \cap \operatorname{Ker} D$, it follows by (2.18) that $\mathscr{V}^M \cap \mathscr{N} \subset \mathscr{V}_0$, and (2.16) is proved. For (2.17), we have first

$$\begin{aligned} A\mathscr{V}_1^M &\subset A\mathscr{V}^M \subset \mathscr{V}^M + \mathscr{B} = \mathscr{V}^M \cap \mathscr{N} + \mathscr{V}_1^M + \mathscr{B} \\ &= \mathscr{V}_0 + \mathscr{B} + \mathscr{V}_1^M, \end{aligned}$$

and therefore, $\mathscr{V}_1^M \subset \mathscr{V}_1$. Now, $\mathscr{V}_1 \cap \mathscr{N} = 0$, as $\mathscr{V}_1 \subset \mathscr{W}$ and $\mathscr{W} \cap \mathscr{N} = 0$. Hence, the subspace

$$(\mathscr{V}_0 \oplus \mathscr{V}_1) \cap \mathscr{N} = \mathscr{V}_0$$

is A-invariant. Since also $\mathscr{V}_0 \oplus \mathscr{V}_1 \subset \operatorname{Ker} D$, and

$$A(\mathscr{V}_0 \oplus \mathscr{V}_1) \subset \mathscr{V}_0 + \mathscr{V}_1 + \mathscr{B},$$

there results $\mathscr{V}_0 \oplus \mathscr{V}_1 \in \mathfrak{V}$. But since

$$\begin{aligned} \mathscr{V}_0 \oplus \mathscr{V}_1 &\supset \mathscr{V}_0 \oplus \mathscr{V}_1^M \\ &= \mathscr{V}^M, \end{aligned} \tag{2.19}$$

the maximality of $\mathscr{V}^M$ implies that equality holds in (2.19), and because $\mathscr{V}_1 \supset \mathscr{V}_1^M$, (2.17) results. □

6.3 Extended Regulator Problem (ERP)

To exploit the advantages of dynamic compensation we bring in auxiliary dynamic elements (integrators), according to the equations

$$\dot{x}_a = B_a u_a, \qquad y_a = C_a x_a,$$

where $u_a \in \mathscr{U}_a$, $x_a \in \mathscr{X}_a$, $y_a \in \mathscr{Y}_a$. Here, $B_a: \mathscr{U}_a \simeq \mathscr{X}_a$, and $C_a: \mathscr{X}_a \simeq \mathscr{Y}_a$ are arbitrary isomorphisms. Introduce the external direct sums

$$\mathscr{U}_e = \mathscr{U} \oplus \mathscr{U}_a, \qquad \mathscr{X}_e = \mathscr{X} \oplus \mathscr{X}_a, \qquad \mathscr{Y}_e = \mathscr{Y} \oplus \mathscr{Y}_a$$

for the extended control, state and (observed) output spaces, and accordingly define the extended maps

$$\begin{aligned} &A_e: \mathscr{X}_e \to \mathscr{X}_e, &&x \oplus x_a \mapsto Ax \\ &B_e: \mathscr{U}_e \to \mathscr{X}_e, &&u \oplus u_a \mapsto Bu \oplus B_a u_a \\ &C_e: \mathscr{X}_e \to \mathscr{Y}_e, &&x \oplus x_a \mapsto Cx \oplus C_a x_a \\ &D_e: \mathscr{X}_e \to \mathscr{Z}, &&x \oplus x_a \mapsto Dx. \end{aligned}$$

The corresponding matrices are

$$A_e = \begin{bmatrix} A & 0 \\ 0 & 0 \end{bmatrix}, \qquad B_e = \begin{bmatrix} B & 0 \\ 0 & B_a \end{bmatrix},$$

$$C_e = \begin{bmatrix} C & 0 \\ 0 & C_a \end{bmatrix}, \qquad D_e = [D \quad 0].$$

Writing $d(\mathscr{X}_e) = n_e$, we have

$$\mathscr{N}_e = \bigcap_{i=1}^{n_e} \operatorname{Ker}(C_e A_e^{i-1}) = \mathscr{N};$$

also

$$\operatorname{Ker} D_e = \operatorname{Ker} D \oplus \mathscr{X}_a.$$

We define the **Extended Regulator Problem (ERP)** as that of finding suitable $\mathscr{X}_a$ (that is, $d(\mathscr{X}_a)$) and then $F_e : \mathscr{X}_e \to \mathscr{U}_e$, such that

$$\operatorname{Ker} F_e \supset \mathscr{N}$$

and

$$\mathscr{X}_e^+(A_e + B_e F_e) \subset \operatorname{Ker} D \oplus \mathscr{X}_a.$$

The main result of this chapter is the following.

Theorem 6.2. *Let RRP be defined as in Section 6.1. Then ERP is solvable if and only if, for RRP,*

$$\mathscr{X}^+(A) \cap \mathscr{N} \subset \operatorname{Ker} D \tag{3.1}$$

and

$$\mathscr{X}^+(A) \subset \langle A | \mathscr{B} \rangle + \mathscr{V}^*, \tag{3.2}$$

where

$$\mathscr{V}^* := \sup \mathfrak{I}(A, B; \operatorname{Ker} D).$$

Furthermore, if ERP is solvable, it is possible to take

$$d(\mathscr{X}_a) \le d\left[\frac{\mathscr{N} \cap \mathscr{V}^*}{\mathscr{N} \cap \bigcap_{i=1}^{n} A^{-i+1}\mathscr{V}^*}\right]. \tag{3.3}$$

The theorem can by paraphrased by saying that output regulation is achievable, at least with dynamic compensation, if and only if (i) any unstable, unobservable modes of the system are nulled at the regulated output, and (ii) output regulation is possible when the constraint of partial observability is dropped, i.e. full state feedback is permitted.

From the proof we shall need

Lemma 6.3. *Suppose $\mathscr{V} \subset \mathscr{X}$ and $\mathscr{N} \subset \mathscr{X}$ with $A\mathscr{N} \subset \mathscr{N}$. Define extended spaces and maps as above, with*

$$\mathscr{X}_a \simeq \frac{\mathscr{N} \cap \mathscr{V}}{\mathscr{N} \cap \bigcap_{i=1}^{n} A^{-i+1}\mathscr{V}}. \tag{3.4}$$

There exists a map $E: \mathscr{X}_e \to \mathscr{X}_e$ with $\operatorname{Im} E = \mathscr{X}_a$, such that the subspace $\mathscr{V}_e = (1 + E)\mathscr{V}$ has the property

$$\mathscr{N} \cap \bigcap_{i=1}^{n} A^{-i+1}\mathscr{V} + A_e(\mathscr{V}_e \cap \mathscr{N}) \subset \mathscr{V}_e.$$

Proof. Write

$$\mathscr{V}_0 = \mathscr{N} \cap \bigcap_{i=1}^{n} A^{-i+1}\mathscr{V}$$

and let

$$\mathscr{V} \cap \mathscr{N} = \mathscr{V}_0 \oplus \mathscr{V}_1, \qquad \mathscr{V} = \mathscr{V} \cap \mathscr{N} \oplus \mathscr{V}_2.$$

Let $E: \mathscr{X}_e \to \mathscr{X}_e$ be any map, such that

$$\operatorname{Ker} E \supset \mathscr{V}_0 \oplus \mathscr{V}_2,$$
$$\operatorname{Ker} E \cap \mathscr{V}_1 = 0,$$
$$E\mathscr{V}_1 = \mathscr{X}_a.$$

Such a map exists by (3.4). Now,

$$\mathscr{V}_e \cap \mathscr{N} = [\mathscr{V}_0 \oplus (1 + E)\mathscr{V}_1 \oplus \mathscr{V}_2] \cap \mathscr{N} = \mathscr{V}_0,$$

so that

$$A_e(\mathscr{V}_e \cap \mathscr{N}) = A\mathscr{V}_0 \subset \mathscr{V}_0 = \mathscr{V}_e \cap \mathscr{N}$$

as required. □

Proof (of Theorem 6.2). (If) Choose $\mathscr{X}_a$ according to (3.4) (with $\mathscr{V}^*$ in place of $\mathscr{V}$). Construct $\mathscr{V}_e^* = (1 + E)\mathscr{V}^*$ as in Lemma 6.3. Thus,

$$\mathscr{N} \cap \bigcap_{i=1}^{n} A^{-i+1}\mathscr{V}^* + A_e(\mathscr{V}_e^* \cap \mathscr{N}) \subset \mathscr{V}_e^*. \tag{3.5}$$

We shall verify that the conditions of Theorem 6.1 hold for the extended problem. Now,

$$\mathscr{V}_e^* \subset \mathscr{V}^* \oplus \mathscr{X}_a \subset \operatorname{Ker} D \oplus \mathscr{X}_a = \operatorname{Ker} D_e$$

and

$$A_e\mathscr{V}_e^* = A\mathscr{V}^* \subset \mathscr{V}^* + \mathscr{B}$$
$$\subset (1+E)\mathscr{V}^* + \mathscr{B} + \mathscr{X}_a = \mathscr{V}_e^* + \mathscr{B}_e.$$

Next,

$$\mathscr{X}_e^+(A_e) \cap \mathscr{N} = [\mathscr{X}^+(A) \oplus \mathscr{X}_a] \cap \mathscr{N} = \mathscr{X}^+(A) \cap \mathscr{N}.$$

As $\mathscr{X}^+(A) \cap \mathscr{N}$ is A-invariant and by (3.1) belongs to Ker D, we get

$$\mathscr{X}_e^+(A_e) \cap \mathscr{N} \subset \mathscr{N} \cap \bigcap_{i=1}^{n} A^{-i+1}\mathscr{V}^* \subset \mathscr{V}_e^* \quad \text{(by (3.5))}$$

and this verifies the extended version of (2.2). Finally,

$$\mathscr{X}_e^+(A_e) = \mathscr{X}^+(A) \oplus \mathscr{X}_a \subset \langle A|\mathscr{B}\rangle + \mathscr{V}^* + \mathscr{X}_a$$
$$= \langle A|\mathscr{B}\rangle + (1+E)\mathscr{V}^* + \mathscr{X}_a = \langle A_e|\mathscr{B}_e\rangle + \mathscr{V}_e^*,$$

which verifies the extended version of (2.3).

(Only if) Let Q be the projection on $\mathscr{X}$ along $\mathscr{X}_a$. Applied to ERP, Theorem 6.1 provides a subspace $\mathscr{V}_e \subset \mathscr{X}_e$ which satisfies the extended version of (2.1)–(2.3). In particular, (2.2) implies

$$\text{Ker } D_e \supset \mathscr{X}_e^+(A_e) \cap \mathscr{N} = [\mathscr{X}^+(A) \oplus \mathscr{X}_a] \cap \mathscr{N} = \mathscr{X}^+(A) \cap \mathscr{N},$$

so that

$$\mathscr{X}^+(A) \cap \mathscr{N} \subset Q \text{ Ker } D_e = \text{Ker } D,$$

proving (3.1). Next, (2.3) applied to ERP yields

$$\mathscr{X}_e^+(A_e) \subset \langle A_e|\mathscr{B}_e\rangle + \mathscr{V}_e$$

and so, with $\mathscr{V} = Q\mathscr{V}_e$,

$$\mathscr{X}^+(A) = Q\mathscr{X}_e^+(A_e) \subset \langle A|\mathscr{B}\rangle + \mathscr{V}. \tag{3.6}$$

Finally, we note that $\mathscr{V}_e \subset$ Ker D_e implies $\mathscr{V} \subset$ Ker D, and $A_e\mathscr{V}_e \subset \mathscr{V}_e + \mathscr{B}_e$ implies

$$A\mathscr{V} = AQ\mathscr{V}_e = QA_e\mathscr{V}_e \subset Q(\mathscr{V}_e + \mathscr{B}_e) = \mathscr{V} + \mathscr{B}.$$

Hence $\mathscr{V} \subset \mathscr{V}^*$, and (3.2) follows from (3.6). □

If $\mathscr{V}_e$ is a solution of ERP then by (2.2)

$$A_e(\mathscr{V}_e \cap \mathscr{N}) \subset \mathscr{V}_e; \tag{3.7}$$

but it is not true, in general, that $A(\mathscr{V} \cap \mathscr{N}) \subset \mathscr{V}$ with $\mathscr{V} = Q\mathscr{V}_e$. Deduction of the last-written inclusion from (3.7) would be immediate if

$$Q(\mathscr{V}_e \cap \mathscr{N}) = Q\mathscr{V}_e \cap Q\mathscr{N};$$

and, as Ker $Q = \mathscr{X}_a$ and $\mathscr{N} \cap \mathscr{X}_a = 0$, this would be true if and only if

$$(\mathscr{V}_e + \mathscr{N}) \cap \mathscr{X}_a = \mathscr{V}_e \cap \mathscr{X}_a. \tag{3.8}$$

But in general (3.8) fails; indeed, the construction of Lemma 6.3 has

$$(\mathscr{V}_e + \mathscr{N}) \cap \mathscr{X}_a = \mathscr{X}_a, \qquad \mathscr{V}_e \cap \mathscr{X}_a = 0.$$

This heuristic reasoning suggests that in some cases ERP is solvable when RRP is not, a conjecture borne out by the following example.

6.4 Example

Let

$$A = \begin{bmatrix} 0 & 1 & 0 \\ -2 & -3 & 0 \\ 0 & 0 & 0 \end{bmatrix}, \qquad B = \begin{bmatrix} 0 \\ 1 \\ 0 \end{bmatrix},$$

$$C = [0 \quad 0 \quad 1], \qquad D = [0 \quad 1 \quad -1].$$

Then,

$$\mathscr{N} = \operatorname{Im} \begin{bmatrix} 1 & 0 \\ 0 & 1 \\ 0 & 0 \end{bmatrix}, \qquad \operatorname{Ker} D = \operatorname{Im} \begin{bmatrix} 1 & 0 \\ 0 & 1 \\ 0 & 1 \end{bmatrix}$$

$$\langle A \mid \mathscr{B} \rangle = \operatorname{Im} \begin{bmatrix} 1 & 0 \\ 0 & 1 \\ 0 & 0 \end{bmatrix}, \qquad \mathscr{X}^+(A) = \operatorname{Im} \begin{bmatrix} 0 \\ 0 \\ 1 \end{bmatrix}.$$

The algorithm (2.7) yields $\mathscr{V}^* = \operatorname{Ker} D$, so that

$$\mathscr{X}^+(A) \subset \mathscr{X} = \langle A \mid \mathscr{B} \rangle + \mathscr{V}^*.$$

Since in addition $\mathscr{X}^+(A) \cap \mathscr{N} = 0$, Theorem 6.2 asserts that ERP is solvable. We find

$$\mathscr{N} \cap \bigcap_{i=1}^{3} A^{-i+1} \mathscr{V}^* = \mathscr{N} \cap \bigcap_{i=1}^{3} \operatorname{Ker}(DA^{i-1}) = 0,$$

so we can take

$$d(\mathscr{X}_a) = d(\mathscr{N} \cap \mathscr{V}^*) = d\left(\operatorname{Im} \begin{bmatrix} 1 \\ 0 \\ 0 \end{bmatrix}\right) = 1.$$

Write explicitly,

$$E \operatorname{col}[1 \quad 0 \quad 0 \quad 0] = \operatorname{col}[1 \quad 0 \quad 0 \quad 1];$$

then,

$$\mathscr{V}_e^* = (1 + E)\mathscr{V}^* = \operatorname{Im} \begin{bmatrix} 1 & 0 \\ 0 & 1 \\ 0 & 1 \\ 1 & 0 \end{bmatrix}.$$

ERP is now essentially solved: it only remains to compute a feedback map F_e, such that Ker $F_e \supset \mathcal{N}$, $(A_e + B_e F_e)\mathcal{V}_e^* \subset \mathcal{V}_e^*$, and the induced map $\overline{A_e + B_e F_e}$ on $\mathcal{X}_e/\mathcal{V}_e^*$ is stable. In this example, these requirements lead to the unique result

$$F_e = \begin{bmatrix} 0 & 0 & 3 & 2 \\ 0 & 0 & 1 & 0 \end{bmatrix}.$$

Controller structure is evident from the signal flow graph, Fig. 6.2. The given system has two stable, unobservable "modes" (with state variables x_1, x_2) and one unstable, observable mode (x_3). Only x_1, x_2 are controllable.

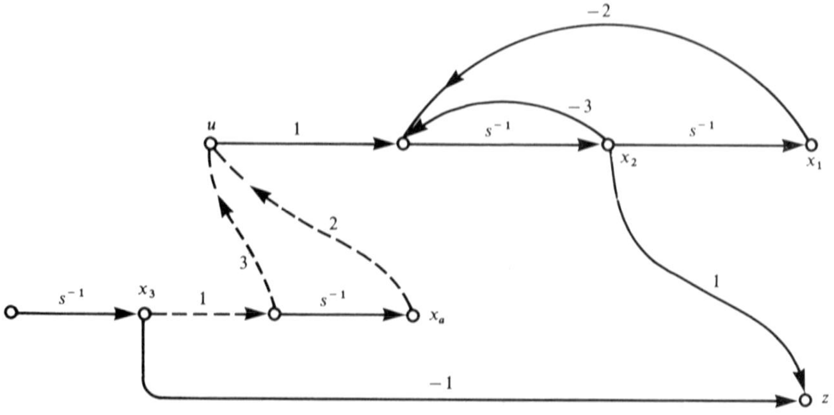

Figure 6.2 Signal Flow Graph: Example, Section 6.4. x_a simulates the variable $w = 2x_1 + f_2 x_2$.

The regulated variable is $z = x_2 - x_3$. To interpret the role of dynamic compensation in state-space terms, suppose first that all states had been observable and consider the corresponding RRP, with $\mathcal{N} = 0$. A simple computation shows that output regulation is achieved, with state feedback $u = f_1 x_1 + f_2 x_2 + f_3 x_3$, only if $f_1 = 2$. But then, $(A + BF)|\langle A|\mathcal{B}\rangle$ has the spectrum $\{0, f_2 - 3\}$, i.e. the controllable subsystem is destabilized. Returning to the example, we see that $\langle A|\mathcal{B}\rangle$ is C-unobservable. The dynamic compensator must, therefore, simulate the variable

$$w := f_1 x_1 + f_2 x_2 = 2x_1 + f_2 x_2,$$

according to an equation of form

$$\dot{w} = \alpha w + \beta x_3,$$

since only x_3 is C-observable. This requirement leads at once to the result that $\alpha = 0$, as displayed in Fig. 6.2. In conventional, frequency terms, the compensator supplies a pole to cancel the plant transmission zero at $s = 0$, as shown in Fig. 6.3. Of course, this cancellation results in internal instability.

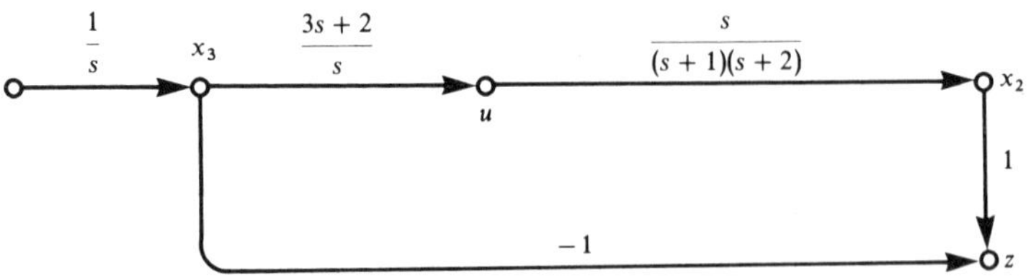

Figure 6.3 Condensed Signal Flow Graph: Example, Section 6.4. Precompensator pole cancels plant transmission zero, producing internal instability.

This example exhibits output regulation as a distinct, fundamental role of linear state feedback and compensation. A second role, observer action, is inoperative by virtue of our initial assumption Ker $C = \mathcal{N}$; a third, pole-shifting, is not possible here, because the controllable observable state subspace is zero.

To verify that RRP is not solvable, we apply Proposition 6.1. Clearly,

$$\mathcal{N} \cap \text{Ker } D = \text{Im} \begin{bmatrix} 1 \\ 0 \\ 0 \end{bmatrix}$$

is not A-invariant, so $\mathcal{V}_0 = 0$. Every subspace $\mathcal{W}$, such that $\mathcal{W} \oplus \mathcal{N} \cap \text{Ker } D = \text{Ker } D$, can be written

$$\mathcal{W}_\mu = \text{Im} \begin{bmatrix} \mu \\ 1 \\ 1 \end{bmatrix}, \qquad \mu \in \mathbb{R}.$$

Application of (2.7), with $\mathcal{W}_\mu$ in place of Ker D, yields

$$\mathcal{V}_\mu^M = 0. \tag{4.1}$$

To satisfy the condition (2.5) of Corollary 6.1 would require that

$$\mathcal{X}^+(A) = \text{Im} \begin{bmatrix} 0 \\ 0 \\ 1 \end{bmatrix} \subset \langle A | \mathcal{B} \rangle + \mathcal{V}_\mu^M = \text{Im} \begin{bmatrix} 1 & 0 \\ 0 & 1 \\ 0 & 0 \end{bmatrix} + \mathcal{V}_\mu^M$$

and this is clearly incompatible with (4.1).

6.5 Concluding Remarks

To conclude the general discussion we remark that there was nothing special about the partition $\mathbb{C} = \mathbb{C}^+ \cup \mathbb{C}^-$. In practice, one might well require that convergence in (0.2) take place with exponents in some "good" subset $\mathbb{C}_g \subset \mathbb{C}^-$, where $\mathbb{C}_g$ is symmetric with respect to the real axis. Writing $\mathbb{C}_b$ for the "bad" complement $\mathbb{C} - \mathbb{C}_g$, and replacing $\mathcal{X}^+$ (resp. $\mathcal{X}^-$) by the corresponding subspaces $\mathscr{X}_b$ (resp. $\mathscr{X}_g$), one obtains the corresponding results by exactly the same procedure.

The main result of this chapter (Theorem 6.2) is constructive and offers significant structural insight. It has, however, the shortcoming that no condition has been imposed to ensure internal stability, that is, stability of controllable and observable modes. Indeed, the example of Section 6.4 demonstrates that output regulation and internal stability are not always compatible. In Chapter 7, we solve the more realistic problem of achieving both these desirable system properties.

6.6 Exercises

6.1. For the example in Section 6.4, work out the Laplace transform $\hat{z}(s)$ as an explicit function of the initial values $x_1(0+)$, $x_2(0+)$, $x_3(0+)$ and $x_a(0+)$ and verify directly that $z(t) \to 0$ as $t \to \infty$.

6.2. Say that the system structure specified by (0.1) *admits indirect feedback control* if the observations $y(\cdot)$ permit reconstruction of $z(\cdot)$, i.e. $\mathscr{N} \subset \operatorname{Ker} D$. In that case, what can be said about the solvability of RRP? of ERP? In particular, show that dynamic compensation is necessary only if $\mathscr{N} \not\subset \operatorname{Ker} D$, and so the system must in this sense operate partially "open-loop."

6.3. Verify in detail that ERP is solvable when RRP is not, only if $\mathscr{X}^-(A) \cap \mathscr{N} \neq 0$.

6.4. ***RRP with direct control feedthrough.*** For the system with direct control feedthrough, i.e. $z = Dx + Eu$, check that RRP should be modified to read as follows: Given A, B, C, D, E, and $\mathscr{N} \subset \mathscr{X}$ with $A\mathscr{N} \subset \mathscr{N}$, find $F: \mathscr{X} \to \mathscr{U}$, such that $\operatorname{Ker} F \supset \mathscr{N}$ and $\mathscr{X}^+(A + BF) \subset \operatorname{Ker}(D + EF)$. Using the results of Exercise 4.6, prove the following modification of Theorem 6.1: RRP is solvable if and only if there exists $\mathscr{V} \subset \mathscr{X}$, such that

$$\mathscr{V} \in \mathfrak{I}(A, B; D, E), \qquad \mathscr{V} \cap \mathscr{N} \subset \operatorname{Ker} D,$$

$$\mathscr{X}^+(A) \cap \mathscr{N} + A(\mathscr{V} \cap \mathscr{N}) \subset \mathscr{V},$$

$$\mathscr{X}^+(A) \subset \langle A \mid \mathscr{B} \rangle + \mathscr{V}.$$

Finally, obtain the appropriate modification of Theorem 6.2.

6.7 Notes and References

The discussion in this chapter closely follows Wonham [7]. The use of dynamic models to represent the structure of reference and disturbance signals is fairly standard; see, for instance, Johnson [1].

7 Tracking and Regulation II: Output Regulation with Internal Stability

In this chapter, we continue the investigation begun in Chapter 6 on output regulation for the system

$$\dot{x}(t) = Ax(t) + Bu(t) \tag{0.1a}$$

$$y(t) = Cx(t) \tag{0.1b}$$

$$z(t) = Dx(t). \tag{0.1c}$$

As in Chapter 6, we shall explore the basic geometric structure, while in Chapter 8, we turn to simple and direct procedures that lend themselves to computation.

To keep the present discussion reasonably self-contained, we shall rely only on Section 6.1 of Chapter 6, concerning the use of an observer, together with Lemmas 6.1 and 6.2 in Section 6.2.

As in Chapter 6, we regard $y(\cdot)$ as the measured variable and $z(\cdot)$ as the variable to be regulated. We postulate also that a dynamic observer is utilized; equivalently, as shown in Section 6.1, we can replace C by any map C_{new}, such that

$$\operatorname{Ker} C_{\text{new}} = \mathscr{N} := \bigcap_{i \geq 1} \operatorname{Ker}(CA^{i-1}).$$

For regulation of $z(\cdot)$ it is required to find a feedback map $F: \mathscr{X} \to \mathscr{U}$, such that

$$\mathscr{X}^+(A + BF) \subset \operatorname{Ker} D. \tag{0.2}$$

Of course, the entire discussion would remain unchanged if $\mathscr{X}^+$ were replaced throughout by $\mathscr{X}_b$, corresponding to some symmetric partition $\mathbb{C}_g \cup \mathbb{C}_b$ of the complex plane into "good" and "bad" subsets, respectively.

To respect the observability constraint, F must satisfy the condition

$$\operatorname{Ker} F \supset \mathscr{N}. \tag{0.3}$$

Necessary and sufficient conditions for the existence of F subject to (0.2) and (0.3) were given in Chapter 6.

In this chapter we impose the additional requirement that F stabilize all the unstable modes of A which are both controllable and observable. Precisely, regard $\mathscr{X}/\mathscr{N}$ as the state space of the system (0.1a,b) made observable by projection mod $\mathscr{N}$. The controllable, observable subspace is then, $(\langle A \mid \mathscr{B} \rangle + \mathscr{N})/\mathscr{N}$. We require that the map induced on $(\langle A \mid \mathscr{B} \rangle + \mathscr{N})/\mathscr{N}$, by the closed-loop system map, be stable. Equivalently, any observable, unstable modes of $A + BF$ must be uncontrollable; that is,

$$\frac{\mathscr{X}^+(A + BF) + \mathscr{N}}{\mathscr{N}} \cap \frac{\langle A \mid \mathscr{B} \rangle + \mathscr{N}}{\mathscr{N}} = 0. \tag{0.4}$$

It is natural to call a system in which F has been chosen to satisfy (0.4), *internally stable*. In this way, we are led to formulate the

Regulator Problem with Internal Stabilization (RPIS). *Given the maps* $A\colon \mathscr{X} \to \mathscr{X}$, $B\colon \mathscr{U} \to \mathscr{X}$, $D\colon \mathscr{X} \to \mathscr{Z}$, *and a subspace* $\mathscr{N} \subset \mathscr{X}$ *with* $A\mathscr{N} \subset \mathscr{N}$, *find* $F\colon \mathscr{X} \to \mathscr{U}$, *such that*

$$\operatorname{Ker} F \supset \mathscr{N} \tag{0.5a}$$

$$\mathscr{X}^+(A + BF) \cap (\langle A \mid \mathscr{B} \rangle + \mathscr{N}) \subset \mathscr{N} \tag{0.5b}$$

and

$$\mathscr{X}^+(A + BF) \subset \operatorname{Ker} D. \tag{0.5c}$$

Here, it is easily checked that (0.5b) is equivalent to (0.4).

The "restricted regulator problem" (RRP) defined in Chapter 6 is identical to RPIS except that the internal stability requirement (0.5b) is absent. We recall that RRP may be solvable when RPIS is not: that is, internal stability need not be compatible with output regulation, as shown by the example in Section 6.4.

In Section 7.1 we provide a preliminary set of necessary and sufficient conditions that RPIS be solvable. While not constructive, they are exploited to show that in this problem dynamic compensation in the sense of Section 6.3 is redundant. In Section 7.2 we give constructive necessary and sufficient conditions in the case $\mathscr{N} = 0$, and in Section 7.3 extend them to the general case. In Section 7.4 the theory is applied to step disturbance rejection, and in Section 7.5 to steady-state decoupling with step inputs. Numerical examples are presented in Sections 7.6 and 7.7.

7.1 Solvability of RPIS: General Considerations

Theorem 7.1. *RPIS is solvable if and only if there exists a subspace* $\mathscr{V} \subset \mathscr{X}$, *such that*

$$\mathscr{V} \subset \operatorname{Ker} D \cap A^{-1}(\mathscr{V} + \mathscr{B}) \tag{1.1a}$$

$$\mathscr{X}^+(A) \cap \mathscr{N} + A(\mathscr{V} \cap \mathscr{N}) \subset \mathscr{V} \tag{1.1b}$$

$$\mathscr{V} \cap (\langle A \mid \mathscr{B} \rangle + \mathscr{N}) \subset \mathscr{N} \tag{1.1c}$$

and

$$\mathscr{X}^+(A) \subset \langle A \mid \mathscr{B} \rangle + \mathscr{V}. \tag{1.1d}$$

We observe that conditions (1.1a,b,d) are equivalent to solvability of RRP, as shown by Theorem 6.1; only condition (1.1c) is new. As in the counterpart Theorem 6.1, $\mathscr{V}$ may be thought of as $\mathscr{X}^+(A + BF)$ or a subspace containing it. Thus, (1.1a,d) simply recast the condition of output regulation, (1.1c) embodies internal stability, while (1.1b) is a technical condition related to the observability constraint.

Although the proof of Theorem 7.1 follows exactly the same lines as that of Theorem 6.1, it will be given here in full.

PROOF. Suppose RPIS is solvable and put $\mathscr{V} = \mathscr{X}^+(A + BF)$. Since $A\mathscr{N} \subset \mathscr{N}$ and $\operatorname{Ker} F \supset \mathscr{N}$, we have by Lemma 6.1,

$$\mathscr{X}^+(A + BF) \cap \mathscr{N} = \mathscr{X}^+(A) \cap \mathscr{N}.$$

Then, (1.1a,b,c) follow immediately from (0.5), and (1.1d) follows from the general identity (Lemma 6.2)

$$\mathscr{X}^+(A + BF) + \langle A \mid \mathscr{B} \rangle = \mathscr{X}^+(A) + \langle A \mid \mathscr{B} \rangle. \tag{1.2}$$

Conversely, let $\mathscr{V}$ have the properties (1.1). In particular,

$$A\mathscr{V} \subset \mathscr{V} + \mathscr{B}, \qquad A(\mathscr{V} \cap \mathscr{N}) \subset \mathscr{V}.$$

Writing

$$\mathscr{V} + \mathscr{N} = \hat{\mathscr{V}} \oplus \mathscr{V} \cap \mathscr{N} \oplus \hat{\mathscr{N}},$$

where $\hat{\mathscr{V}} \subset \mathscr{V}$ and $\hat{\mathscr{N}} \subset \mathscr{N}$, we can easily construct $F_0 \colon \mathscr{X} \to \mathscr{U}$, such that $\operatorname{Ker} F_0 \supset \mathscr{N}$ and $(A + BF_0)\mathscr{V} \subset \mathscr{V}$. Write $A_0 = A + BF_0$. By (1.1d) and (1.2),

$$\mathscr{X}^+(A_0) \subset \langle A_0 \mid \mathscr{B} \rangle + \mathscr{V}. \tag{1.3}$$

Again by Lemma 6.1, and using (1.1b),

$$\mathscr{X}^+(A_0) \cap \mathscr{N} = \mathscr{X}^+(A) \cap \mathscr{N} \subset \mathscr{V}. \tag{1.4}$$

It will be shown that (1.3) and (1.4) imply the existence of $F_1 \colon \mathscr{X} \to \mathscr{U}$, such that $\operatorname{Ker} F_1 = \mathscr{N}$ and $\mathscr{X}^+(A_0 + BF_1) \subset \mathscr{V}$. Then, $F = F_0 + F_1$ will have all the properties required.

Let $P: \mathscr{X} \to \bar{\mathscr{X}} := \mathscr{X}/\mathscr{V}$. Then, in obvious notation (1.3) yields

$$\bar{\mathscr{X}}^+(\bar{A}_0) = P\mathscr{X}^+(A_0) \subset \langle \bar{A}_0 | \bar{\mathscr{B}} \rangle,$$

where $\bar{\mathscr{B}} = \operatorname{Im} \bar{B}$, $\bar{B} = PB$. Thus, $(\bar{A}_0, \bar{B})$ is stabilizable, and there exists $\bar{F}_1: \bar{\mathscr{X}} \to \mathscr{U}$, such that $\bar{\mathscr{X}}^+(\bar{A}_0 + \bar{B}\bar{F}_1) = 0$. It is easy to see that $\bar{F}_1$ can also be chosen so that $\bar{F}_1 | \bar{\mathscr{X}}^-(\bar{A}_0) = 0$. Let $F_1 := \bar{F}_1 P$. Then, $P(A_0 + BF_1) = (\bar{A}_0 + \bar{B}\bar{F}_1)P$, so that $\overline{A_0 + BF_1} = (\bar{A}_0 + \bar{B}\bar{F}_1)$. Therefore,

$$\mathscr{X}^+(A_0 + BF_1) \subset P^{-1}\bar{\mathscr{X}}^+(\bar{A}_0 + \bar{B}\bar{F}_1) = \operatorname{Ker} P = \mathscr{V},$$

as required. It remains to show that $\operatorname{Ker} F_1 \supset \mathscr{N}$. Now, $A_0\mathscr{N} \subset \mathscr{N}$, so

$$\mathscr{N} = \mathscr{N} \cap \mathscr{X}^+(A_0) \oplus \mathscr{N} \cap \mathscr{X}^-(A_0).$$

By (1.4), $P(\mathscr{N} \cap \mathscr{X}^+(A_0)) = 0$, so that

$$\bar{\mathscr{N}} := P\mathscr{N} \subset \bar{\mathscr{X}}^-(\bar{A}_0).$$

This means that $\bar{F}_1\bar{\mathscr{N}} = 0$, or $F_1\mathscr{N} = 0$ as claimed. □

While Theorem 7.1 does not indicate how to find a suitable $\mathscr{V}$ if one exists, it is well suited to showing that if RPIS is not solvable, then no solution can be obtained by broadening the assumptions to include the possibility of state space extension, that is, dynamic compensation. We may interpret this result as a "deterministic separation theorem," which asserts that, after insertion of a dynamic observer, no further *dynamic* signal processing is required to achieve the stated design objectives, if these objectives can be achieved at all.[1]

To be precise, introduce extended spaces and maps exactly as in Section 6.3. Namely, we adjoin the auxiliary equations

$$\dot{x}_a = B_a u_a, \qquad y_a = C_a x_a,$$

where $u_a \in \mathscr{U}_a$, $x_a \in \mathscr{X}_a$, $y_a \in \mathscr{Y}_a$. Here, $B_a: \mathscr{U}_a \simeq \mathscr{X}_a$ and $C_a: \mathscr{X}_a \simeq \mathscr{Y}_a$ are arbitrary isomorphisms. Introduce the external direct sums

$$\mathscr{U}_e := \mathscr{U} \oplus \mathscr{U}_a, \qquad \mathscr{X}_e := \mathscr{X} \oplus \mathscr{X}_a, \qquad \mathscr{Y}_e := \mathscr{Y} \oplus \mathscr{Y}_a$$

for the extended control, state and (observed) output spaces, and accordingly define the extended maps

$$\begin{aligned}
A_e&: \mathscr{X}_e \to \mathscr{X}_e, & x \oplus x_a &\mapsto Ax\\
B_e&: \mathscr{U}_e \to \mathscr{X}_e, & u \oplus u_a &\mapsto Bu \oplus B_a u_a\\
C_e&: \mathscr{X}_e \to \mathscr{Y}_e, & x \oplus x_a &\mapsto Cx \oplus C_a x_a\\
D_e&: \mathscr{X}_e \to \mathscr{Z}, & x \oplus x_a &\mapsto Dx.
\end{aligned}$$

[1] It should be borne in mind that as yet our problem formulation takes no account of the sensitivity of the synthesis to parameter variations. If this is done, additional dynamic elements may be used to advantage, as will be shown in Chapter 8.

The corresponding matrices are

$$A_e = \begin{bmatrix} A & 0 \\ 0 & 0 \end{bmatrix}, \qquad B_e = \begin{bmatrix} B & 0 \\ 0 & B_a \end{bmatrix},$$

$$C_e = \begin{bmatrix} C & 0 \\ 0 & C_a \end{bmatrix}, \qquad D_e = [D \quad 0].$$

In this notation, we now define the **Extended Regulator Problem with Internal Stability (ERPIS)** as that of finding suitable $\mathscr{X}_a$ (that is, $d(\mathscr{X}_a)$) and then $F_e: \mathscr{X}_e \to \mathscr{U}_e$, such that

$$\text{Ker } F_e \supset \mathscr{N}$$

$$\mathscr{X}_e^+(A_e + B_e F_e) \cap (\langle A_e | \mathscr{B}_e \rangle + \mathscr{N}) \subset \mathscr{N}$$

and

$$\mathscr{X}_e^+(A_e + B_e F_e) \subset \text{Ker } D \oplus \mathscr{X}_a.$$

Theorem 7.2. *ERPIS is solvable only if RPIS is solvable.*

PROOF. If ERPIS is solvable Theorem 7.1 implies the existence of $\mathscr{V}_e \subset \mathscr{X}_e$, such that

$$\mathscr{V}_e \subset (\text{Ker } D \oplus \mathscr{X}_a) \cap A_e^{-1}(\mathscr{V}_e + \mathscr{B} + \mathscr{B}_a), \tag{1.5a}$$

$$\mathscr{X}_e^+(A_e) \cap \mathscr{N} + A_e(\mathscr{V}_e \cap \mathscr{N}) \subset \mathscr{V}_e, \tag{1.5b}$$

$$\mathscr{V}_e \cap (\langle A | \mathscr{B} \rangle + \mathscr{X}_a + \mathscr{N}) \subset \mathscr{N}, \tag{1.5c}$$

and

$$\mathscr{X}_e^+(A_e) \subset \langle A | \mathscr{B} \rangle + \mathscr{X}_a + \mathscr{V}_e. \tag{1.5d}$$

Here, we have used the evident facts that $\mathscr{N}_e = \mathscr{N}$, $\text{Ker } D_e = \text{Ker } D \oplus \mathscr{X}_a$, and

$$\langle A_e | \mathscr{B}_e \rangle = \langle A | \mathscr{B} \rangle \oplus \mathscr{X}_a.$$

Let $P: \mathscr{X}_e \to \mathscr{X}_e$ be the projection on $\mathscr{X}$ along $\mathscr{X}_a$ and define $\mathscr{V} = P\mathscr{V}_e$. It is enough to show that $\mathscr{V}$ has the properties (1.1), and this requires only the application of P to both sides of the corresponding relations (1.5). By definition of P and A_e, $PA_e = A_e P$ and $A_e | \mathscr{X} = A$. Using these facts and rewriting (1.5a) as

$$\mathscr{V}_e \subset \text{Ker } D \oplus \mathscr{X}_a, \qquad A_e \mathscr{V}_e \subset \mathscr{V}_e + \mathscr{B} + \mathscr{B}_a,$$

there follows

$$\mathscr{V} \subset \text{Ker } D, \qquad A\mathscr{V} \subset \mathscr{V} + \mathscr{B},$$

which is equivalent to (1.1a). Next, the obvious relation

$$\mathscr{X}_e^+(A_e) = \mathscr{X}^+(A) \oplus \mathscr{X}_a \tag{1.6}$$

together with (1.5d), establishes (1.1d). To verify (1.1c) from (1.5c), we use the following general result for a map P and subspaces $\mathcal{R}$, $\mathcal{S}$ (0.4.2), (0.4.3):

$$P(\mathcal{R} \cap \mathcal{S}) = (P\mathcal{R}) \cap (P\mathcal{S}) \tag{1.7}$$

if and only if

$$(\mathcal{R} + \mathcal{S}) \cap \text{Ker } P = \mathcal{R} \cap \text{Ker } P + \mathcal{S} \cap \text{Ker } P. \tag{1.8}$$

With Ker $P = \mathcal{X}_a$, $\mathcal{R} = \mathcal{V}_e$, and $\mathcal{S} = \langle A | \mathcal{B} \rangle + \mathcal{X}_a + \mathcal{N}$, (1.8) follows at once, and then (1.7) applied to (1.5c) yields (1.1c). It remains to check (1.1b) from (1.5b). By (1.6)

$$\mathcal{X}^+(A_e) \cap \mathcal{N} = \mathcal{X}^+(A) \cap \mathcal{N},$$

and so

$$\mathcal{X}^+(A) \cap \mathcal{N} \subset P\mathcal{V}_e = \mathcal{V}. \tag{1.9}$$

Also, by (1.7), (1.8) we shall have

$$P(\mathcal{V}_e \cap \mathcal{N}) = \mathcal{V} \cap \mathcal{N} \tag{1.10}$$

provided

$$(\mathcal{V}_e + \mathcal{N}) \cap \mathcal{X}_a = \mathcal{V}_e \cap \mathcal{X}_a + \mathcal{N} \cap \mathcal{X}_a. \tag{1.11}$$

It is in proving (1.11) that the condition of internal stability (1.5c) plays a crucial role. Let

$$x_a = v_e + n \in (\mathcal{V}_e + \mathcal{N}) \cap \mathcal{X}_a$$

with $v_e \in \mathcal{V}_e$ and $n \in \mathcal{N}$. Then,

$$v_e = x_a - n \in \mathcal{X}_a + \mathcal{N}$$

and by (1.5c), $v_e \in \mathcal{N}$. Therefore, $x_a \in \mathcal{N} \cap \mathcal{X}_a = 0$, that is,

$$\mathcal{X}_a \cap (\mathcal{V}_e + \mathcal{N}) = 0$$

proving (1.11). Then (1.10) is true, and (1.5b) yields

$$\mathcal{V} \supset PA_e(\mathcal{V}_e \cap \mathcal{N}) = A(\mathcal{V} \cap \mathcal{N}). \tag{1.12}$$

Finally, (1.1b) results from (1.9) and (1.12). □

7.2 Constructive Solution of RPIS: $\mathcal{N} = 0$

Let

$$\mathcal{V}^* := \sup \mathfrak{I}(A, B; \text{Ker } D)$$

and

$$\mathcal{R}^* := \sup \mathfrak{C}(A, B; \text{Ker } D).$$

Recall from Theorem 5.5 that $\mathscr{R}^* \subset \mathscr{V}^*$, and from Corollary 5.1 that $\mathbf{F}(A, B; \mathscr{V}^*) \subset \mathbf{F}(A, B; \mathscr{R}^*)$. If $F \in \mathbf{F}(\mathscr{V}^*)$ $(:= \mathbf{F}(A, B; \mathscr{V}^*))$ and $A_F := A + BF$, let $\bar{A}_F$ denote the map induced in $\mathscr{X}/\mathscr{R}^*$ by A_F. Then (Theorem 5.7) the restriction $\bar{A}_F \mid (\mathscr{V}^*/\mathscr{R}^*)$ is independent of the choice of $F \in \mathbf{F}(\mathscr{V}^*)$.

Let $A\mathscr{T} \subset \mathscr{T}$ and $A\mathscr{R} \subset \mathscr{R} \subset \mathscr{T}$. Recall the definition that the subspace $\mathscr{R}$ *decomposes* $\mathscr{T}$ *relative to* A if there exists a subspace $\mathscr{S}$, such that $A\mathscr{S} \subset \mathscr{S}$ and $\mathscr{R} \oplus \mathscr{S} = \mathscr{T}$. An elementary and constructive criterion that $\mathscr{R}$ decompose $\mathscr{T}$ was given in Section 0.11; it amounts to the fact that decomposability is equivalent to the existence of a solution to Sylvester's linear matrix equation. Let $\bar{A}$ be the map induced by A on $\mathscr{T}/\mathscr{R}$. A simple sufficient condition for decomposability is that $\sigma(A \mid \mathscr{R})$ and $\sigma(\bar{A})$ be disjoint. From a systemic viewpoint, decomposability can be related to the topology of signal flow: for illustrations see Exercises 7.8 and 7.9.

The main result of this chapter is the following.

Theorem 7.3. *Let $\mathscr{N} = 0$. Then, RPIS is solvable if and only if*

$$\mathscr{X}^+(A) \subset \langle A \mid \mathscr{B} \rangle + \mathscr{V}^*, \tag{2.1}$$

and in $\mathscr{X}/\mathscr{R}^*$, with $F \in \mathbf{F}(\mathscr{V}^*)$, the subspace

$$\frac{\mathscr{V}^* \cap \mathscr{X}^+(A_F) \cap \langle A \mid \mathscr{B} \rangle + \mathscr{R}^*}{\mathscr{R}^*}$$

decomposes the subspace

$$\frac{\mathscr{V}^* \cap \mathscr{X}^+(A_F) + \mathscr{R}^*}{\mathscr{R}^*}$$

relative to the map induced by A_F in $\mathscr{V}^*/\mathscr{R}^*$.

We emphasize that the decomposability condition need be checked only for a single, arbitrarily selected map $F \in \mathbf{F}(\mathscr{V}^*)$. Thus, it is clear that the solvability criterion is constructive, as $\mathscr{V}^*$ and $\mathscr{R}^*$ are computable by simple algorithms (Theorems 4.3 and 5.6), an $F \in \mathbf{F}(\mathscr{V}^*)$ is readily constructed (Lemma 4.2) and decomposability is verifiable in the way already described. Intuitively, the first condition of the theorem is just the familiar criterion for output regulation (cf. Theorems 4.4 and 6.2); while decomposability ensures that any "internal" (i.e. controllable) unstable modes can be "split off" from unstable modes of the exosystem: once they are split off, such modes admit stabilization by feedback that preserves output regulation. (Exercise 7.8 provides a simple illustration of how this process can fail.) Based on the property of spectral disjointness already mentioned, a sufficient condition for decomposability that is almost always effective in practice is given below in Corollary 7.2.

Proof. (If) We shall construct a subspace $\mathcal{V}$ which satisfies the conditions of Theorem 7.1. While the details are somewhat intricate, the construction is readily visualized with the aid of the lattice diagrams of Fig. 7.1. For simplicity, the reader might first consider what happens when $\mathcal{R}^*$ is zero.

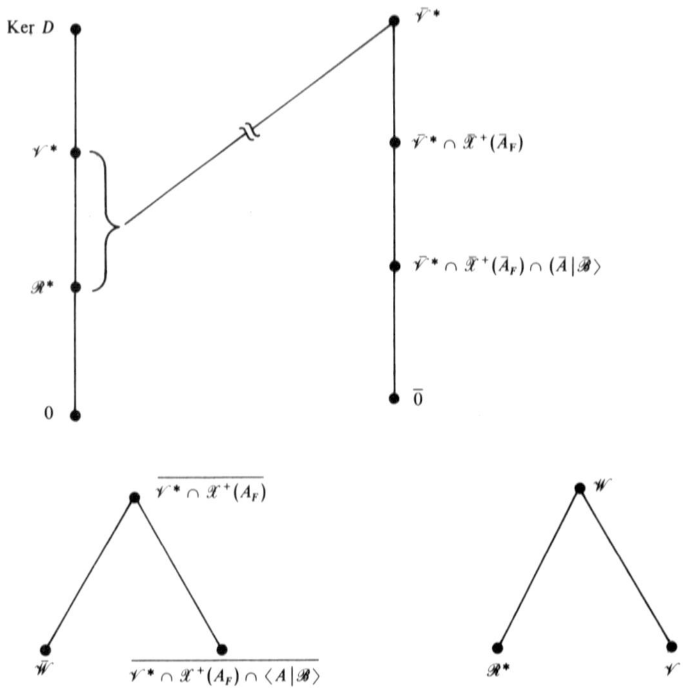

Figure 7.1 Lattice Diagrams: Construction of Subspace $\mathcal{V}$. (Bars denote projections mod $\mathcal{R}^*$.)

Let $F \in \mathbf{F}(\mathcal{V}^*)$. By (2.1) and Lemma 6.4,

$$\mathcal{X}^+(A_F) \subset \langle A | \mathcal{B} \rangle + \mathcal{V}^*.$$

As the subspaces on the right are A_F-invariant, we have

$$\mathcal{X}^+(A_F) = \langle A | \mathcal{B} \rangle \cap \mathcal{X}^+(A_F) + \mathcal{V}^* \cap \mathcal{X}^+(A_F). \tag{2.2}$$

Also, by the assumption of decomposability, there exists a subspace $\mathcal{W} \subset \mathcal{X}$, such that

$$A_F \mathcal{W} \subset \mathcal{W} \tag{2.3}$$

$$\mathcal{R}^* \subset \mathcal{W} \subset \mathcal{V}^* \cap \mathcal{X}^+(A_F) + \mathcal{R}^* \tag{2.4}$$

and

$$\frac{\mathcal{V}^* \cap \mathcal{X}^+(A_F) + \mathcal{R}^*}{\mathcal{R}^*} = \frac{\mathcal{V}^* \cap \langle A | \mathcal{B} \rangle \cap \mathcal{X}^+(A_F) + \mathcal{R}^*}{\mathcal{R}^*} \oplus \frac{\mathcal{W}}{\mathcal{R}^*}. \tag{2.5}$$

We remark that with $\mathscr{W}$ fixed, (2.3)–(2.5) hold for all $F \in \mathbf{F}(\mathscr{V}^*)$. By (2.5),

$$\mathscr{V}^* \cap \mathscr{X}^+(A_F) \subset \mathscr{V}^* \cap \langle A \mid \mathscr{B} \rangle \cap \mathscr{X}^+(A_F) + \mathscr{W} \tag{2.6}$$

and by (2.2) and (2.6),

$$\mathscr{X}^+(A_F) \subset \langle A \mid \mathscr{B} \rangle + \mathscr{V}^* \cap \mathscr{X}^+(A_F) \subset \langle A \mid \mathscr{B} \rangle + \mathscr{W}. \tag{2.7}$$

Clearly,

$$\mathscr{W} \subset \mathscr{V}^*. \tag{2.8}$$

Also, by (2.5)

$$\frac{\mathscr{W}}{\mathscr{R}^*} \cap \frac{\langle A \mid \mathscr{B} \rangle}{\mathscr{R}^*} = 0,$$

so that

$$\mathscr{W} \cap \langle A \mid \mathscr{B} \rangle = \mathscr{R}^*. \tag{2.9}$$

Finally, let $\bar{A}_F$ denote the map induced by A_F in $\mathscr{X}/\mathscr{R}^*$, and choose $F \in \mathbf{F}(\mathscr{V}^*)$, such that

$$\sigma(A_F \mid \mathscr{R}^*) \cap \sigma(\bar{A}_F \mid \mathscr{W}/\mathscr{R}^*) = \varnothing.$$

Since the two spectra are now disjoint, $\mathscr{R}^*$ certainly decomposes $\mathscr{W}$ relative to A_F; that is, there exists $\mathscr{V} \subset \mathscr{W}$ (depending on F), such that

$$A_F \mathscr{V} \subset \mathscr{V} \tag{2.10}$$

and

$$\mathscr{R}^* \oplus \mathscr{V} = \mathscr{W}. \tag{2.11}$$

From (2.7)–(2.11), we conclude that

$$\mathscr{V} \subset \operatorname{Ker} D \cap A^{-1}(\mathscr{V} + \mathscr{B}),$$
$$\mathscr{V} \cap \langle A \mid \mathscr{B} \rangle = 0,$$
$$\mathscr{X}^+(A) \subset \langle A \mid \mathscr{B} \rangle + \mathscr{V},$$

and it follows by Theorem 7.1 that RPIS is solvable.

(Only if) If RPIS is solvable, Theorem 7.1 supplies a subspace $\mathscr{V}$, such that

$$\mathscr{V} \subset \operatorname{Ker} D \cap A^{-1}(\mathscr{V} + \mathscr{B})$$

and

$$\mathscr{X}^+(A) \subset \langle A \mid \mathscr{B} \rangle \oplus \mathscr{V}. \tag{2.12}$$

Since $\mathscr{V} \cap \langle A \mid \mathscr{B} \rangle = 0$, we have $\mathscr{V} \cap \mathscr{R}^* = 0$, so that $\mathscr{V}^* \supset \mathscr{R}^* \oplus \mathscr{V}$. From this it is clear that $F \in \mathbf{F}(\mathscr{V}) \cap \mathbf{F}(\mathscr{V}^*)$ can be chosen such that $A_F \mid \mathscr{R}^*$ is stable. By (2.12) and Lemma 6.2, we have

$$\mathscr{X}^+(A_F) \subset \langle A \mid \mathscr{B} \rangle \oplus \mathscr{V} \subset \langle A \mid \mathscr{B} \rangle + \mathscr{V}^*.$$

Since all the subspaces here are A_F-invariant there follows

$$\begin{aligned}\mathcal{X}^+(A_F) &\subset \langle A \,|\, \mathcal{B}\rangle \cap \mathcal{X}^+(A_F) \oplus \mathcal{V} \cap \mathcal{X}^+(A_F)\\ &\subset \langle A \,|\, \mathcal{B}\rangle \cap \mathcal{X}^+(A_F) + \mathcal{V}^* \cap \mathcal{X}^+(A_F)\\ &\subset \mathcal{X}^+(A_F),\end{aligned}$$

and therefore,

$$\mathcal{V}^* \cap \mathcal{X}^+(A_F) \subset \langle A \,|\, \mathcal{B}\rangle \cap \mathcal{X}^+(A_F) \oplus \mathcal{V} \cap \mathcal{X}^+(A_F).$$

Intersecting both sides with $\mathcal{V}^*$ and using $\mathcal{V}^* \supset \mathcal{V}$, we obtain

$$\begin{aligned}\mathcal{V}^* \cap \mathcal{X}^+(A_F) &= \langle A \,|\, \mathcal{B}\rangle \cap \mathcal{X}^+(A_F) \cap \mathcal{V}^* \oplus \mathcal{V} \cap \mathcal{X}^+(A_F)\\ &= \mathcal{S} \oplus \mathcal{T}, \text{ say.}\end{aligned} \tag{2.13}$$

Let $P\colon \mathcal{X} \to \mathcal{X}/\mathcal{R}^*$ be the canonical projection. By the stability of $A_F \,|\, \mathcal{R}^*$, we have

$$\begin{aligned}(\mathcal{S} \oplus \mathcal{T}) \cap \operatorname{Ker} P = (\mathcal{S} \oplus \mathcal{T}) \cap \mathcal{R}^* &\subset \mathcal{X}^+(A_F) \cap \mathcal{R}^*\\ &= 0\\ &= \mathcal{S} \cap \operatorname{Ker} P \oplus \mathcal{T} \cap \operatorname{Ker} P\end{aligned}$$

and therefore,

$$(P\mathcal{S}) \cap (P\mathcal{T}) = P(\mathcal{S} \cap \mathcal{T}) = 0. \tag{2.14}$$

By (2.13), (2.14), we have finally,

$$\begin{aligned}P[\mathcal{V}^* \cap \mathcal{X}^+(A_F)] = P[&\langle A \,|\, \mathcal{B}\rangle \cap \mathcal{X}^+(A_F) \cap \mathcal{V}^*]\\ &\oplus P[\mathcal{V} \cap \mathcal{X}^+(A_F)],\end{aligned}$$

a decomposition of the type required. To complete the proof it suffices to remark that the condition just derived must hold for all $F \in \mathbf{F}(\mathcal{V}^*)$ if it holds for one. □

Remark 1. The foregoing proof of sufficiency made no essential use of the fact that $\mathcal{V}^*$ is actually the supremal element of the family of subspaces

$$\mathfrak{B} = \{\tilde{\mathcal{V}} : \tilde{\mathcal{V}} \subset \operatorname{Ker} D \cap A^{-1}(\tilde{\mathcal{V}} + \mathcal{B})\}.$$

The sole reason for stating Theorem 7.3 in terms of $\mathcal{V}^*$ is that this element of $\mathfrak{B}$ is readily computable algorithmically and so the obtained conditions are constructive. It is clear from the proof that the conclusion of Theorem 7.3 is valid provided the stated conditions hold for *some* element $\tilde{\mathcal{V}} \in \mathfrak{B}$, with $\mathcal{R}^*$ replaced by the largest c.s. $\tilde{\mathcal{R}} \subset \tilde{\mathcal{V}}$.

Remark 2. In applications it is often true that the map $\tilde{A}$ (say), induced by A in $\tilde{\mathcal{X}} := \mathcal{X}/\langle A \,|\, \mathcal{B}\rangle$, is completely unstable, i.e. $\sigma(\tilde{A}) \subset \mathbb{C}^+$. This merely reflects the fact that $\tilde{A}$ embodies the structure of the exosystem, or dynamic model of the disturbance and reference signals external to the plant. Under this condition Theorem 7.3 can be stated more simply as follows.

Corollary 7.1. *Let* $\mathscr{N} = 0$ *and assume that the map induced by* A *in* $\mathscr{X}/\langle A \mid \mathscr{B}\rangle$ *has its spectrum in* $\mathbb{C}^+$. *Then, RPIS is solvable if and only if*

$$\langle A \mid \mathscr{B}\rangle + \mathscr{V}^* = \mathscr{X},$$

and in $\mathscr{X}/\mathscr{R}^*$, *with* $F \in \mathbf{F}(\mathscr{V}^*)$, *the subspace* $(\mathscr{V}^* \cap \langle A \mid \mathscr{B}\rangle)/\mathscr{R}^*$ *decomposes the subspace* $\mathscr{V}^*/\mathscr{R}^*$ *relative to the map induced by* A_F *in* $\mathscr{V}^*/\mathscr{R}^*$.

We remark again that the decomposability condition need be checked only for a single, arbitrarily selected map $F \in \mathbf{F}(\mathscr{V}^*)$.

The proof is left as Exercise 7.6.

Corollary 7.1 leads to a simple sufficient condition that is usually satisfied when RPIS is, in fact, solvable. For brevity write (D_1, A_1, B_1) for the restriction of the triple (D, A, B) to $\langle A \mid \mathscr{B}\rangle$:

$$D_1 := D \mid \langle A \mid \mathscr{B}\rangle, \qquad A_1 := A \mid \langle A \mid \mathscr{B}\rangle, \qquad B_1 := \langle A \mid \mathscr{B}\rangle \mid B.$$

It is easy to check (Exercise 7.7) that

$$\mathscr{V}_1^* := \sup \mathfrak{J}(A_1, B_1; \operatorname{Ker} D_1) = \mathscr{V}^* \cap \langle A \mid \mathscr{B}\rangle$$

and that, correspondingly, $\mathscr{R}_1^* = \mathscr{R}^*$ (considered as a subspace of $\langle A \mid \mathscr{B}\rangle$). As in Section 5.4, write

$$\sigma_1^*(A_1, B_1; \operatorname{Ker} D_1)$$

for the spectrum of the map induced by $A_1 + B_1 F_1$ on $\mathscr{V}_1^*/\mathscr{R}_1^*$, with $F_1 \in \mathbf{F}(\mathscr{V}_1^*)$; by Corollary 5.2 this spectrum is, of course, independent of the choice of such F_1. Now the decomposability condition of Corollary 7.1 means that the factor

$$\frac{\mathscr{V}^*/\mathscr{R}^*}{(\mathscr{V}^* \cap \langle A \mid \mathscr{B}\rangle)/\mathscr{R}^*} \simeq \frac{\mathscr{V}^* + \langle A \mid \mathscr{B}\rangle}{\langle A \mid \mathscr{B}\rangle} = \frac{\mathscr{X}}{\langle A \mid \mathscr{B}\rangle}$$

is representable as an $\bar{A}_F$-invariant subspace of $\mathscr{V}^*/\mathscr{R}^*$. Since $\mathscr{X}/\langle A \mid \mathscr{B}\rangle$ can be thought of as the state space of the exosystem, we have the following.

Corollary 7.2. *Let* $\mathscr{N} = 0$ *and assume that the map* $\bar{A}$ *induced by* A *in* $\mathscr{X}/\langle A \mid \mathscr{B}\rangle$ *has its spectrum in* $\mathbb{C}^+$. *Then, RPIS is solvable provided*

$$\langle A \mid \mathscr{B}\rangle + \mathscr{V}^* = \mathscr{X}$$

and

$$\sigma_1^*(A_1, B_1; \operatorname{Ker} D_1) \cap \sigma(\bar{A}) = \varnothing.$$

A frequency interpretation of Corollary 7.2 can be based on the fact (cf. Section 6.4) that an element $\lambda \in \sigma_1^*$ corresponds to a "zero" which may block the passage of error corrective signals at the corresponding complex frequency. Indeed, in the special case where (D_1, A_1) is observable, σ_1^* is precisely the set of transmission zeros (Section 5.5) of the plant transfer matrix $D_1(\lambda 1_1 - A_1)^{-1}B_1$. If such complex frequencies occur among the signals from the exosystem, error correction may fail unless the zero is cancelled by (say) a precompensator pole. But with a typically unstable exosystem, such a pole would introduce instability of a precompensator mode, and these modes are both controllable and observable, hence are subject to "internal stabilization." Corollary 7.2 states the converse: namely, if no plant transmission zero coincides with an exosystem pole, then error correction is always possible without illegal pole-zero cancellation and the internal instability that would ensue.

7.3 Constructive Solution of RPIS: $\mathcal{N}$ Arbitrary

It is not difficult to extend Theorem 7.3 to the general case. Suppose first that RPIS is solvable with the map F. Since Ker $F \supset \mathcal{N}$, we have by Lemma 6.2 that

$$\mathcal{X}^+(A) \cap \mathcal{N} = \mathcal{X}^+(A + BF) \cap \mathcal{N} \tag{3.1}$$

is $(A + BF)$-invariant. Let

$$P: \mathcal{X} \to \bar{\mathcal{X}} = \frac{\mathcal{X}}{\mathcal{X}^+(A) \cap \mathcal{N}}$$

be the canonical projection, and let bars designate the maps induced in $\bar{\mathcal{X}}$. As Ker $F \supset$ Ker P, $\bar{F}: \bar{\mathcal{X}} \to \mathcal{U}$ exists uniquely such that $\bar{F}P = F$, and it is easily seen that P Ker $F =$ Ker $\bar{F}$. Similarly, by (0.5c) and (3.1), $\bar{D}: \bar{\mathcal{X}} \to \mathcal{Z}$ exists uniquely such that $\bar{D}P = D$, and P Ker $D =$ Ker $\bar{D}$. Finally, define $\bar{B}: \mathcal{U} \to \bar{\mathcal{X}}$ by $\bar{B} = PB$.

Now $\overline{A + BF} = \bar{A} + \bar{B}\bar{F}$, so (by Lemma 4.6 applied to $A + BF$)

$$P\mathcal{X}^+(A + BF) = \bar{\mathcal{X}}^+(\bar{A} + \bar{B}\bar{F}).$$

Also,

$$\begin{aligned}[\mathcal{X}^+(A + BF) + \langle A \mid \mathcal{B}\rangle + \mathcal{N}] \cap \operatorname{Ker} P &= \mathcal{X}^+(A) \cap \mathcal{N} \\ &= \mathcal{X}^+(A + BF) \cap \operatorname{Ker} P + (\langle A \mid \mathcal{B}\rangle + \mathcal{N}) \cap \operatorname{Ker} P.\end{aligned}$$

With these observations we may project both sides of (0.5) to obtain

$$\operatorname{Ker} \bar{F} \supset \bar{\mathcal{N}} \tag{3.2a}$$

$$\bar{\mathcal{X}}^+(\bar{A} + \bar{B}\bar{F}) \cap (\langle \bar{A} \mid \bar{\mathcal{B}}\rangle + \bar{\mathcal{N}}) \subset \bar{\mathcal{N}} \tag{3.2b}$$

$$\bar{\mathcal{X}}^+(\bar{A} + \bar{B}\bar{F}) \subset \operatorname{Ker} \bar{D}. \tag{3.2c}$$

Automatically,

$$\bar{\mathcal{X}}^+(\bar{A}) \cap \bar{\mathcal{N}} = 0$$

or equivalently,

$$\bar{\mathcal{N}} \subset \bar{\mathcal{X}}^-(\bar{A}). \tag{3.3}$$

We have shown that if **RPIS** is solvable, so is the reduced problem (3.2) in $\bar{\mathcal{X}}$, and (3.3) is true as well. Conversely, suppose

$$\mathcal{X}^+(A) \cap \mathcal{N} \subset \operatorname{Ker} D \tag{3.4}$$

and that $\bar{F}: \bar{\mathcal{X}} \to \mathcal{U}$ exists, such that (3.2) are true. Define $F = \bar{F}P$. By reversing the steps which led to (3.2) it is routine to verify that (0.5) are true, that is, RPIS is solvable. We therefore have

Lemma 7.1. *RPIS is solvable if and only if the reduced problem* (3.2) *is solvable under the assumption* (3.4).

Next, we show that in (3.2b) we may set $\bar{\mathcal{N}} = 0$.

Lemma 7.2. *If* (3.2)–(3.4) *are true, then*

$$\bar{\mathcal{X}}^+(\bar{A} + \bar{B}\bar{F}) \cap \langle \bar{A} | \bar{\mathcal{B}} \rangle = 0. \tag{3.5}$$

Conversely, if (3.5) *holds, so does* (3.2b).

PROOF. By (3.2b)

$$\begin{aligned}\bar{\mathcal{X}}^+(\bar{A} + \bar{B}\bar{F}) \cap (\langle \bar{A} | \bar{\mathcal{B}} \rangle + \bar{\mathcal{N}}) &\subset \bar{\mathcal{N}} \cap \bar{\mathcal{X}}^+(\bar{A} + \bar{B}\bar{F}) \\ &= \bar{\mathcal{N}} \cap \bar{\mathcal{X}}^+(\bar{A}) \quad \text{(by (3.2a))} \\ &= 0.\end{aligned}$$

Conversely, the left side of (3.2b) can be written

$$\begin{aligned}&\bar{\mathcal{X}}^+(\bar{A} + \bar{B}\bar{F}) \cap [\langle \bar{A} | \bar{\mathcal{B}} \rangle \cap \bar{\mathcal{X}}^+(\bar{A} + \bar{B}\bar{F}) \\ &\qquad\qquad + \langle \bar{A} | \bar{\mathcal{B}} \rangle \cap \bar{\mathcal{X}}^-(\bar{A} + \bar{B}\bar{F}) + \bar{\mathcal{N}} \cap \bar{\mathcal{X}}^-(\bar{A})] \\ &= \bar{\mathcal{X}}^+(\bar{A} + \bar{B}\bar{F}) \cap \langle \bar{A} | \bar{\mathcal{B}} \rangle \\ &= 0.\end{aligned}$$

□

By Lemmas 7.1 and 7.2 the solvability of **RPIS** is equivalent to solvability of the reduced problem (3.2a), (3.2c), (3.5) under the assumption (3.4). Our next result implies that the condition (3.2a) is redundant. For simplicity of notation we temporarily drop bars.

Lemma 7.3. *Let* $F_0: \mathcal{X} \to \mathcal{U}$, *be such that*

$$\mathcal{X}^+(A + BF_0) \cap \langle A | \mathcal{B} \rangle = 0.$$

There exists $F_1: \mathcal{X} \to \mathcal{U}$, *such that*

$$\operatorname{Ker} F_1 \supset \mathcal{X}^-(A) \tag{3.6}$$

and

$$\mathcal{X}^+(A + BF_1) = \mathcal{X}^+(A + BF_0). \tag{3.7}$$

PROOF. We first establish the decomposition

$$\mathcal{X} = \mathcal{X}^-(A) \oplus \langle A \mid \mathcal{B}\rangle \cap \mathcal{X}^+(A) \oplus \mathcal{X}^+(A + BF_0). \tag{3.8}$$

For this let $P: \mathcal{X} \to \mathcal{X}/\langle A \mid \mathcal{B}\rangle =: \bar{\mathcal{X}}$ and note that

$$\bar{\mathcal{X}} = \bar{\mathcal{X}}^+(\bar{A}) + \bar{\mathcal{X}}^-(\bar{A}) = \bar{\mathcal{X}}^+(\overline{A + BF_0}) + \bar{\mathcal{X}}^-(\bar{A}),$$

so that

$$\mathcal{X} = \mathcal{X}^+(A + BF_0) + \mathcal{X}^-(A) + \langle A \mid \mathcal{B}\rangle,$$

which equals the span of the right side of (3.8). For the independence in (3.8) note that

$$\begin{aligned}
&P\{\mathcal{X}^+(A + BF_0) \cap [\mathcal{X}^-(A) \oplus \langle A \mid \mathcal{B}\rangle \cap \mathcal{X}^+(A)]\} \\
&\quad \subset P\mathcal{X}^+(A + BF_0) \cap P\mathcal{X}^-(A) \\
&\quad = \bar{\mathcal{X}}^+(\bar{A}) \cap \bar{\mathcal{X}}^-(\bar{A}) \\
&\quad = 0,
\end{aligned}$$

so that

$$\mathcal{X}^+(A + BF_0) \cap [\mathcal{X}^-(A) \oplus \langle A \mid \mathcal{B}\rangle \cap \mathcal{X}^+(A)] \subset \langle A \mid \mathcal{B}\rangle.$$

Intersecting the left side with $\langle A \mid \mathcal{B}\rangle$ and using the hypothesis yields

$$\mathcal{X}^+(A + BF_0) \cap [\mathcal{X}^-(A) \oplus \langle A \mid \mathcal{B}\rangle \cap \mathcal{X}^+(A)] = 0$$

as required.

Next, let Q^+ be the projection on $\mathcal{X}^+(A) \cap \langle A \mid \mathcal{B}\rangle$ along $\mathcal{X}^-(A) \cap \langle A \mid \mathcal{B}\rangle$ and consider

$$A^+ := A \mid [\langle A \mid \mathcal{B}\rangle \cap \mathcal{X}^+(A)], \qquad B^+ = Q^+ B.$$

Since controllability is preserved in a modal decomposition (Exercise 1.5), we have that (A^+, B^+) is controllable, hence there exists $F^+: \langle A \mid \mathcal{B}\rangle \cap \mathcal{X}^+(A) \to \mathcal{U}$, such that $A^+ + B^+F^+$ is stable.

Now define $F_1: \mathcal{X} \to \mathcal{U}$ according to

$$F_1 \mid \mathcal{X}^-(A) = 0 \tag{3.9a}$$

$$F_1 \mid [\langle A \mid \mathcal{B}\rangle \cap \mathcal{X}^+(A)] = F^+ \tag{3.9b}$$

$$F_1 \mid \mathcal{X}^+(A + BF_0) = F_0 \mid \mathcal{X}^+(A + BF_0). \tag{3.9c}$$

Since

$$(F_1 - F_0) \mid \mathcal{X}^+(A + BF_0) = 0,$$

there follows

$$\mathcal{X}^+(A + BF_1) = \mathcal{X}^+(A + BF_0 + B(F_1 - F_0)) \supset \mathcal{X}^+(A + BF_0), \tag{3.10}$$

by an application of Lemma 6.1. But

$$\mathscr{X}^+(A + BF_1) \cap \langle A | \mathscr{B} \rangle = 0$$

by (3.9a,b), and as

$$\bar{\mathscr{X}}^+(\overline{A + BF_1}) = \bar{\mathscr{X}}^+(\bar{A}) = \bar{\mathscr{X}}^+(\overline{A + BF_0}),$$

there results

$$\mathscr{X}^+(A + BF_1) \oplus \langle A | \mathscr{B} \rangle = \mathscr{X}^+(A + BF_0) \oplus \langle A | \mathscr{B} \rangle.$$

Therefore, $\mathscr{X}^+(A + BF_1) \simeq \mathscr{X}^+(A + BF_0)$; from this and (3.10), we get

$$\mathscr{X}^+(A + BF_1) = \mathscr{X}^+(A + BF_0),$$

as required. □

By (3.9c) our construction also achieved that

$$(A + BF_1) | \mathscr{X}^+(A + BF_0) = (A + BF_0) | \mathscr{X}^+(A + BF_0).$$

It is now easy to prove our main result. For this, we revert to the notation introduced at the beginning of this section.

Theorem 7.4. *In the general case* $\mathscr{N} \neq 0$, *RPIS is solvable if and only if*

i. $\mathscr{X}^+(A) \cap \mathscr{N} \subset \operatorname{Ker} D$ (3.11)

and

ii. *In the factor space* $\bar{\mathscr{X}} = \mathscr{X} / [\mathscr{X}^+(A) \cap \mathscr{N}]$ *the reduced problem is solvable: that is, there exists* $\bar{F}_0 \colon \bar{\mathscr{X}} \to \mathscr{U}$, *such that*

$$\bar{\mathscr{X}}^+(\bar{A} + \bar{B}\bar{F}_0) \subset \operatorname{Ker} \bar{D} \tag{3.12a}$$

and

$$\bar{\mathscr{X}}^+(\bar{A} + \bar{B}\bar{F}_0) \cap \langle \bar{A} | \bar{\mathscr{B}} \rangle = 0. \tag{3.12b}$$

Of course, the reduced problem (ii) is identical in form to the one solved by Theorem 7.3.

Proof. (If) Suppose the reduced problem (RP) defined by (3.12) is solvable. Lemma 7.3 applied to RP yields a map $\bar{F} \colon \bar{\mathscr{X}} \to \mathscr{U}$, such that

$$\operatorname{Ker} \bar{F} \supset \bar{\mathscr{X}}^-(\bar{A})$$

and $\bar{F}$ satisfies (3.2c) and (3.5). Since $\bar{\mathscr{X}}^-(\bar{A}) \supset \bar{\mathscr{N}}$, we have that (3.2a) is true as well. As already noted, Lemmas 7.1 and 7.2 now imply that RPIS is solvable.

(Only if) The necessity of (3.11) is immediate from (1.1a,b); and that of (3.12) follows by Lemmas 7.1 and 7.2. □

7.4 Application: Regulation Against Step Disturbances

As a simple application of Theorem 7.3, consider the system

$$\dot{x}_1 = A_1 x_1 + A_3 x_2 + B_1 u$$
$$\dot{x}_2 = 0$$
$$z = D_1 x_1 + D_2 x_2.$$

We assume that $y = x$ and (A_1, B_1) is controllable. The equations represent a controllable plant subjected to step disturbances which enter both dynamically and directly at the regulated output, a situation common in industrial process control.

In basis-free terms our assumptions amount to the following:

$$\mathcal{N} = 0, \tag{4.1a}$$
$$\operatorname{Im} A \subset \langle A \mid \mathcal{B} \rangle. \tag{4.1b}$$

We now have

Theorem 7.5. *Subject to the assumptions* (4.1), *RPIS is solvable if and only if*

$$\langle A \mid \mathcal{B} \rangle + \operatorname{Ker} D \cap A^{-1}\mathcal{B} = \mathcal{X}. \tag{4.2}$$

PROOF. (If) Exploiting the remark after the proof of Theorem 7.3, let

$$\tilde{\mathcal{V}} = \operatorname{Ker} D \cap A^{-1}\mathcal{B}. \tag{4.3}$$

From (4.2), (4.3) it is clear, first, that

$$\mathcal{X}^+(A) \subset \langle A \mid \mathcal{B} \rangle + \tilde{\mathcal{V}}.$$

Also, as $A\tilde{\mathcal{V}} \subset \mathcal{B}$ there exists $F \in \mathbf{F}(\tilde{\mathcal{V}})$, such that $A_F \tilde{\mathcal{V}} = 0$, where $A_F = A + BF$. Then,

$$\tilde{\mathcal{V}} \subset \operatorname{Ker} A_F \subset \mathcal{X}^+(A_F)$$

so

$$\tilde{\mathcal{V}} \cap \mathcal{X}^+(A_F) \cap \langle A \mid \mathcal{B} \rangle = \tilde{\mathcal{V}} \cap \langle A \mid \mathcal{B} \rangle.$$

According to Theorem 5.5 the supremal c.s. $\tilde{\mathcal{R}}$ in $\tilde{\mathcal{V}}$ is given by

$$\tilde{\mathcal{R}} = \langle A_F \mid \mathcal{B} \cap \tilde{\mathcal{V}} \rangle = \mathcal{B} \cap \tilde{\mathcal{V}}.$$

The second condition of Theorem 7.3 (with $\tilde{\mathcal{V}}$ in place of $\mathcal{V}^*$) will thus be satisfied if

$$\frac{\tilde{\mathcal{V}} \cap \langle A \mid \mathcal{B} \rangle}{\mathcal{B} \cap \tilde{\mathcal{V}}}$$

decomposes $\tilde{\mathscr{V}}/(\mathscr{B} \cap \tilde{\mathscr{V}})$ relative to the map induced by A_F in $\tilde{\mathscr{V}}/(\mathscr{B} \cap \tilde{\mathscr{V}})$. Since $A_F \mid \tilde{\mathscr{V}} = 0$ this is trivial, and the result follows.

(Only if) Let $\bar{\mathscr{X}} := \mathscr{X}/\langle A \mid \mathscr{B} \rangle$ and now use bars for subspaces and induced maps in $\bar{\mathscr{X}}$. By (4.1b), $\bar{A} = 0$ and, since $\bar{\mathscr{X}}^-(\bar{A}) = 0$, we have $\mathscr{X}^-(A) \subset \langle A \mid \mathscr{B} \rangle$. Let F solve RPIS. Since $\bar{A}_F = \bar{A} = 0$ for all F, and since $\mathscr{X}^+(A_F) \cap \langle A \mid \mathscr{B} \rangle = 0$, we have $\mathscr{X}^+(A_F) = \text{Ker } A_F$. Now, Ker $A_F \subset A^{-1}\mathscr{B}$ for any F, so

$$\mathscr{X}^+(A_F) \subset \text{Ker } D \cap A^{-1}\mathscr{B}. \tag{4.4}$$

By application of Lemma 6.3 to (4.4) there results

$$\mathscr{X}^+(A) \subset \langle A \mid \mathscr{B} \rangle + \text{Ker } D \cap A^{-1}\mathscr{B},$$

and therefore,

$$\begin{aligned} \mathscr{X} &= \mathscr{X}^-(A) \oplus \mathscr{X}^+(A) \subset \langle A \mid \mathscr{B} \rangle + \mathscr{X}^+(A) \\ &\subset \langle A \mid \mathscr{B} \rangle + \text{Ker } D \cap A^{-1}\mathscr{B} \subset \mathscr{X}. \end{aligned}$$

□

For the more general case with $\mathscr{N} \neq 0$, the reader may verify that RPIS is solvable if and only if

$$\mathscr{X}^+(A) \cap \mathscr{N} \subset \text{Ker } D$$

and

$$\langle A \mid \mathscr{B} \rangle + \text{Ker } D \cap A^{-1}[\mathscr{B} + \mathscr{X}^+(A) \cap \mathscr{N}] = \mathscr{X}.$$

7.5 Application: Static Decoupling

Many control processes call for the occasional resetting of scalar output variables to new values, which are then held constant for time intervals long compared to the time constants of the process. It is convenient to associate a reset control v_i with each such output w_i so that, if v_i is given a step change at $t = 0$, then

$$w_i(t) \to v_i(0+), \qquad t \to \infty.$$

In addition, it is required that the remaining variables return to their initial values, possibly after an intervening transient, i.e.

$$w_j(t) \to w_j(0-), \qquad j \neq i, \ t \to \infty.$$

It is straightforward to treat this situation with the methods already developed. We have the plant equation

$$\dot{x}_1 = A_1 x_1 + B_1 u$$

and assume (A_1, B_1) controllable. The output equation is

$$w = D_1 x_1.$$

Denote the reset control vector by x_2, so that

$$\dot{x}_2(t) = 0, \qquad t \geq 0.$$

For simplicity, we assume that both the plant state x_1 and (reasonably enough) the reset control x_2 are observable. We require state feedback F_1 and reset gain F_2, such that, if

$$u = F_1 x_1 + F_2 x_2,$$

then $w(t) - x_2(t) \to 0$ $(t \to \infty)$. In addition, we ask for internal (plant) stabilization. Thus, defining

$$z = D_1 x_1 - x_2$$

we obtain a problem of the type solved in Section 7.4.

7.6 Example 1: RPIS Unsolvable

It is instructive to return to the example in Section 6.4 and verify that the conditions of Theorem 7.4 fail. We had

$$A = \begin{bmatrix} 0 & 1 & 0 \\ -2 & -3 & 0 \\ 0 & 0 & 0 \end{bmatrix}, \qquad B = \begin{bmatrix} 0 \\ 1 \\ 0 \end{bmatrix},$$

$$C = [0 \quad 0 \quad 1], \qquad D = [0 \quad 1 \quad -1].$$

This yields

$$\mathcal{N} = \operatorname{Im} \begin{bmatrix} 1 & 0 \\ 0 & 1 \\ 0 & 0 \end{bmatrix}, \qquad \operatorname{Ker} D = \operatorname{Im} \begin{bmatrix} 1 & 0 \\ 0 & 1 \\ 0 & 1 \end{bmatrix},$$

$$\langle A \mid \mathscr{B} \rangle = \operatorname{Im} \begin{bmatrix} 1 & 0 \\ 0 & 1 \\ 0 & 0 \end{bmatrix}, \qquad \mathscr{X}^+(A) = \operatorname{Im} \begin{bmatrix} 0 \\ 0 \\ 1 \end{bmatrix}.$$

Since $\mathscr{X}^+(A) \cap \mathcal{N} = 0$, the "reduced problem" of Theorem 7.4 is simply the problem given, with $\mathcal{N}$ replaced by zero. Now,

$$A \operatorname{Ker} D \subset \operatorname{Ker} D + \mathscr{B},$$

hence, $\mathscr{V}^* = \operatorname{Ker} D$; and $\mathscr{B} \cap \mathscr{V}^* = 0$ implies $\mathscr{R}^* = 0$. The map

$$F = [2 \quad 3 \quad 0]$$

is in $\mathbf{F}(\mathscr{V}^*)$;

$$A_F = \begin{bmatrix} 0 & 1 & 0 \\ 0 & 0 & 0 \\ 0 & 0 & 0 \end{bmatrix};$$

and $\mathscr{X}^+(A_F) = \mathscr{X}$.

Since now $\mathscr{N} = 0$, we use Theorem 7.3. Clearly,

$$\langle A \mid \mathscr{B} \rangle + \mathscr{V}^* = \mathscr{X}$$

and so condition (2.1) holds. Since $\mathscr{R}^* = 0$, we must check (for the decomposability condition) whether

$$\mathscr{V}^* \cap \langle A \mid \mathscr{B} \rangle \cap \mathscr{X}^+(A_F) = \operatorname{Im} \begin{bmatrix} 1 \\ 0 \\ 0 \end{bmatrix}$$

decomposes

$$\mathscr{V}^* \cap \mathscr{X}^+(A_F) = \operatorname{Im} \begin{bmatrix} 1 & 0 \\ 0 & 1 \\ 0 & 1 \end{bmatrix}$$

relative to the map $A_F \mid \mathscr{V}^*$. For this, let

$$e_1 = \begin{bmatrix} 1 \\ 0 \\ 0 \end{bmatrix}, \qquad e_2 = \begin{bmatrix} 0 \\ 1 \\ 1 \end{bmatrix}$$

and write $\mathscr{e}_1$ for the span of e_1. In the basis $\{e_1, e_2\}$,

$$A_F \mid \mathscr{V}^* = \begin{bmatrix} 0 & 1 \\ 0 & 0 \end{bmatrix} =: A_F^*, \text{ say.} \tag{6.1}$$

Since $\sigma(A_F^*) = \{0, 0\}$, it is clear that $\mathscr{e}_1$ decomposes $\mathscr{V}^*$ relative to A_F^* if and only if there is a vector of form $\left[\begin{smallmatrix}\alpha \\ 1\end{smallmatrix}\right]$, such that

$$A_F^* \begin{bmatrix} \alpha \\ 1 \end{bmatrix} = 0.$$

A trivial computation shows that no such vector exists, hence decomposability fails, and RPIS cannot be solvable.

Referring to Proposition 0.5, we could alternatively check the condition on elementary divisors. By inspection of (6.1), we see that $A_F^* \mid \mathscr{e}_1$ has e.d. λ, as does the induced map $\bar{A}_F^*$ on $\mathscr{V}^*/\mathscr{e}_1$; but A_F^* has the single e.d. λ^2, which shows again that decomposability fails.

7.7 Example 2: Servo-Regulator

We shall design a controller for the single-input, single-output system with the signal-flow graph of Fig. 7.2. The state and output equations are:

$$\dot{x}_1 = x_2$$
$$\dot{x}_2 = -x_2 + x_3 + u$$
$$\dot{x}_3 = 0$$
$$\dot{x}_4 = x_5$$
$$\dot{x}_5 = 0$$
$$y = z = -x_1 + x_4.$$

The system represents a second-order plant (state variables x_1, x_2) subject to a step load disturbance x_3, to be designed to track a ramp input x_4. The tracking error $x_4 - x_1$ is assumed to be the only variable accessible to direct measurement. What is required is a suitable compensator $T(s)$.

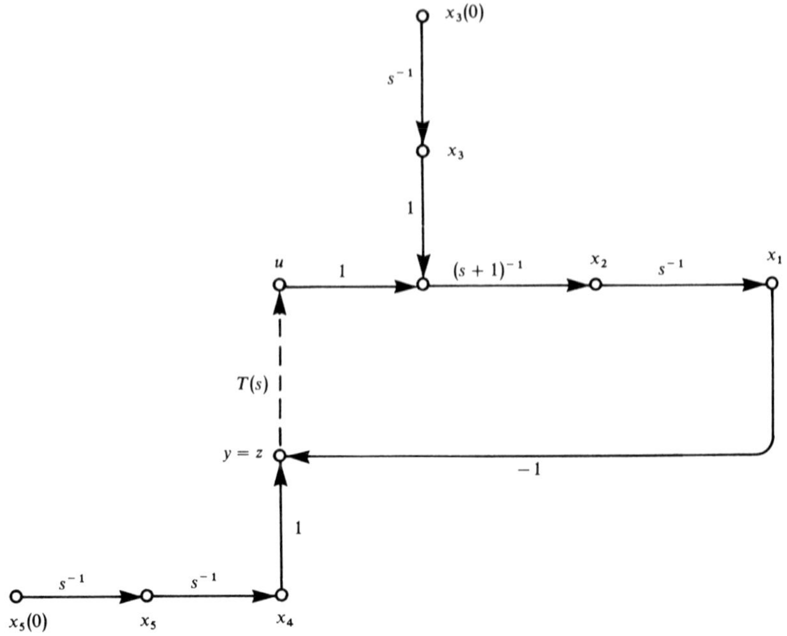

Figure 7.2 Signal Flow Graph: Servo-Regulator, Section 7.7. $T(s)$ combines observer-compensator.

We have

$$A = \begin{bmatrix} 0 & 1 & 0 & 0 & 0 \\ 0 & -1 & 1 & 0 & 0 \\ 0 & 0 & 0 & 0 & 0 \\ 0 & 0 & 0 & 0 & 1 \\ 0 & 0 & 0 & 0 & 0 \end{bmatrix}, \qquad B = \begin{bmatrix} 0 \\ 1 \\ 0 \\ 0 \\ 0 \end{bmatrix},$$

$$C = D = [-1 \quad 0 \quad 0 \quad 1 \quad 0].$$

There follows

$$\mathcal{N} = \operatorname{Im} \begin{bmatrix} 1 & 0 \\ 0 & 1 \\ 0 & 1 \\ 1 & 0 \\ 0 & 1 \end{bmatrix}, \qquad \mathcal{X}^+(A) = \operatorname{Im} \begin{bmatrix} 1 & 0 & 0 & 0 \\ 0 & 1 & 0 & 0 \\ 0 & 1 & 0 & 0 \\ 0 & 0 & 1 & 0 \\ 0 & 0 & 0 & 1 \end{bmatrix}.$$

Thus, $\mathcal{N} \cap \mathcal{X}^+(A) = \mathcal{N} \subset \operatorname{Ker} D$, as required by the first condition (3.11) of Theorem 7.4. To examine the "reduced" problem in $\mathcal{X}/\mathcal{N}$, write $\mathcal{X} = \mathcal{M} \oplus \mathcal{N}$, with

$$\mathcal{M} = \operatorname{Im} \begin{bmatrix} 1 & 0 & 0 \\ 0 & 1 & 0 \\ 0 & 0 & 1 \\ 0 & 0 & 0 \\ 0 & 0 & 0 \end{bmatrix}.$$

The projection $P\colon \mathcal{X} \to \mathcal{X}/\mathcal{N}$ is represented by the natural projection $P\colon \mathcal{M} \oplus \mathcal{N} \to \mathcal{M}$ defined by $P|\mathcal{M} = 1$, $P|\mathcal{N} = 0$. This gives

$$P = \begin{bmatrix} 1 & 0 & 0 & -1 & 0 \\ 0 & 1 & 0 & 0 & -1 \\ 0 & 0 & 1 & 0 & -1 \end{bmatrix}.$$

The induced maps

$$\bar{A}\colon \mathcal{M} \to \mathcal{M}, \qquad \bar{B}\colon \mathcal{U} \to \mathcal{M}, \qquad \bar{D}\colon \mathcal{M} \to \mathcal{Z},$$

determined by

$$\bar{A}P = PA, \qquad \bar{B} = PB, \qquad \bar{D}P = D,$$

are then

$$\bar{A} = \begin{bmatrix} 0 & 1 & 0 \\ 0 & -1 & 1 \\ 0 & 0 & 0 \end{bmatrix}, \qquad \bar{B} = \begin{bmatrix} 0 \\ 1 \\ 0 \end{bmatrix}, \qquad \bar{D} = [-1 \quad 0 \quad 0]. \tag{7.1}$$

The reduced problem is solvable if the conditions of Theorem 7.3 are satisfied by the triple (7.1). We have by simple computations

$$\bar{\mathscr{X}}^+(\bar{A}) = \text{Im}\begin{bmatrix} 1 & 0 \\ 0 & 1 \\ 0 & 1 \end{bmatrix}, \qquad \langle \bar{A} \,|\, \bar{\mathscr{B}} \rangle = \text{Im}\begin{bmatrix} 1 & 0 \\ 0 & 1 \\ 0 & 0 \end{bmatrix}, \qquad \bar{\mathscr{V}}^* = \text{Im}\begin{bmatrix} 0 \\ 0 \\ 1 \end{bmatrix}, \tag{7.2}$$

so (2.1) holds. The condition of decomposability is trivially satisfied, as $\bar{\mathscr{V}}^* \cap \langle \bar{A} \,|\, \bar{\mathscr{B}} \rangle = 0$.

Having verified that RPIS is solvable, we construct a solution in three stages: first a controller for the reduced problem with data (7.1), second an observer to generate $\bar{x} = Px \in \mathscr{M}$, and third the compensator $T(s)$ in which controller and observer are combined.

1. For the controller, we could follow the constructive procedure in the proof (sufficiency half) of Theorem 7.3, or alternatively look directly for a subspace $\bar{\mathscr{V}}$ with properties (1.1) of Theorem 7.1. Since our problem is of low dimension, the latter method is quicker. Referring to (7.2), we see that $\bar{\mathscr{X}}^+(\bar{A}) \not\subset \langle \bar{A} \,|\, \bar{\mathscr{B}} \rangle$, hence to satisfy (1.1c,d) (with $\mathscr{N} = 0$) $\bar{\mathscr{V}}$ must be of the form

$$\bar{\mathscr{V}} = \text{Im}\begin{bmatrix} \alpha \\ \beta \\ 1 \end{bmatrix} \tag{7.3}$$

for some α, β. Applying (1.1a), we require

$$\bar{A}\bar{\mathscr{V}} \subset \bar{\mathscr{V}} + \bar{\mathscr{B}}, \qquad \bar{\mathscr{V}} \subset \text{Ker } \bar{D}. \tag{7.4}$$

From (7.1), (7.3), (7.4), there results

$$\bar{\mathscr{V}} = \text{Im}\begin{bmatrix} 0 \\ 0 \\ 1 \end{bmatrix};$$

thus, $\bar{\mathscr{V}} = \bar{\mathscr{V}}^*$, as was also clear from (7.2) and (7.3). Next, choose $\bar{F}_0 \in \mathbf{F}(\bar{\mathscr{V}})$ arbitrarily, e.g. $\bar{F}_0 = [0 \quad 0 \quad -1]$; then,

$$\bar{A}_0 = \bar{A} + \bar{B}\bar{F} = \begin{bmatrix} 0 & 1 & 0 \\ 0 & -1 & 0 \\ 0 & 0 & 0 \end{bmatrix}.$$

It remains to choose $\bar{F}_1$, such that $\bar{\mathscr{X}}^+(\bar{A}_0 + \bar{B}\bar{F}_1) \subset \bar{\mathscr{V}}$ and $(\bar{A}_0 + \bar{B}\bar{F}_1) | \langle \bar{A} \,|\, \bar{\mathscr{B}} \rangle$ is stable. Write $\bar{F}_1 = [\gamma, \delta, \epsilon]$, so

$$\bar{A}_0 + \bar{B}\bar{F}_1 = \begin{bmatrix} 0 & 1 & 0 \\ \gamma & -1+\delta & \epsilon \\ 0 & 0 & 0 \end{bmatrix}$$

and

$$(\bar{A}_0 + \bar{B}\bar{F}_1)|\langle \bar{A} | \bar{\mathscr{B}} \rangle = \begin{bmatrix} 0 & 1 \\ \gamma & -1+\delta \end{bmatrix}.$$

Assigning the spectrum on $\langle \bar{A} | \bar{\mathscr{B}} \rangle$ as $\{-2, -2\}$, we get $\gamma = -4, \delta = -3$; then,

$$\bar{\mathscr{X}}^+(\bar{A}_0 + \bar{B}\bar{F}_1) = \operatorname{Im} \begin{bmatrix} \epsilon \\ 0 \\ 4 \end{bmatrix},$$

which belongs to $\bar{\mathscr{V}}$ if $\epsilon = 0$. Finally,

$$\bar{F} = \bar{F}_0 + \bar{F}_1 = [-4 \quad -3 \quad -1]. \tag{7.5}$$

2. For the observer design we utilize again the decomposition $\mathscr{X} = \mathscr{M} \oplus \mathscr{N}$, the pair $(\bar{A}, \bar{B})$ of (7.1), and measured output matrix

$$\bar{C} = [-1 \quad 0 \quad 0]$$

determined by $\bar{C}P = C$. We adopt an observer of minimal order 2 with spectrum $\{-4, -4\}$. Applied to $(\bar{C}, \bar{A}, \bar{B})$ the procedure of Exercise 3.5 yields the observer equation

$$\dot{\bar{w}} = \begin{bmatrix} -8 & 1 \\ -16 & 0 \end{bmatrix} \bar{w} + \begin{bmatrix} 40 \\ 112 \end{bmatrix} y + \begin{bmatrix} 1 \\ 0 \end{bmatrix} u \tag{7.6}$$

and asymptotic evaluation

$$\bar{x} - \begin{bmatrix} -1 & 0 & 0 \\ -7 & 1 & 0 \\ -16 & 0 & 1 \end{bmatrix} \begin{bmatrix} y \\ \bar{w}_1 \\ \bar{w}_2 \end{bmatrix}. \tag{7.7}$$

From (7.5) and (7.7) the control is given by

$$u = \bar{F}\bar{x} = 41y - 3\bar{w}_1 - \bar{w}_2. \tag{7.8}$$

3. The compensator $T(s)$ is now obtained as the transfer function from y to u determined by (7.6) and (7.8). The result is

$$T(s) = \frac{41s^2 + 96s + 64}{s(s+11)}.$$

A straightforward computation from the signal flow graph yields, as a check,

$$\hat{z}(s) = \frac{s(s+11)[-\hat{x}_3(s) + s(s+1)\hat{x}_4(s)]}{(s+2)^2(s+4)^2}. \tag{7.9}$$

It is clear from (7.9) that the system is internally stable and that the tracking error $z(t) \to 0$ $(t \to \infty)$ in the presence of step disturbances

$$\hat{x}_3(s) = x_3(0+)s^{-1}$$

and ramp reference signals

$$\hat{x}_4(s) = x_4(0+)s^{-1} + x_5(0+)s^{-2}.$$

7.8 Exercises

The first four exercises are directed to programming a solution F of RPIS when a solution exists. It is assumed that the results of Exercises 0.6, 2.1, 4.2, and 5.5 are available as subprocedures.

7.1. *Regulator synthesis.* Given A, B, $\mathscr{V}$, $\mathscr{N}$ with the properties

$$A\mathscr{V} \subset \mathscr{V} + \mathscr{B}, \qquad \langle A \mid \mathscr{B} \rangle \oplus \mathscr{V} = \mathscr{X},$$
$$A\mathscr{N} \subset \mathscr{N} \subset \langle A \mid \mathscr{B} \rangle \cap \mathscr{X}^-(A),$$

compute $F \in \mathbf{F}(\mathscr{V})$, such that $\operatorname{Ker} F \supset \mathscr{N}$ and $(A + BF) \mid \langle A \mid \mathscr{B} \rangle$ is stable. Hint:

1. Choose a basis adapted to the decomposition

$$\mathscr{X} = \langle A \mid \mathscr{B} \rangle \cap \mathscr{X}^-(A) \oplus \langle A \mid \mathscr{B} \rangle \cap \mathscr{X}^+(A) \oplus \mathscr{V}.$$

2. In this basis, compute

$$A = \left[\begin{array}{cc|c} A_1^- & 0 & A_3^- \\ 0 & A_1^+ & A_3^+ \\ \hline 0 & 0 & A_2 \end{array}\right], \qquad B = \left[\begin{array}{c} B_1^- \\ B_1^+ \\ \hline 0 \end{array}\right].$$

3. Compute $F_0 = [0 \quad 0 \quad F_2]$, such that $(A + BF_0)\mathscr{V} \subset \mathscr{V}$.
4. Compute $F_1 = [0 \quad F_1^+ \quad 0]$, such that $A_1^+ + B_1^+ F_1^+$ is stable.
5. Set $F := F_0 + F_1$.

7.2. *Decomposition.* Given A, $\mathscr{R}$, $\mathscr{T}$ with $A\mathscr{T} \subset \mathscr{T}$ and $A\mathscr{R} \subset \mathscr{R} \subset \mathscr{T}$, compute $\mathscr{S}$, such that $A\mathscr{S} \subset \mathscr{S}$ and $\mathscr{R} \oplus \mathscr{S} = \mathscr{T}$. Hint:

1. Compute any $\hat{\mathscr{S}}$, such that $\mathscr{T} = \mathscr{R} \oplus \hat{\mathscr{S}}$.
2. Compute $A \mid \mathscr{T}$ in a basis adapted to the decomposition in 1, so that

$$A \mid \mathscr{T} = \begin{bmatrix} A_1 & A_3 \\ 0 & A_2 \end{bmatrix}.$$

3. Compute any solution Q of

$$A_1 Q - Q A_2 - A_3 = 0.$$

If no solution exists, $\mathscr{R}$ does not decompose $\mathscr{T}$ relative to A.

4. Represent $\mathscr{S}$ as

$$\mathscr{S} = \operatorname{Ker}[1 \quad Q] = \operatorname{Im}\begin{bmatrix} -Q \\ 1 \end{bmatrix}.$$

7.3. *Solution of reduced RPIS.* Given A, B, C, D, such that

$$\mathscr{N} \subset \mathscr{X}^-(A) \subset \langle A \mid \mathscr{B} \rangle, \tag{8.1}$$

compute (if one exists) a solution F of RPIS. Hint:

1. Compute $\langle A \mid \mathscr{B} \rangle$, $\mathscr{V}^*$, $\mathscr{V}^* \cap \langle A \mid \mathscr{B} \rangle$, $\mathscr{R}^*$ and any $\mathscr{T}$, such that $\mathscr{R}^* \oplus \mathscr{T} = \mathscr{V}^*$ [$\mathscr{T}$ coordinatizes $\mathscr{V}^*/\mathscr{R}^*$].
2. Check $\langle A \mid \mathscr{B} \rangle + \mathscr{V}^* = \mathscr{X}$. If this condition fails, RPIS is not solvable.
3. Compute arbitrary $F_0 \in \mathbf{F}(\mathscr{V}^*)$ [without regard to the constraint $\operatorname{Ker} F \supset \mathscr{N}$], and set $A_0 := A + BF_0$.

4. Compute $A_0|\mathscr{V}^*$, $\mathscr{B} \cap \mathscr{V}^*$ in a basis adapted to the decomposition $\mathscr{V}^* = \mathscr{R}^* \oplus \mathscr{T}$:

$$\hat{A}_0 := A_0|\mathscr{V}^* = \begin{bmatrix} A_{01} & A_{03} \\ 0 & A_{02} \end{bmatrix}, \qquad \mathscr{B} \cap \mathscr{V}^* = \operatorname{Im}\begin{bmatrix} B_1 \\ 0 \end{bmatrix}. \tag{8.2}$$

5. Check the condition $\sigma(A_{01}) \cap \sigma(A_{02}) = \emptyset$. If it fails, achieve it by replacing A_{01} with $A_{01} + B_1 F_{01}$ for suitable (random!) F_{01}: i.e. exploit controllability of (A_{01}, B_1).
6. Redefine $\mathscr{T}$ as the $\hat{A}_0$-invariant complement of $\mathscr{R}^*$ in $\mathscr{V}^*$: namely, if π_2 is the ch.p. of A_{02}, then $\mathscr{T} := \operatorname{Ker} \pi_2(\hat{A}_0)$. The result of Steps 5 and 6 is to ensure $A_{03} = 0$ in (8.2).
7. Compute $\mathscr{T}_1 := \mathscr{T} \cap \langle A|\mathscr{B}\rangle$: its representation in $\mathscr{V}^*$ is of form $\operatorname{Im}[{0 \atop T_1}]$, and in $\mathscr{T}$ is $\operatorname{Im} T_1$.
8. Using Exercise 7.2 compute $\mathscr{T}_2 = \operatorname{Im} T_2 \subset \mathscr{T}$, such that $A_{02}\mathscr{T}_2 \subset \mathscr{T}_2$ and $\mathscr{T}_1 \oplus \mathscr{T}_2 = \mathscr{T}$. If this step fails, RPIS is not solvable.
9. Compute any $\mathscr{S}$, such that $\mathscr{X} = \mathscr{R}^* \oplus \mathscr{T} \oplus \mathscr{S}$ and set

$$\mathscr{V} := \operatorname{Im}\begin{bmatrix} 0 \\ T_2 \\ 0 \end{bmatrix}.$$

10. Compute a solution F by Exercise 7.1 applied to A, B, $\mathscr{V}$, $\mathscr{N}$.

7.4. *Reduction of general RPIS*. Given A, B, C, D compute reduced versions $(\bar{A}, \bar{B}, \bar{C}, \bar{D})$ for which (8.1) holds. Check solvability of RPIS for the original data. Hint:

1. The first condition of (8.1) is equivalent to $\bar{\mathscr{N}} \cap \bar{\mathscr{X}}^+(\bar{A}) = 0$. To achieve it, first check that $\mathscr{N} \cap \mathscr{X}^+(A) \subset \operatorname{Ker} D$. If this condition fails, RPIS is not solvable.
2. Compute arbitrary $\hat{\mathscr{X}}$, such that $\mathscr{N} \cap \mathscr{X}^+(A) \oplus \hat{\mathscr{X}} = \mathscr{X}$, then $P: \mathscr{X} \to \mathscr{X}$ to satisfy $P(\mathscr{N} \cap \mathscr{X}^+(A)) = 0$, $P|\hat{\mathscr{X}} = 1_{\hat{\mathscr{X}}}$. With $\hat{\mathscr{X}}$ as a representation of $\bar{\mathscr{X}}$, compute the induced maps $\bar{A}$ etc. according to $\bar{A}P = PA$, $\bar{B} = PB$, $\bar{C}P = C$, and $\bar{D}P = D$.
3. The second condition of (8.1) states that the map induced by A on $\mathscr{X}/\langle A|\mathscr{B}\rangle$ is completely unstable. To achieve this, set $\bar{\mathscr{X}}_1 := \langle \bar{A}|\bar{\mathscr{B}}\rangle$, compute arbitrary $\bar{\mathscr{X}}_2$, such that $\bar{\mathscr{X}}_1 \oplus \bar{\mathscr{X}}_2 = \bar{\mathscr{X}}$, and compute $\bar{A}$ in a compatible basis as

$$\bar{A} = \begin{bmatrix} \bar{A}_1 & \bar{A}_3 \\ 0 & \bar{A}_2 \end{bmatrix}.$$

4. Split $\bar{\mathscr{X}}_2$ according to

$$\bar{\mathscr{X}}_2 = \bar{\mathscr{X}}_2^+(\bar{A}_2) \oplus \bar{\mathscr{X}}_2^-(\bar{A}_2).$$

5. With a compatible sub-basis for $\bar{\mathscr{X}}_2$ the maps are now

$$\bar{A} = \begin{bmatrix} \bar{A}_1 & \bar{A}_3^+ & \bar{A}_3^- \\ 0 & \bar{A}_2^+ & 0 \\ 0 & 0 & \bar{A}_2^- \end{bmatrix}, \qquad \bar{B} = \begin{bmatrix} \bar{B}_1 \\ 0 \\ 0 \end{bmatrix},$$

$$\bar{C} = [\bar{C}_1 \quad \bar{C}_2^+ \quad \bar{C}_2^-], \qquad \bar{D} = [\bar{D}_1 \quad \bar{D}_2^+ \quad \bar{D}_2^-].$$

To obtain the final versions of $\bar{A}$ etc. delete the third (block) row and column of $\bar{A}$, the third row of $\bar{B}$, and third column of $\bar{C}$ and $\bar{D}$.

7.5. Apply the procedures of Exercises 7.1–7.4 to solve a realistic multivariable example with at least two inputs to be tracked and at least one disturbance to be rejected. The physical origin of the example should be plausible, and the parameter values representative of the application. Check the final design by simulation. Hint: The example to follow illustrates the main steps, in a contrived situation of minimal complexity (Fig. 7.3). To make the problem more impressive, disguise it by a random change of basis in $\mathscr{U}, \mathscr{X}, \mathscr{Y}, \mathscr{Z}$. For the graph illustrated, we have

$$A = \begin{bmatrix} 1 & 1 & 0 & 0 & 0 & 0 & 0 & 0 \\ 0 & 0 & 0 & 0 & 0 & 0 & 0 & 0 \\ 0 & 0 & 2 & 0 & 0 & 0 & 0 & 0 \\ 0 & 0 & 0 & 1 & 1 & 0 & 0 & 0 \\ 0 & 0 & 0 & 0 & -2 & 1 & 0 & 0 \\ 0 & 0 & 0 & 0 & 0 & 0 & 0 & 0 \\ 0 & 0 & 0 & 0 & 0 & 0 & 0 & 0 \\ 0 & 0 & 0 & 0 & 0 & 0 & 0 & -3 \end{bmatrix}, \qquad B = \begin{bmatrix} 0 & 0 & 0 \\ 1 & 0 & 0 \\ 0 & 1 & 0 \\ 0 & 0 & 0 \\ 0 & 0 & 1 \\ 0 & 0 & 0 \\ 0 & 0 & 0 \\ 0 & 0 & 0 \end{bmatrix},$$

$$C = \begin{bmatrix} 1 & 0 & 0 & 0 & 0 & 0 & 0 & 0 \\ 0 & 0 & 1 & 0 & 0 & 0 & 0 & 0 \\ 0 & 0 & 0 & -1 & 0 & 0 & 1 & 0 \end{bmatrix},$$

$$D = \begin{bmatrix} 0 & 1 & 0 & 0 & 0 & 0 & 0 & 0 \\ 0 & 0 & 0 & -1 & 0 & 0 & 1 & 0 \\ 0 & 0 & 0 & 0 & 0 & 0 & 0 & 1 \end{bmatrix}.$$

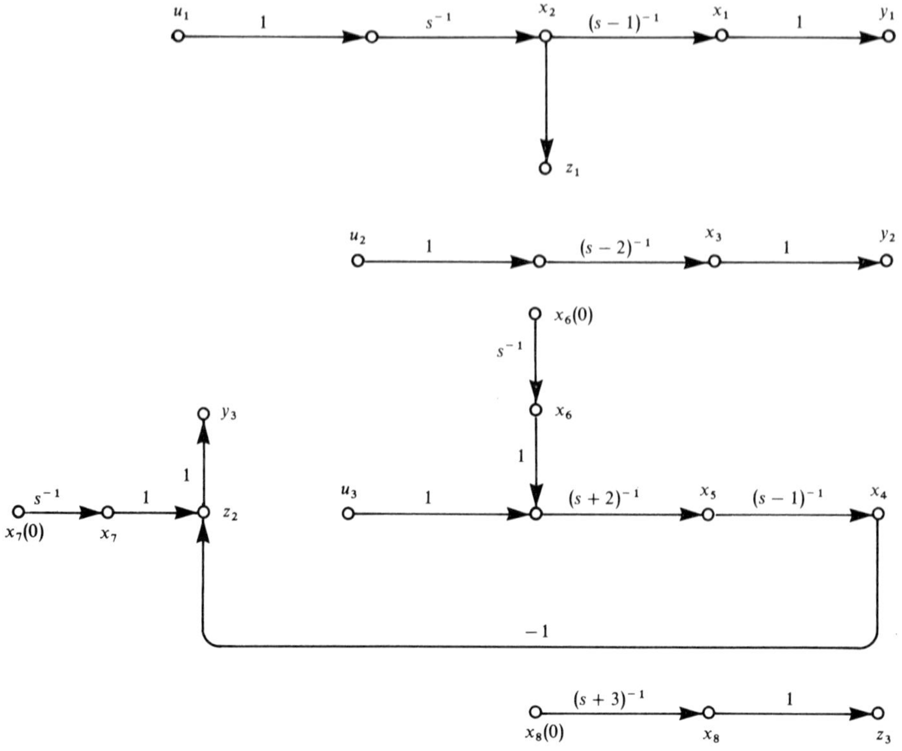

Figure 7.3 Signal Flow Graph: Exercise 7.5.

Exercise 7.4 yields

$$\mathscr{N} = \operatorname{Im}\begin{bmatrix} 0 & 0 \\ 0 & 0 \\ 0 & 0 \\ 1 & 0 \\ -1 & 0 \\ -2 & 0 \\ 1 & 0 \\ 0 & 1 \end{bmatrix} = \operatorname{Im}[n_1, n_2], \text{ say,}$$

where $An_1 = 0$, $An_2 = -3n_2$. Thus,

$$\mathscr{N} \cap \mathscr{X}^-(A) = \operatorname{Span}\{n_2\}$$

$$\mathscr{N} \cap \mathscr{X}^+(A) = \operatorname{Span}\{n_1\} \subset \operatorname{Ker} D.$$

Writing e_i ($i \in \mathbf{8}$) for the unit vectors in $\mathbb{R}^8$, and taking

$$\hat{\mathscr{X}} = \operatorname{Im}[e_1 \quad e_2 \quad e_3 \quad e_4 \quad e_5 \quad e_6 \quad e_8],$$

one finds for $P: \mathscr{N} \cap \mathscr{X}^+(A) \oplus \hat{\mathscr{X}} \to \hat{\mathscr{X}}$ the matrix

$$P = [0^{7 \times 1} \quad 1^{7 \times 7}][n_1 \quad e_1 \quad e_2 \quad e_3 \quad e_4 \quad e_5 \quad e_6 \quad e_8]^{-1}$$

$$= \begin{bmatrix} 1 & 0 & 0 & 0 & 0 & 0 & 0 & 0 \\ 0 & 1 & 0 & 0 & 0 & 0 & 0 & 0 \\ 0 & 0 & 1 & 0 & 0 & 0 & 0 & 0 \\ 0 & 0 & 0 & 1 & 0 & 0 & -1 & 0 \\ 0 & 0 & 0 & 0 & 1 & 0 & 1 & 0 \\ 0 & 0 & 0 & 0 & 0 & 1 & 2 & 0 \\ 0 & 0 & 0 & 0 & 0 & 0 & 0 & 1 \end{bmatrix}.$$

From this (Exercise 7.4, Step 2)

$$\bar{A} = \begin{bmatrix} 1 & 1 & 0 & 0 & 0 & 0 & 0 \\ 0 & 0 & 0 & 0 & 0 & 0 & 0 \\ 0 & 0 & 2 & 0 & 0 & 0 & 0 \\ 0 & 0 & 0 & 1 & 1 & 0 & 0 \\ 0 & 0 & 0 & 0 & -2 & 1 & 0 \\ 0 & 0 & 0 & 0 & 0 & 0 & 0 \\ 0 & 0 & 0 & 0 & 0 & 0 & -3 \end{bmatrix}, \qquad \bar{B} = \begin{bmatrix} 0 & 0 & 0 \\ 1 & 0 & 0 \\ 0 & 1 & 0 \\ 0 & 0 & 0 \\ 0 & 0 & 1 \\ 0 & 0 & 0 \\ 0 & 0 & 0 \end{bmatrix}, \tag{8.3a}$$

$$\bar{C} = \begin{bmatrix} 1 & 0 & 0 & 0 & 0 & 0 & 0 \\ 0 & 0 & 1 & 0 & 0 & 0 & 0 \\ 0 & 0 & 0 & -1 & 0 & 0 & 0 \end{bmatrix},$$

$$\bar{D} = \begin{bmatrix} 0 & 1 & 0 & 0 & 0 & 0 & 0 \\ 0 & 0 & 0 & -1 & 0 & 0 & 0 \\ 0 & 0 & 0 & 0 & 0 & 0 & 1 \end{bmatrix}. \tag{8.3b}$$

To carry out Steps 3–5, note that

$$\langle \bar{A} \mid \bar{\mathscr{B}} \rangle = \operatorname{Im}[\bar{e}_1 \cdots \bar{e}_5],$$

where $\bar{e}_i$ ($i \in \mathbf{7}$) are the unit vectors in $\mathbb{R}^7$. By inspection,

$$\bar{\mathscr{X}}_2 = \operatorname{Im}[\bar{e}_6 \quad \bar{e}_7] = \bar{\mathscr{X}}_2^+(\bar{A}_2) \oplus \bar{\mathscr{X}}_2^-(\bar{A}_2)$$
$$= \operatorname{Span}\{\bar{e}_6\} \oplus \operatorname{Span}\{\bar{e}_7\}.$$

The final versions, say $\tilde{A}$, etc., of $\bar{A}$, etc. are obtained by deleting the 7th row and column of $\bar{A}$, the 7th row of $\bar{B}$ and 7th column of $\bar{C}$ and $\bar{D}$. This yields

$$\tilde{A} = \begin{bmatrix} 1 & 1 & 0 & 0 & 0 & 0 \\ 0 & 0 & 0 & 0 & 0 & 0 \\ 0 & 0 & 2 & 0 & 0 & 0 \\ 0 & 0 & 0 & 1 & 1 & 0 \\ 0 & 0 & 0 & 0 & -2 & 1 \\ 0 & 0 & 0 & 0 & 0 & 0 \end{bmatrix}, \qquad \tilde{B} = \begin{bmatrix} 0 & 0 & 0 \\ 1 & 0 & 0 \\ 0 & 1 & 0 \\ 0 & 0 & 0 \\ 0 & 0 & 1 \\ 0 & 0 & 0 \end{bmatrix}, \tag{8.4a}$$

$$\tilde{C} = \begin{bmatrix} 1 & 0 & 0 & 0 & 0 & 0 \\ 0 & 0 & 1 & 0 & 0 & 0 \\ 0 & 0 & 0 & -1 & 0 & 0 \end{bmatrix},$$

$$\tilde{D} = \begin{bmatrix} 0 & 1 & 0 & 0 & 0 & 0 \\ 0 & 0 & 0 & -1 & 0 & 0 \end{bmatrix}. \tag{8.4b}$$

The meaning of these steps should be transparent from Figure 7.3.

Next, we carry out Exercise 7.3 with the data (8.4). Writing $\tilde{e}_i$ ($i \in \mathbf{6}$) for the unit vectors in $\mathbb{R}^6$, we have

$$\langle \tilde{A} \mid \tilde{\mathscr{B}} \rangle = \operatorname{Im}[\tilde{e}_1 \quad \tilde{e}_2 \quad \tilde{e}_3 \quad \tilde{e}_4 \quad \tilde{e}_5],$$

and after a short computation,

$$\tilde{\mathscr{V}}^* = \operatorname{Im}[\tilde{e}_1 \quad \tilde{e}_3 \quad \tilde{e}_6].$$

For $\tilde{F}_0 \in \tilde{\mathbf{F}}(\tilde{\mathscr{V}}^*)$, we may take

$$\tilde{F}_0 = \begin{bmatrix} 0^{3 \times 5} & \begin{matrix} 0 \\ 0 \\ -1 \end{matrix} \end{bmatrix}, \tag{8.5}$$

and then

$$\tilde{A}_0 = \tilde{A} + \tilde{B}\tilde{F}_0 = \begin{bmatrix} 1 & 1 & 0 & 0 & 0 & 0 \\ 0 & 0 & 0 & 0 & 0 & 0 \\ 0 & 0 & 2 & 0 & 0 & 0 \\ 0 & 0 & 0 & 1 & 1 & 0 \\ 0 & 0 & 0 & 0 & -2 & 0 \\ 0 & 0 & 0 & 0 & 0 & 0 \end{bmatrix}.$$

Next,

$$\tilde{\mathscr{V}}^* = \tilde{\mathscr{R}}^* \oplus \tilde{\mathscr{T}} = \operatorname{Span}\{\tilde{e}_3\} \oplus \operatorname{Span}\{\tilde{e}_1, \tilde{e}_6\},$$

say. Since

$$\tilde{\mathscr{T}}_1 = \tilde{\mathscr{T}} \cap \langle \tilde{A} \mid \tilde{\mathscr{B}} \rangle = \operatorname{Span}\{\tilde{e}_1\},$$

and since

$$\tilde{A}_0 \tilde{e}_1 = \tilde{e}_1, \qquad \tilde{A}_0 \tilde{e}_6 = 0,$$

the decomposition of $\tilde{\mathscr{T}}$ yields

$$\tilde{\mathscr{V}} = \tilde{\mathscr{T}}_2 = \mathrm{Span}\{\tilde{e}_6\}. \tag{8.6}$$

To complete the solution we go to Exercise 7.1 with data (8.4)–(8.6); here $\tilde{\mathscr{N}} = 0$. The details are quite straightforward. Choosing $\tilde{F}_1$, such that $\tilde{F}_1\tilde{\mathscr{V}} = 0$ and $(\tilde{A}_0 + \tilde{B}\tilde{F}_1)|\langle\tilde{A}|\tilde{\mathscr{B}}\rangle$ has spectrum at -1, we get, for instance,

$$\tilde{F}_1 = \begin{bmatrix} -4 & -3 & 0 & 0 & 0 & 0 \\ 0 & 0 & -3 & 0 & 0 & 0 \\ 0 & 0 & 0 & -4 & -1 & 0 \end{bmatrix},$$

and then

$$\tilde{F} = \tilde{F}_0 + \tilde{F}_1 = \begin{bmatrix} -4 & -3 & 0 & 0 & 0 & 0 \\ 0 & 0 & -3 & 0 & 0 & 0 \\ 0 & 0 & 0 & -4 & -1 & -1 \end{bmatrix}.$$

Writing $Q: \bar{\mathscr{X}} \to \tilde{\mathscr{X}}$ for the projection employed above, namely

$$Q = [1^{6\times 6} \quad 0^{6\times 1}],$$

we have that $\bar{F} = \tilde{F}Q$, and finally

$$F = \bar{F}P = \begin{bmatrix} -4 & -3 & 0 & 0 & 0 & 0 & 0 & 0 \\ 0 & 0 & -3 & 0 & 0 & 0 & 0 & 0 \\ 0 & 0 & 0 & -4 & -1 & -1 & 1 & 0 \end{bmatrix}.$$

It is readily checked that F is indeed a solution of RPIS. Implementation of $u = Fx$ with an observer may be left to the reader; its order would be

$$d\left(\frac{\mathrm{Ker}\ C}{\mathscr{N}}\right) = 3.$$

7.6. Prove Corollary 7.1. Hint: With $A\mathscr{R} \subset \mathscr{R} \subset \mathscr{X}$, note that $\mathscr{R}$ decomposes $\mathscr{X}$ relative to A if and only if $\mathscr{R} \cap \mathscr{X}^{\pm}(A)$ decomposes $\mathscr{X}^{\pm}(A)$ relative to A.

7.7. Referring to the discussion preceding Corollary 7.2, prove that $\mathscr{V}_1^* = \mathscr{V}^* \cap \langle A|\mathscr{B}\rangle$.

7.8. For the signal flow graph of Fig. 7.4, we have

$$A = \begin{bmatrix} 0 & 0 & 0 \\ 1 & 0 & 1 \\ 0 & 0 & 0 \end{bmatrix}, \quad B = \begin{bmatrix} 1 \\ 0 \\ 0 \end{bmatrix}, \quad D = [1 \quad 0 \quad 0].$$

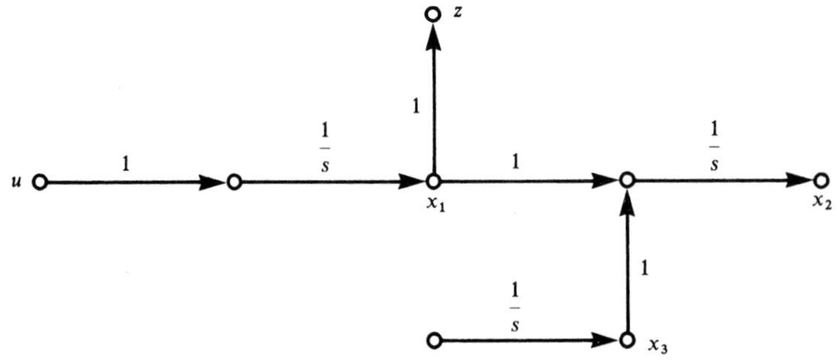

Figure 7.4 Signal Flow Graph: Exercise 7.8.

Assuming $C = 1$, show that RPIS is unsolvable. Explain the failure of the decomposability condition in terms of the "topology" of the signal flow. Hint: Check that $\mathscr{R}^* = 0$ and compute the matrix of $A \mid \mathscr{V}^*$. What would be the effect of (internally) stabilizing state feedback on the regulated output z?

7.9. "*Network-style*" *interpretation of decomposability.* Let $\mathscr{X} = \mathscr{X}_1 \oplus \mathscr{X}_2$ and $A: \mathscr{X} \to \mathscr{X}$, with

$$A = \begin{bmatrix} A_1 & A_3 \\ 0 & A_2 \end{bmatrix}.$$

For the system $\dot{x} = Ax$, the signal flow (Fig. 7.5a) exhibits the subsystems S_1, S_2 (say), with S_1 driven by S_2 through the binding map A_3. Show that $\mathscr{X}_1$ decomposes $\mathscr{X}$ relative to A if and only if the response of S_1 to S_2 can be "tuned out" by a constant matrix gain T, in the network of Fig. 7.5b. The tuning condition is that $\hat{x}_1(s) \equiv 0$ when $x_{10} = 0$ and x_{20} is arbitrary. Reconsider Exercise 7.8 in the light of this result.

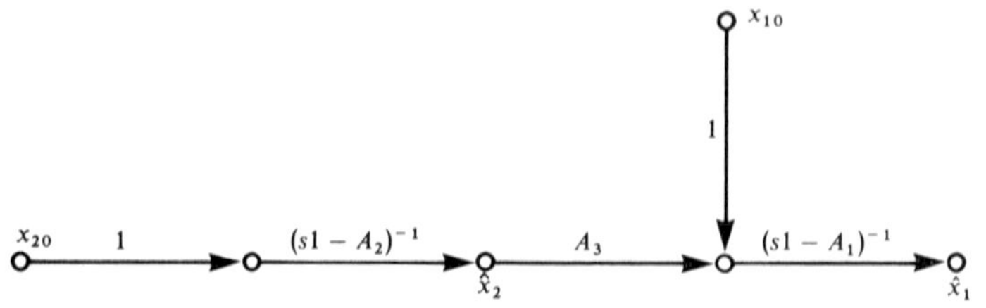

Figure 7.5a

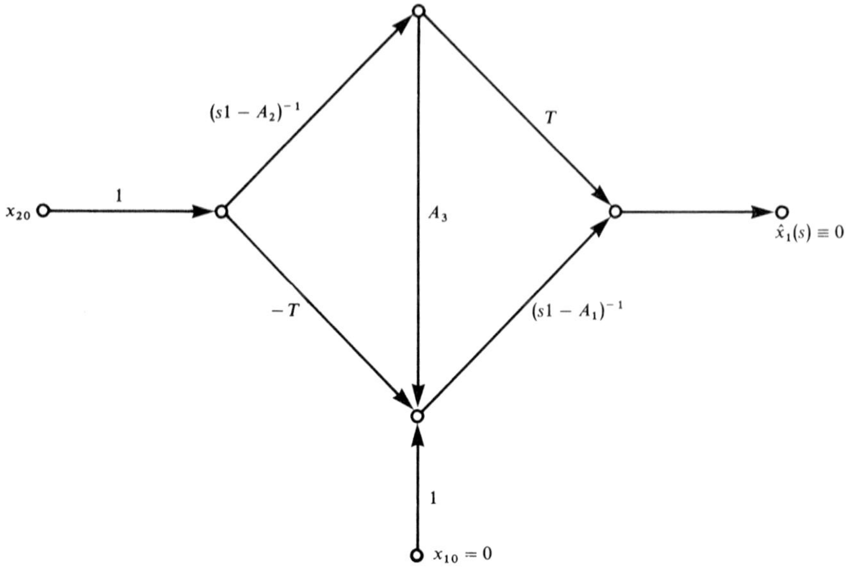

Figure 7.5b Signal Flow Graphs: Exercise 7.9.

7.10. Carry out the routine verification indicated just before Lemma 7.1.

7.11. Starting from the decompositions

$$\mathscr{X} = \mathscr{X}^+(A) \oplus \mathscr{X}^-(A)$$
$$\langle A \mid \mathscr{B} \rangle = \langle A \mid \mathscr{B} \rangle \cap \mathscr{X}^+(A) \oplus \langle A \mid \mathscr{B} \rangle \cap \mathscr{X}^-(A),$$

give a matrix-style proof of Lemma 7.3. The theory of Sylvester's equation (Section 0.11) will help.

7.12. Verify the extension of Theorem 7.5 given at the end of Section 7.4.

7.13. In the system shown in Fig. 7.6,

$$\sigma(A_1) \subset \mathbb{C}^+ := \{\lambda : \Re e\ \lambda \geq 0\}$$
$$\sigma(A_2) \subset \mathbb{C}^- := \{\lambda : \Re e\ \lambda < 0\}$$

and B_2, F_2 are arbitrary. Show that there exists F_1 such that, for all $x_1(0)$ and $x_2(0)$,

$$z(t) := F_1 x_1(t) - F_2 x_2(t) \to 0, \qquad t \to \infty.$$

Hint: Let

$$Q(t) := \int_0^t e^{sA_2} B_2 e^{-sA_1}\, ds.$$

Show that

$$Q_\infty := \lim Q(t), \qquad t \to \infty,$$

exists, and is determined uniquely by

$$A_2 Q_\infty - Q_\infty A_1 + B_2 = 0.$$

Then let $F_1 := F_2 Q_\infty$. Can you interpret Lemma 7.3 in the light of this result?

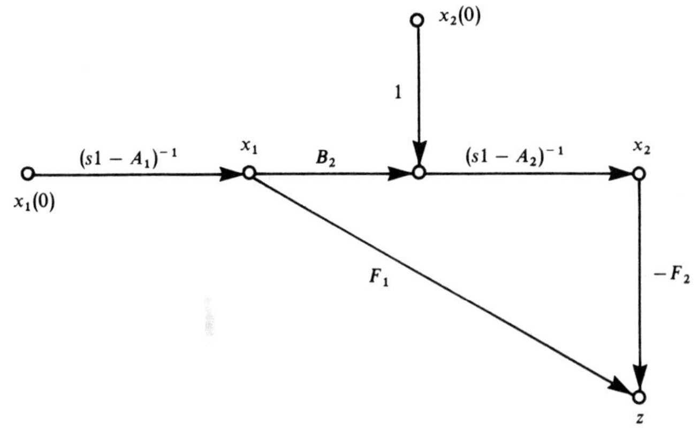

Figure 7.6 Signal Flow Graph: Exercise 7.13.

7.14. *RPIS with direct control feedthrough.* Verify the following. If the output equation (0.1c) is modified to read $z = Dx + Eu$, then RPIS is modified only by the replacement of the regulation condition (0.5c) by the condition $\mathscr{X}^+(A + BF) \subset \operatorname{Ker}(D + EF)$. Theorem 7.1 remains valid provided the condition $\mathscr{V} \in \mathfrak{I}(A, B; \operatorname{Ker} D)$ [i.e. (1.1a)] is replaced by the two conditions

$$\mathscr{V} \cap \mathscr{N} \subset \operatorname{Ker} D, \qquad \mathscr{V} \in \mathfrak{I}(A, B; D, E),$$

where $\mathfrak{I}(A, B; D, E)$ is defined in Exercise 4.6.

With the latter notational change, Theorem 7.2 remains as before. In Theorem 7.3 it is required only to replace $\mathscr{V}^*$ by $\mathscr{V}^\Delta$ (see Exercise 4.6) and $\mathscr{R}^*$ by $\mathscr{R}^\Delta$ (see Exercise 5.9). Theorem 7.4 remains exactly as stated, except for the obvious replacement of (3.12b) by $\bar{\mathscr{X}}^+(\bar{A} + \bar{B}\bar{F}_0) \subset \operatorname{Ker}(\bar{D} + \bar{E}\bar{F}_0)$.

7.9 Notes and References

In one form or another, the problem of tracking and regulation with internal stability is perhaps the oldest and most central issue of control theory. While a full bibliography would be voluminous we mention, among the approaches closest to the spirit of this chapter, the earlier work of Johnson [1], Young and Willems [1] and Smith and Davison [1]. The present formulation and results are due to Wonham and Pearson [1]. For an interesting discussion of poles and zeros in multivariable systems the reader is referred to MacFarlane and Karcanias [1].

8 Tracking and Regulation III: Structurally Stable Synthesis

In this chapter we investigate the regulator problem with internal stability (RPIS) discussed in Chapter 7, from the viewpoint of well-posedness and genericity in the sense of Section 0.16, and of structurally stable implementation. Subject to mild restrictions it is shown that, if and only if RPIS is well-posed, a controller can be synthesized which preserves output regulation and loop stability in the presence of small parameter variations, of a specified type, in controller and plant. Synthesis is achieved by means of a feedback configuration which, in general, incorporates an invariant, and suitably redundant, copy of the exosystem, namely the dynamic model adopted for the exogenous reference and disturbance signals which the system is required to process. The geometric idea underlying these results is transversality, or the intersection of subspaces in general position.

8.1 Preliminaries

As in Chapter 7 we consider the system

$$\dot{x} = Ax + Bu, \qquad y = Cx, \qquad z = Dx, \tag{1.1}$$

where y is the measured vector and z the vector to be regulated.[1] The system pair (A, B) describes the plant (controllable subsystem) together with the exogenous reference and disturbance signals (e.g. steps, ramps, ...) with respect to which control is needed. As before, our spaces $\mathscr{X}, \mathscr{U}, \ldots$ and maps A,

[1] As usual, the case of direct control feedthrough, $z = Dx + Eu$, merely complicates the notation while requiring no fresh ideas. The results are summarized in Exercise 8.8.

$B, \ldots$ are defined initially over the field $\mathbb{R}$ but we sometimes adopt, without comment, the natural complexifications of $\mathscr{X}$, etc. Finally, it will be convenient to use the formalism of tensor products as summarized in Section 0.13.

In the following, we shall assume without essential loss of generality that the exogenous signals are completely unstable, namely the map $\bar{A}$ induced by A in $\bar{\mathscr{X}} := \mathscr{X}/\langle A \mid \mathscr{B}\rangle$ satisfies

$$\sigma(\bar{A}) \subset \mathbb{C}^+ := \{\lambda: \Re e\, \lambda \geq 0\}. \tag{1.2}$$

It is also natural to assume that the pair (C, A) is detectable, namely

$$\mathscr{X}^+(A) \cap \mathscr{N} = 0, \tag{1.3}$$

since otherwise, by Theorem 7.4, RPIS can be reformulated in the factor space $\mathscr{X}/\mathscr{X}^+(A) \cap \mathscr{N}$. We then have the following result, as an immediate formal simplification of Theorem 7.1. From this result we shall obtain the simple computational criterion of Corollary 8.1, below.

Theorem 8.1. *Subject to assumptions* (1.2) *and* (1.3), *RPIS is solvable if and only if there exists a subspace* $\mathscr{V} \subset \mathscr{X}$, *such that*

$$\mathscr{V} \subset \operatorname{Ker} D \cap A^{-1}(\mathscr{V} + \mathscr{B}), \tag{1.4a}$$

$$\mathscr{V} \cap \langle A \mid \mathscr{B}\rangle = 0, \tag{1.4b}$$

and

$$\langle A \mid \mathscr{B}\rangle + \mathscr{V} = \mathscr{X}. \tag{1.4c}$$

We may clearly assume that $D: \mathscr{X} \to \mathscr{Z}$ is epic (otherwise replace $\mathscr{Z}$ by $\operatorname{Im} D$); then, (1.4a,c) imply

$$D\langle A \mid \mathscr{B}\rangle = \mathscr{Z}. \tag{1.5}$$

From now on we adopt (1.5) as a standing assumption.

To recall the interpretation of Theorem 8.1, let $F: \mathscr{X} \to \mathscr{U}$ solve RPIS, so that $(A + BF)|\langle A \mid \mathscr{B}\rangle$ is stable and, because of (1.2),

$$\langle A \mid \mathscr{B}\rangle \oplus \mathscr{X}^+(A + BF) = \mathscr{X}. \tag{1.6}$$

Since $\mathscr{X}^+(A + BF) \subset \operatorname{Ker} D$ it follows that $\mathscr{V} := \mathscr{X}^+(A + BF)$ satisfies (1.4). Conversely, if $\mathscr{V}$ satisfies (1.4), and if $F \in \mathbf{F}(\mathscr{V})$ is such that $(A + BF)|\langle A \mid \mathscr{B}\rangle$ is stable, then it is straightforward to check that $\mathscr{V} = \mathscr{X}^+(A + BF)$. Fix such F, let $P: \mathscr{X} \to \bar{\mathscr{X}}$ be the canonical projection and write $\bar{x} = Px$. By the decomposition (1.6), $x = x^- + x^+$ with $x^- \in \langle A \mid \mathscr{B}\rangle$ and $x^+ \in \mathscr{X}^+(A + BF)$ uniquely determined. Define $V: \bar{\mathscr{X}} \to \mathscr{X}$ according to $V\bar{x} = x^+$. Then, $V\bar{x} = 0$ implies that $x \in \langle A \mid \mathscr{B}\rangle$, that is $\bar{x} = 0$, so V is monic. Since $\bar{\mathscr{X}} \simeq \mathscr{X}^+(A + BF)$, there follows $\operatorname{Im} V = \mathscr{X}^+(A + BF)$ and $PV = 1$. Also,

$$\begin{aligned}(A + BF)V\bar{x} &= (A + BF)x^+ = V\overline{(A + BF)x^+} \\ &= VP(A + BF)x^+ = V\bar{A}\bar{x}\end{aligned}$$

for all $\bar{x} \in \bar{\mathscr{X}}$. To summarize, if F solves RPIS then there exists $V: \bar{\mathscr{X}} \to \mathscr{X}$, such that the diagram below commutes. Furthermore, $\operatorname{Im} V = \mathscr{X}^+(A + BF) \subset \operatorname{Ker} D$.

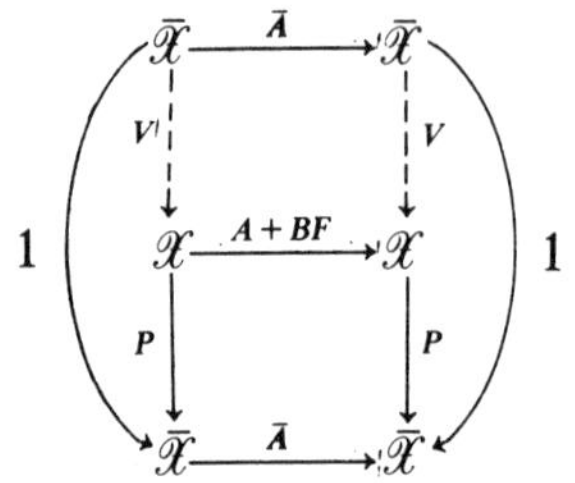

Conversely, suppose V, F are such that the diagram commutes and also $\operatorname{Im} V \subset \operatorname{Ker} D$. Since $PV = 1$, we have

$$\langle A | \mathscr{B} \rangle \cap \operatorname{Im} V = \operatorname{Ker} P \cap \operatorname{Im} V = 0.$$

Therefore, $F | \langle A | \mathscr{B} \rangle$ can be modified, if necessary, to ensure that $F\mathscr{N} = 0$ and $(A + BF) | \langle A | \mathscr{B} \rangle$ is stable, and the diagram will still commute. But then, $\operatorname{Im} V = \mathscr{X}^+(A + BF)$ and F solves RPIS.

It is convenient to set $K := FV$. Then, the diagram (top square) yields

$$AV - V\bar{A} + BK = 0.$$

On the other hand, if this equation holds and $PV = 1$, the diagram is recovered on setting $F = KP$. Thus, we have the following.

Corollary 8.1. *Subject to assumptions* (1.2), (1.3) *and* (1.5), *RPIS is solvable if and only if there exist maps* $V: \bar{\mathscr{X}} \to \mathscr{X}$ *and* $K: \bar{\mathscr{X}} \to \mathscr{U}$, *such that*

$$AV - V\bar{A} + BK = 0 \tag{1.7a}$$

$$DV = 0 \tag{1.7b}$$

$$PV = 1. \tag{1.7c}$$

This result will be our point of departure for the synthesis described later. It is well to note that when these equations are used to compute a solution of RPIS, the correct procedure is first to select F_0, such that $F_0\mathscr{N} = 0$ and $(A + BF_0) | \langle A | \mathscr{B} \rangle$ is stable; then solve (1.7) with A replaced by $A_0 := A + BF_0$; obtain $F_1 := KP$; and finally, set $F := F_0 + F_1$. Here the first step is feasible because, of course, $\mathscr{N} = \mathscr{N} \cap \mathscr{X}^-(A) \subset \langle A | \mathscr{B} \rangle$.

8.2 Example 1: Structural Stability

To motivate further developments, we point out here that naive application of the results of Section 8.1 may lead to systems which are somewhat unsatisfactory from a practical viewpoint. Consider the trivial RPIS defined by the following:

$$\dot{x}_1 = -ax_1 + u, \qquad a > 0$$
$$\dot{x}_2 = 0,$$
$$y = (x_1, x_2), \qquad z = x_2 - x_1,$$

where x_1, x_2 are scalars. Certainly a solution is furnished by

$$u = f_1 x_1 + f_2 x_2; \qquad f_1 = 0, f_2 = a;$$

with signal flow shown in Fig. 8.1. However, if the parameter a is not precisely known to the designer, who takes instead (say) $f_2 = a + \epsilon$, then

$$\dot{z} = ax_1 - f_2 x_2 = -az - \epsilon x_2.$$

But now,

$$z(t) \to \left(-\frac{\epsilon}{a}\right) x_2(0+), \qquad t \to \infty,$$

and output regulation fails.

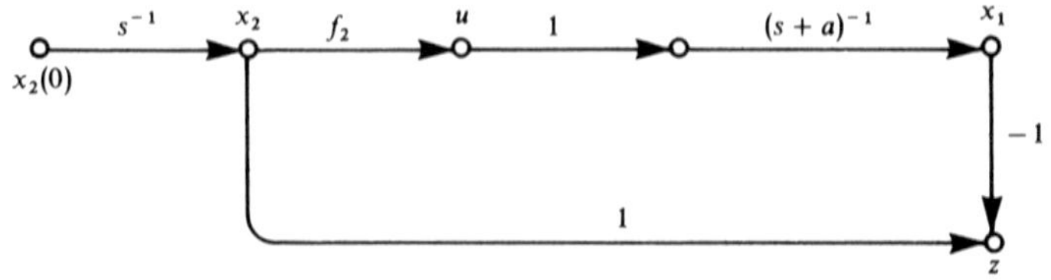

Figure 8.1 "Naive" Solution of RPIS. Open-loop control: $f_2 = a$.

Every control engineer knows that the cure in this example is to adopt feedback together with "integral" compensation, according to Fig. 8.2. A short computation yields

$$\hat{z}(s) = \frac{s(s + a)}{s^2 + (a + f_3)s + f_4} \hat{x}_2(s).$$

Then, with $\hat{x}_2(s) = x_2(0+)/s$, we shall have internal (loop) stability and output regulation, provided $a + f_3 > 0$ and $f_4 > 0$. The new design is *structurally stable* in the sense that internal stability and output regulation are

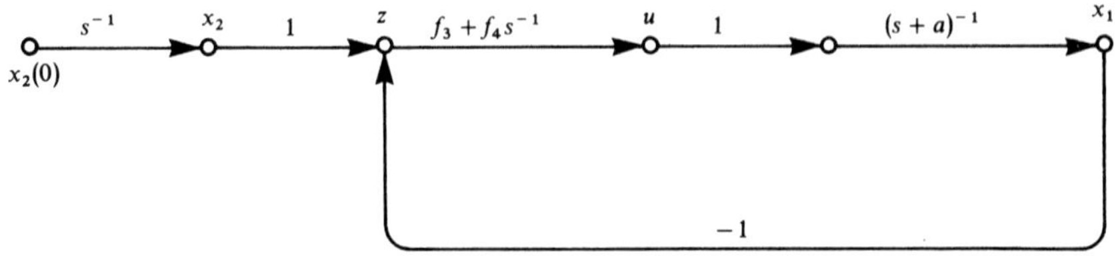

Figure 8.2 Structurally Stable Solution of RPIS. Closed-loop control: integral compensation, $f_3 > -a, f_4 > 0$ arbitrary.

both preserved in the presence of arbitrary variations in the parameters (a, f_3, f_4) of plant and controller, at least in a sufficiently small neighborhood of their nominal values. For the approach to succeed, of course, we must ensure that the loop integration s^{-1}, as well as the comparator which generates the error signal z, remain precisely fixed. In practice this usually poses no serious problem, as such information-processing elements can be constructed to be highly accurate over extended periods of time.

The crucial features of the second design are the feedback loop topology of the signal flow, together with the presence in the loop of an "internal model" (s^{-1}) of the dynamic system which generates the exogenous signal x_2 which the system is required to track. Now RPIS can always be formulated, as in the example of Section 7.7, to build in feedback from scratch: just assume $y = z$. The technique will work for the present example too, because the loop integration is supplied by the observer. But this simplistic approach is unduly restrictive and, it turns out, in the multivariable case may actually fail. To see why, and to develop a structurally stable synthesis in the general case, we place RPIS in a simple topological setting, along lines already sketched in Section 0.16.

8.3 Well-Posedness and Genericity

Write $\mathscr{X}_1 := \langle A | \mathscr{B} \rangle$, $\mathscr{X} = \mathscr{X}_1 \oplus \mathscr{X}_2$, and fix a compatible basis to obtain matrix representations

$$A = \begin{bmatrix} A_1 & A_3 \\ 0 & A_2 \end{bmatrix}, \qquad B = \begin{bmatrix} B_1 \\ 0 \end{bmatrix}, \tag{3.1}$$
$$C = [C_1 \quad C_2], \qquad D = [D_1 \quad D_2],$$

with dimensions $A_1: n_1 \times n_1$, $A_2: n_2 \times n_2$, $B_1: n_1 \times m$, $C_i: p \times n_i$, and $D_i: q \times n_i$. Here,

$$(A_1, B_1) \text{ is controllable} \tag{3.2a}$$

and A_2 represents the map $\bar{A}$ defined above, so that

$$\sigma(A_2) \subset \mathbb{C}^+. \tag{3.2b}$$

Also, in view of (1.5), we shall assume at the outset

$$\operatorname{Rank} D_1 = q. \tag{3.3}$$

Now fix (A_2, C, D) and, after listing the matrix elements in some arbitrary order, regard (A_1, A_3, B_1) as a data point $\mathbf{p}$ in $\mathbb{R}^N$, $N = n_1^2 + n_1 n_2 + n_1 m$. In the spirit of Section 0.16 we shall say that RPIS is *generically solvable* if it is solvable at all $\mathbf{p}$ in the complement of a proper algebraic variety in $\mathbb{R}^N$, and that RPIS is *well-posed at* $\mathbf{p}$ if it is solvable at all points in some neighborhood of $\mathbf{p}$ in $\mathbb{R}^N$.

Of course, many other definitions of well-posedness are possible, the idea being to allow for uncertainty about the precise numerical values of various system parameters. We elect to fix C and D because "typically" the variables which can be measured ($y = Cx$) or which must be regulated ($z = Dx$) are defined by the problem coordinatization in a way which is independent of small parameter variations, for instance, when y is a position measurement and z a tracking error. Here we neglect any sensor or comparator imprecision. The matrix A_2 describes the exosystem, namely, it embodies the dynamic structure of disturbance and/or reference signals external to the plant: as such it is normally fixed by *a priori* specification of the class of exogenous signals which the regulator is to be designed to handle: steps, ramps, and the like. Similarly, the controllable subspace $\mathscr{X}_1$ (i.e. the plant state space) is or can be fixed by coordinatization: since (A_1, B_1) is controllable, so are all pairs (A_1', B_1') in some neighborhood of (A_1, B_1) in $\mathbb{R}^{n_1^2+n_1 m}$. On the other hand, by allowing completely free small variations in (A_1, B_1), we tacitly ignore special features of plant structure often fixed by coordinatization, like definitional relations of the type $\xi_2 = \dot{\xi}_1$. More conservatively still, we allow free variations in the map A_3 which binds the exogenous signals into the plant. While somewhat unrealistic, such latitude is technically convenient, and it will turn out that our results are not unduly restrictive from a practical point of view.

The main results of this section are the following. Here, a prime denotes duality (or matrix transpose) and 1_1, $1_2'$ are the identities on $\mathscr{X}_1$, $\mathscr{X}_2'$.

Theorem 8.2. *Subject to assumptions* (1.2), (1.3) *and* (1.5) [*equivalently* (3.2), (1.3) *and* (3.3)], *RPIS is well-posed at* (A_1, A_3, B_1) *if and only if*

$$(A_1 \otimes 1_2' - 1_1 \otimes A_2') \operatorname{Ker}(D_1 \otimes 1_2') + \operatorname{Im}(B_1 \otimes 1_2') = \mathscr{X}_1 \otimes \mathscr{X}_2'. \quad (3.4)$$

Corollary 8.2. *Under the assumptions of Theorem 8.2, RPIS is generically solvable if and only if*

$$m := d(\mathscr{B}) \geq d(\mathscr{Z}) =: q.$$

If $m < q$, no data point is well-posed.

Notice that the criterion of Theorem 8.2 does not involve A_3. Several equivalent criteria together with their systemic interpretations are given in Section 8.4. The result of Corollary 8.2 is pleasingly simple: it says that our regulation problem is almost always solvable provided the number of independent controls is at least as great as the number of output variables to be regulated, and that our problem is almost never solvable otherwise.

PROOF. The proof follows on rewriting Corollary 8.1 by means of (3.1). Setting $P = [0 \quad 1_2]$, we have from (1.7c) that

$$V = \begin{bmatrix} V_1 \\ 1_2 \end{bmatrix}.$$

Then (1.7a,b) give

$$A_1 V_1 - V_1 A_2 + A_3 + B_1 K = 0$$

$$D_1 V_1 + D_2 = 0. \tag{3.5}$$

Let $\check{D}_1$ be a right inverse of D_1 and set

$$V_1 = -\check{D}_1 D_2 + V_0,$$

where $D_1 V_0 = 0$. Substitution in (3.5) yields

$$A_1 V_0 - V_0 A_2 + B_1 K = -A_3 + A_1 \check{D}_1 D_2 - \check{D}_1 D_2 A_2. \tag{3.6}$$

Thus, RPIS is solvable if and only if (3.6) has a solution (V_0, K) with $D_1 V_0 = 0$. For well-posedness, a solution must exist for all $\hat{A}_3$ in some neighborhood of $A_3 \in \mathbb{R}^{n_1 \times n_2}$. But this is true just when the linear map

$$(V_0, K) \mapsto A_1 V_0 - V_0 A_2 + B_1 K, \tag{3.7a}$$

restricted to the subspace

$$\{(V_0, K): D_1 V_0 = 0\}, \tag{3.7b}$$

is epic. Thus, (3.4) is necessary. Since the map (3.7) remains epic for all $(\hat{A}_1, \hat{B}_1)$ in some neighborhood of (A_1, B_1) if it is epic at (A_1, B_1), (3.4) is also sufficient.

For the corollary, it is clear that (3.4) holds only if

$$d(\text{Ker}(D_1 \otimes 1_2')) + d(\text{Im}(B_1 \otimes 1_2')) \geq d(\mathscr{X}_1 \otimes \mathscr{X}_2'),$$

namely,

$$(n_1 - q)n_2 + mn_2 \geq n_1 n_2,$$

or $m \geq q$. Thus, if $m < q$ no data point is well-posed. Conversely, suppose $m \geq q$. It is clear that if the linear map (3.7) is epic at some particular data point, then it is epic at almost all. To display such a data point, first set $A_1 = -1_1$. Since $\sigma(A_2) \subset \mathbb{C}^+$ it follows that

$$A_1 V_0 - V_0 A_2 = -V_0(1_2 + A_2)$$

with $1_2 + A_2$ invertible. Next, choose B_1 such that $D_1 \text{ Im } B_1 = \text{Im } D_1$, or

$$\text{Ker } D_1 + \text{Im } B_1 = \mathscr{X}_1. \tag{3.8}$$

Since $m \geq q$, such a choice is possible. Denote the right side of (3.6) by R. By (3.8) there exist U, and V_0 with $D_1 V_0 = 0$, such that

$$-V_0 + B_1 U = R(1_2 + A_2)^{-1}.$$

Then (V_0, K) solves (3.6) if $K = U(1_2 + A_2)$, and as R was arbitrary the result follows. □

8.4 Well-Posedness and Transmission Zeros

In this section we provide alternative criteria for well-posedness of **RPIS** that shed further light on its systemic meaning. Our main results are the following; the notation is that of Section 8.3.

Theorem 8.3. *Under the assumptions of Theorem 8.2, RPIS is well-posed at* (A_1, A_3, B_1) *if and only if either of the following conditions holds:*

i. *For all* $\lambda \in \sigma(A_2)$,

$$(A_1 - \lambda 1_1) \operatorname{Ker} D_1 + \mathscr{B}_1 = \mathscr{X}_1 \tag{4.1}$$

or

ii.
$$\sigma^*(A_1, B_1; \operatorname{Ker} D_1) \cap \sigma(A_2) = \varnothing \tag{4.2}$$

and

$$d(\mathscr{B}_1 \cap \mathscr{V}_1^*) = d(\mathscr{B}_1) - d(\mathscr{Y}) \tag{4.3}$$
$$(= m - q).$$

Theorem 8.4. *In addition to the assumptions of Theorem 8.2, suppose* (D_1, A_1) *is observable* [*i.e.* (D_1, A_1, B_1) *is complete*] *and that* $\sigma(A_1) \cap \sigma(A_2) = \varnothing$. *Then RPIS is well-posed at* (A_1, A_3, B_1) *if and only if the plant transfer matrix*

$$H_1(\lambda) := D_1(\lambda 1_1 - A_1)^{-1} B_1$$

is right-invertible (over $\mathbb{C}$*) for every complex number* $\lambda \in \sigma(A_2)$, *i.e. at every point in the spectrum of the exosystem.*

We recall that the condition (4.2) occurred earlier, in Corollary 7.2, as a sufficient condition for the general property of decomposability required by Theorem 7.3 and Corollary 7.1. As might be expected, the spectral disjointness expressed by (4.2) is necessary and sufficient for decomposability to be a well-posed property. In Theorem 8.4, right-invertibility of $H_1(\lambda)$ for some $\lambda \in \mathbb{C}$ is easily seen to imply that $H_1(\lambda)$ is right-invertible as a rational matrix over $\mathbb{R}(\lambda)$, λ indeterminate. Thus, our condition is equivalent to right-invertibility over $\mathbb{R}(\lambda)$ together with the requirement that no plant transmission zero (where right-invertibility over $\mathbb{C}$ would fail) coincide with an eigenvalue of A_2. In either version the condition of well-posedness can be roughly paraphrased by saying that "the plant, as a signal processor, is an invertible operator when restricted to signals of the type generated by the exosystem."

Proof (of Theorem 8.3(i)). The condition (4.1) is merely a "modal decomposition" of the condition (3.4) of Theorem 8.2. To see this let

$$\mathscr{X}_2 = \bigoplus_{\lambda} \mathscr{X}_{2\lambda}$$

be a modal decomposition of $\mathscr{X}_2$ relative to A_2 (as in Section 0.11); in the direct sum, λ ranges over the distinct elements of $\sigma(A_2)$. Then (cf. (0.11.5)),

$$A_{2\lambda} := A_2 | \mathscr{X}_{2\lambda} = \lambda 1_\lambda + N_\lambda,$$

where 1_λ is the identity on $\mathscr{X}_{2\lambda}$ and $N_\lambda: \mathscr{X}_{2\lambda} \to \mathscr{X}_{2\lambda}$ is nilpotent, i.e. $N_\lambda^k = 0$ for some k. With these substitutions (3.4) can be written

$$[(A_1 - \lambda 1_1) \otimes 1'_\lambda - 1_1 \otimes N'_\lambda] \operatorname{Ker}(D_1 \otimes 1'_\lambda) + \operatorname{Im}(B_1 \otimes 1'_\lambda) = \mathscr{X}_1 \otimes \mathscr{X}'_{2\lambda}, \qquad \lambda \in \sigma(A_2). \quad (4.4)$$

Let N_λ have index of nilpotence ν, i.e. $N_\lambda^\nu = 0$, $N_\lambda^{\nu-1} \neq 0$. There is a natural isomorphism

$$\mathscr{X}_\lambda \simeq \bigoplus_{j=1}^{\nu} \mathscr{X}_{\lambda j},$$

where

$$\mathscr{X}_{\lambda j} := \frac{\operatorname{Im} N_\lambda^{\nu-j}}{\operatorname{Im} N_\lambda^{\nu-j+1}}, \qquad j \in \mathbf{v}.$$

Explicitly if

$$x_\lambda = \sum_{j=1}^{\nu} (N_\lambda^{\nu-j} - N_\lambda^{\nu-j+1}) x_\lambda \in \mathscr{X}_\lambda,$$

then

$$x_\lambda \mapsto \bigoplus_{j=1}^{\nu} x_{\lambda j},$$

where (in the notation of cosets, Section 0.5)

$$x_{\lambda j} = N_\lambda^{\nu-j} x_\lambda + \operatorname{Im} N_\lambda^{\nu-j+1}.$$

In this copy of $\mathscr{X}_\lambda$, N_λ is determined by the maps

$$N_{\lambda j}: \mathscr{X}_{\lambda j} \to \mathscr{X}_{\lambda j-1}: x_{\lambda j} \mapsto N_\lambda^{\nu-j+1} x + \operatorname{Im} N_\lambda^{\nu-j+2}, \qquad j \in \mathbf{v}.$$

Note that $N_{\lambda 1} = 0$. In terms of this representation (4.4) is evidently true if and only if the system of equations

$$[(A_1 - \lambda 1_1) \otimes 1'_{\lambda 1}] X_1 + (B_1 \otimes 1'_{\lambda 1}) U_1 = Y_1 \quad (4.5a)$$

$$[(A_1 - \lambda 1_1) \otimes 1'_{\lambda j}] X_j - (1_1 \otimes N'_{\lambda j}) X_{j-1} + (B_1 \otimes 1'_{\lambda j}) U_j = Y_j, \qquad j = 2, \ldots, \nu \quad (4.5b)$$

has a solution

$$X_j \in \operatorname{Ker}(D_1 \otimes 1'_{\lambda j}), \qquad U_j \in \mathscr{U} \otimes \mathscr{X}'_{\lambda j}, \qquad j \in \mathbf{v}$$

for every set of elements

$$Y_j \in \mathscr{X}_1 \otimes \mathscr{X}'_{\lambda j}, \qquad j \in \mathbf{v}.$$

But it is clear from the chain structure of (4.5) that the system is always solvable if and only if (4.5a) is always solvable, and this is true if and only if

$$[(A_1 - \lambda 1_1) \otimes 1'_{\lambda 1}] \operatorname{Ker}(D_1 \otimes 1'_{\lambda 1}) + \operatorname{Im}(B_1 \otimes 1'_{\lambda 1}) = \mathscr{X}_1 \otimes \mathscr{X}'_{\lambda 1}. \quad (4.6)$$

Since (4.6) is equivalent to (4.3), the result follows. □

To prove Theorem 8.3(ii), we need several preliminary results, and shall use the following notation. Since in Lemmas 8.1 and 8.2 the triple (D, A, B) can be arbitrary, for convenience we temporarily drop the subscript 1. Now let

$$\Lambda := \{\lambda : \lambda \in \mathbb{C}, (A - \lambda 1) \operatorname{Ker} D + \mathscr{B} = \mathscr{X}\},$$

$$\tilde{\mathscr{X}} := \frac{\mathscr{X}}{\mathscr{R}^*},$$

$$\tilde{A}_F := \text{map induced on } \tilde{\mathscr{X}} \text{ by } A_F := A + BF, \text{ with } F \in \mathbf{F}(\mathscr{V}^*),$$

$$\sigma^* := \sigma^*(A, B; \operatorname{Ker} D) := \sigma\left(\tilde{A}_F \,\middle|\, \frac{\mathscr{V}^*}{\mathscr{R}^*}\right).$$

Recall that σ^* is fixed for all $F \in \mathbf{F}(\mathscr{V}^*)$.

Lemma 8.1. *Let $F \in \mathbf{F}(\mathscr{V}^*)$. Then for all $\lambda \in \mathbb{C}$,*

$$\mathscr{B} \cap (A_F - \lambda 1) \operatorname{Ker} D \subset \mathscr{B} \cap \mathscr{V}^*,$$

with equality if $\lambda \notin \sigma(A_F \mid \mathscr{V}^*)$.

Proof. Let $b = Bu = (A_F - \lambda 1)x_0$, $x_0 \in \operatorname{Ker} D$. Choose $F_0 : \mathscr{X} \to \mathscr{U}$, such that $F_0 x_0 = -u$. Then,

$$(A_F + BF_0 - \lambda 1)x_0 = 0,$$

so that $\operatorname{Span}\{x_0\} \in \mathfrak{I}(A, B; \operatorname{Ker} D)$. This implies that $x_0 \in \mathscr{V}^*$, hence so is $(A_F - \lambda 1)x_0$, namely $b \in \mathscr{B} \cap \mathscr{V}^*$. Also if $b \in \mathscr{B} \cap \mathscr{V}^*$ and $\lambda \notin \sigma(A_F \mid \mathscr{V}^*)$, then

$$(A_F - \lambda 1)^{-1} b \in \mathscr{V}^* \subset \operatorname{Ker} D,$$

so that $b \in \mathscr{B} \cap (A_F - \lambda 1) \operatorname{Ker} D$. □

Lemma 8.2.

$$\sigma^* \cap \Lambda = \varnothing.$$

Proof. If $\Lambda = \varnothing$ there is nothing to prove, so assume $\lambda \in \Lambda$. Let $F \in \mathbf{F}(\mathscr{V}^*)$ and $P : \mathscr{X} \to \tilde{\mathscr{X}}$ be the canonical projection. Clearly, Λ is an open subset of $\mathbb{C}$, so we can and do assume that $\lambda \notin \sigma(A_F)$. Then,

$$(A_F - \lambda 1) \operatorname{Ker} D \supset (A_F - \lambda 1)\mathscr{V}^* = \mathscr{V}^* \supset \operatorname{Ker} P.$$

Therefore (in obvious notation),

$$\begin{aligned}\tilde{\mathscr{B}} \cap (\tilde{A}_F - \lambda\tilde{1}) \operatorname{Ker} \tilde{D} &= P[\mathscr{B} \cap (A_F - \lambda 1) \operatorname{Ker} D] \quad \text{(using (0.4.3))}\\ &= P(\mathscr{B} \cap \mathscr{V}^*) \quad \text{(by Lemma 8.1)}\\ &= 0,\end{aligned}$$

since $\mathscr{B} \cap \mathscr{V}^* \subset \mathscr{R}^* = \operatorname{Ker} P$. It follows that

$$(\tilde{A}_F - \lambda\tilde{1}) \operatorname{Ker} \tilde{D} \oplus \tilde{\mathscr{B}} = \tilde{\mathscr{X}},$$

and as $\lambda \notin \sigma(\tilde{A}_F)$,

$$d(\operatorname{Ker} \tilde{D}) + d(\tilde{\mathscr{B}}) = d(\tilde{\mathscr{X}}).$$

But if $\mu \in \sigma^*$, then

$$d[(\tilde{A}_F - \mu\tilde{1}) \operatorname{Ker} \tilde{D}] < d(\operatorname{Ker} \tilde{D})$$

and it results, in turn, that

$$\begin{aligned}(\tilde{A}_F - \mu\tilde{1}) \operatorname{Ker} \tilde{D} + \tilde{\mathscr{B}} &\neq \tilde{\mathscr{X}},\\ P(A_F - \mu 1) \operatorname{Ker} D + P\mathscr{B} &\neq P\mathscr{X},\\ (A_F - \mu 1) \operatorname{Ker} D + \mathscr{B} &\neq \mathscr{X},\\ (A - \mu 1) \operatorname{Ker} D + \mathscr{B} &\neq \mathscr{X}.\end{aligned}$$

That is, $\mu \notin \Lambda$, so $\sigma^* \subset \mathbb{C} - \Lambda$, i.e. $\sigma^* \cap \Lambda = \varnothing$. □

In the next three lemmas, let

$$\begin{aligned}\Lambda_1 &:= \{\lambda : \lambda \in \mathbb{C},\ (A_1 - \lambda 1_1) \operatorname{Ker} D_1 + \mathscr{B}_1 = \mathscr{X}_1\},\\ \sigma_1^* &:= \sigma^*(A_1, B_1; \operatorname{Ker} D_1).\end{aligned}$$

Lemma 8.3. *Under the assumptions of Theorem 8.2, RPIS is well-posed only if*

$$\sigma_1^* \cap \sigma(A_2) = \varnothing.$$

Proof. Applying Lemma 8.2 to the triple (D_1, A_1, B_1), we have $\sigma_1^* \cap \Lambda_1 = \varnothing$. By Theorem 8.3(i), RPIS is well-posed only if $\Lambda_1 \supset \sigma(A_2)$, and the result follows. □

Lemma 8.4. *Under the assumptions of Theorem 8.2, RPIS is well-posed only if*

$$d(\mathscr{B}_1 \cap \mathscr{V}_1^*) = m - q.$$

Proof. Choose $F_1 \in \mathbf{F}(\mathscr{V}_1^*)$, such that

$$\sigma(A_1 + B_1 F_1 | \mathscr{R}_1^*) \subset \sigma_1^*.$$

Writing $A_{1F_1} := A_1 + B_1 F_1$ we have, since RPIS is well-posed,

$$(A_{1F_1} - \lambda 1_1) \operatorname{Ker} D_1 + \mathscr{B}_1 = \mathscr{X}_1, \qquad \lambda \in \sigma(A_2).$$

Thus $\Lambda_1 \neq \emptyset$ and as $\Lambda_1 \subset \mathbb{C}$ is open, we can and do select $\lambda_1 \in \Lambda_1 - \sigma_1^*$. By Lemma 8.1,

$$d[(A_{1F_1} - \lambda_1 1_1) \operatorname{Ker} D_1 \cap \mathscr{B}_1] = d(\mathscr{B}_1 \cap \mathscr{V}_1^*) =: m_1, \text{ say.}$$

Also,

$$d[(A_{1F_1} - \lambda_1 1_1) \operatorname{Ker} D_1] = d(\operatorname{Ker} D_1).$$

Therefore,

$$d(\mathscr{X}_1) = d(\operatorname{Ker} D_1) + d(\mathscr{B}_1) - m_1,$$

or

$$n_1 = (n_1 - q) + m - m_1,$$

or

$$m_1 = m - q,$$

as claimed. □

Lemma 8.5. *Under the assumptions of Theorem 8.2, RPIS is well-posed if*

$$\sigma_1^* \cap \sigma(A_2) = \emptyset$$

and

$$d(\mathscr{B}_1 \cap \mathscr{V}_1^*) \leq m - q.$$

Proof. Choose $F_1 \in \mathbf{F}(\mathscr{V}_1^*)$, such that

$$\sigma(A_1 + B_1 F_1 | \mathscr{R}_1^*) \subset \sigma_1^*.$$

Writing $A_{1F_1} := A_1 + B_1 F_1$ we then have

$$d[(A_{1F_1} - \lambda 1_1) \operatorname{Ker} D_1] = d(\operatorname{Ker} D_1)$$

for all $\lambda \notin \sigma_1^*$. Also, by Lemma 8.1 applied to (D_1, A_1, B_1), we have

$$d[\mathscr{B}_1 \cap (A_{1F_1} - \lambda 1_1) \operatorname{Ker} D_1] \leq d(\mathscr{B}_1 \cap \mathscr{V}_1^*) \leq m - q.$$

It follows that, for $\lambda \notin \sigma_1^*$,

$$\begin{aligned} d[(A_{1F_1} - \lambda 1_1) \operatorname{Ker} D_1 + \mathscr{B}_1] &\geq d(\operatorname{Ker} D_1) + d(\mathscr{B}_1) - (m - q) \\ &= (n_1 - q) + m - (m - q) \\ &= n_1, \end{aligned}$$

so that

$$(A_{1F_1} - \lambda 1_1) \operatorname{Ker} D_1 + \mathscr{B}_1 = \mathscr{X}_1, \qquad \lambda \notin \sigma_1^*.$$

Therefore equality holds if $\lambda \in \sigma(A_2)$, and so by Theorem 8.3(i), RPIS is well-posed. □

Proof (of Theorem 8.3(ii)). The result follows directly from Lemmas 8.3–8.5. □

To prove Theorem 8.4 let (D, A, B) be an arbitrary triple (with dimensions q, n, m as usual) and bring in the "polynomial system map"

$$M(\lambda): \mathbb{R}^n(\lambda) \oplus \mathbb{R}^m(\lambda) \to \mathbb{R}^n(\lambda) \oplus \mathbb{R}^q(\lambda)$$

defined by the matrix

$$M(\lambda) := \begin{bmatrix} A - \lambda 1 & B \\ D & 0 \end{bmatrix}.$$

Corresponding to $M(\lambda)$, we have the transfer matrix

$$H(\lambda) := D(\lambda 1 - A)^{-1}B: \mathbb{R}^m(\lambda) \to \mathbb{R}^q(\lambda).$$

Lemma 8.6. *$H(\lambda)$ is epic [i.e. right-invertible over $\mathbb{R}(\lambda)$] if and only if $M(\lambda)$ is epic. If $\lambda_0 \notin \sigma(A)$ then $H(\lambda_0)$ is epic (over $\mathbb{C}$) if and only if $M(\lambda_0)$ is epic (over $\mathbb{C}$).*

PROOF.

$$\text{Rank } M(\lambda) = \text{Rank}\left\{\begin{bmatrix} 1_n & 0 \\ D(\lambda 1 - A)^{-1} & 1_q \end{bmatrix} M(\lambda)\right\}$$

$$= \text{Rank}\begin{bmatrix} A - \lambda 1 & B \\ 0 & H(\lambda) \end{bmatrix} = n + \text{Rank } H(\lambda).$$

Thus, $H(\lambda)$ is epic if and only if Rank $M(\lambda) = n + q$, i.e. if and only if $M(\lambda)$ is epic. Clearly, the argument is valid at $\lambda_0 \in \mathbb{C}$ provided $\lambda_0 \notin \sigma(A)$. □

PROOF (of Theorem 8.4). Let $M_1(\lambda)$, $H_1(\lambda)$ be defined, as above, for the triple (D_1, A_1, B_1). The condition (4.1) is clearly equivalent to the condition that $M_1(\lambda)$ be epic (over $\mathbb{C}$) at all $\lambda \in \sigma(A_2)$. The result follows by application of Lemma 8.6. □

In conclusion we remark that a more detailed investigation reveals that

$$d[\text{Ker } M_1(\lambda)] = d(\mathscr{B}_1 \cap \mathscr{V}_1^*),$$

a fact that completes the link between Theorem 8.4 and Theorem 8.3(ii). Also, the two technical hypotheses of Theorem 8.4 [(D_1, A_1) observable and $\sigma(A_1) \cap \sigma(A_2) = \varnothing$] were imposed only for simplicity in stating and interpreting the result solely in terms of $H_1(\lambda)$.

8.5 Example 2. RPIS Solvable but Ill-Posed

We present in this section a very simple example for which RPIS is solvable, but ill-posed. Let

$$A = \left[\begin{array}{ccc|c} -1 & 0 & 1 & 1 \\ 0 & 0 & 1 & 0 \\ 0 & -2 & -2 & 0 \\ \hline 0 & 0 & 0 & 0 \end{array}\right], \qquad B = \begin{bmatrix} 0 \\ 0 \\ 1 \\ \hline 0 \end{bmatrix},$$

$$C = 1_4, \qquad D = [1 \quad 0 \quad 0 \quad -1].$$

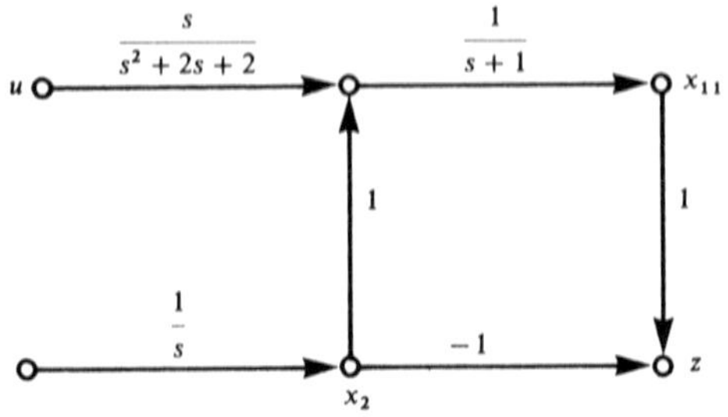

Figure 8.3 Signal Flow Graph: Example, Section 8.5. RPIS is solvable but ill-posed.

The matrices are partitioned in accordance with the decomposition $\mathscr{X} = \mathscr{X}_1 \oplus \mathscr{X}_2$, $d(\mathscr{X}_1) = 3$, $d(\mathscr{X}_2) = 1$. A signal flow graph is shown in Fig. 8.3. By inspection of the graph it is clear that RPIS is trivially solvable with the control $u = 0$. However, on checking the condition of Theorem 8.3(i), we find

$$\sigma(A_2) = \{0\}, \qquad A_1 \operatorname{Ker} D_1 + \mathscr{B}_1 = \operatorname{Im}\begin{bmatrix} 0 & 1 \\ 0 & 1 \\ 1 & -2 \end{bmatrix} \neq \mathscr{X}_1.$$

Thus, RPIS is not well-posed. Similarly, for Theorem 8.3(ii), we find that

$$\mathscr{V}_1^* = \operatorname{Im}\begin{bmatrix} 0 \\ 1 \\ 0 \end{bmatrix}; \qquad F_1 = [f_{11} \quad 2 \quad f_{13}], \quad F_1 \in \mathbf{F}(\mathscr{V}_1^*);$$

$$\mathscr{B}_1 \cap \mathscr{V}_1^* = 0, \quad \mathscr{R}_1^* = 0.$$

Thus, in this case

$$\begin{aligned} \sigma^*(A_1, B_1; \operatorname{Ker} D_1) &= \sigma[(A_1 + B_1 F_1) \mid \mathscr{V}_1^*] \\ &= \{0\} = \sigma(A_2), \end{aligned}$$

and condition (4.2) fails. As to Theorem 8.4, we have for the plant transfer function

$$H_1(\lambda) = \frac{\lambda}{(\lambda + 1)(\lambda^2 + 2\lambda + 2)},$$

which is not invertible at $\lambda = 0$.

Now suppose A is perturbed to A_ϵ by replacing the upper right block,

$$A_3 = \begin{bmatrix} 1 \\ 0 \\ 0 \end{bmatrix}, \quad \text{by } A_{3\epsilon} = \begin{bmatrix} 1 + \epsilon \\ 0 \\ 0 \end{bmatrix}, \qquad \epsilon \neq 0. \tag{5.1}$$

We find that

$$\mathscr{V}^* = \operatorname{Im}\begin{bmatrix} 0 & 1 \\ 1 & 0 \\ 0 & 0 \\ 0 & 1 \end{bmatrix}, \qquad \mathscr{V}_\epsilon^* = \operatorname{Im}\begin{bmatrix} 0 \\ 1 \\ 0 \\ 0 \end{bmatrix}.$$

Checking the first condition of Corollary 7.1, we have

$$\langle A | \mathscr{B} \rangle + \mathscr{V}^* = \mathscr{X}$$

but

$$\langle A_\epsilon | \mathscr{B} \rangle + \mathscr{V}_\epsilon^* \neq \mathscr{X},$$

so that **RPIS** is now unsolvable.

For this example the situation is, of course, transparent by inspection of the signal flow graph. At $\epsilon = 0$ the "feedforward" branch bypassing the plant transmission zero has exactly the gain required for tracking of x_2 by x_{11}. Perturbation of this gain forces the use of feedback to the control node (u), but the corrective signals required are blocked by the plant zero.

8.6 Structurally Stable Synthesis

In this section we show how to synthesize a controller which implements a solution of (well-posed) **RPIS**, and has the desirable property that internal stability and output regulation are preserved, when parameters of the plant and controller undergo small variations. In other words, the controller is "flexible" enough to permit some uncertainty, at the design stage, about values of system parameters, and also to permit (slow) drift (within limits) of these parameters while the system is in operation. It will be no surprise that a feedback configuration is used: less familiar is the fact that the feedback compensator includes a (fixed) model of the external dynamics which, in general, must be reduplicated in a sense made precise below.

To present the main ideas most simply, we shall assume that the measured variables y are precisely the regulated variables z and that the system (1.1) is completely observable from z; thus,

$$C = D, \qquad \mathscr{N}_D := \bigcap_{i=1}^{n} \operatorname{Ker}(DA^{i-1}) = 0. \tag{6.1}$$

As explained in Chapter 7, the control law $u = Fx$ provided by formal solution of **RPIS** can be implemented by means of a dynamic observer. For the present, however, we shall represent these additional dynamics simply as a compensator triple (F_c, A_c, B_c) with state space $\mathscr{X}_c$. The compensator is assumed to be driven by the regulated variable $z(\cdot)$ through the input map B_c. In turn, the control $u(\cdot)$ is obtained both from the compensator output via F_c and directly from $z(\cdot)$ via a feedthrough map K. Thus, the system equations take the form

$$\dot{x}_1 = A_1 x_1 + B_1(F_c x_c + Kz) + A_3 x_2$$
$$\dot{x}_c = A_c x_c + B_c z$$
$$\dot{x}_2 = A_2 x_2.$$

Setting $z = D_1 x_1 + D_2 x_2$, we have

$$\begin{bmatrix} \dot{x}_1 \\ \dot{x}_c \\ \dot{x}_2 \end{bmatrix} = \begin{bmatrix} A_1 + B_1 K D_1 & B_1 F_c & A_3 + B_1 K D_2 \\ B_c D_1 & A_c & B_c D_2 \\ 0 & 0 & A_2 \end{bmatrix} \begin{bmatrix} x_1 \\ x_c \\ x_2 \end{bmatrix}. \tag{6.2}$$

For our purposes there is no loss of generality in taking $K = 0$; otherwise A_1, A_3 may be relabeled accordingly; and we assume this is done.

The signal flow described by (6.2) takes the form of a feedback loop, as shown in Fig. 8.4. Write

$$\mathscr{X}_L := \mathscr{X}_1 \oplus \mathscr{X}_c, \qquad x_L := x_1 \oplus x_c$$

and bring in the loop maps

$$A_L := \begin{bmatrix} A_1 & B_1 F_c \\ B_c D_1 & A_c \end{bmatrix}, \qquad B_L := \begin{bmatrix} A_3 \\ B_c D_2 \end{bmatrix}, \qquad D_L := [D_1 \quad 0]. \tag{6.3}$$

Then,

$$\begin{bmatrix} \dot{x}_L \\ \dot{x}_2 \end{bmatrix} = \begin{bmatrix} A_L & B_L \\ 0 & A_2 \end{bmatrix} \begin{bmatrix} x_L \\ x_2 \end{bmatrix}$$
$$z = D_L x_L + D_2 x_2.$$

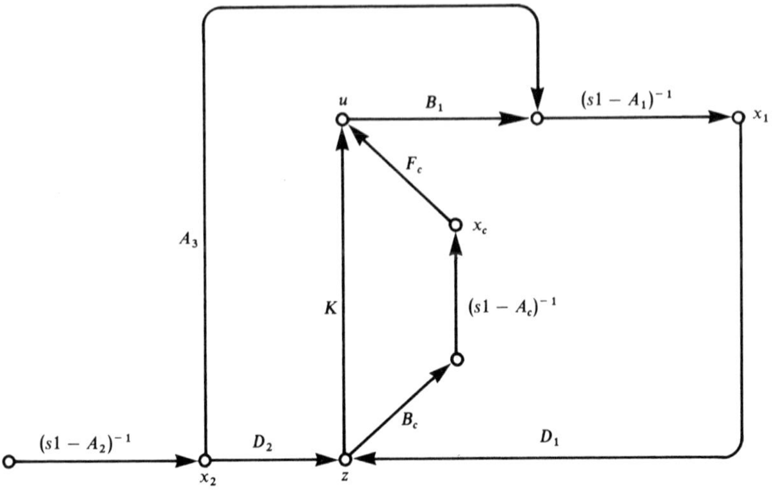

Figure 8.4 Signal Flow in Feedback Regulator.

Finally, introduce the extensions

$$\mathscr{X}_e := \mathscr{X}_L \oplus \mathscr{X}_2, \qquad x_e := x_L \oplus x_2$$

$$A_e := \begin{bmatrix} A_L & B_L \\ 0 & A_2 \end{bmatrix}, \qquad D_e := [D_L \quad D_2].$$

We shall refer to the triple $S_c := (F_c, A_c, B_c)$ as a *synthesis* of RPIS provided

i. (F_c, A_c, B_c) is complete: i.e. (F_c, A_c) is observable and (A_c, B_c) is controllable;
ii. the internal stability condition holds, i.e.

$$A_L \text{ is stable;} \tag{6.4}$$

iii. the output regulation condition holds, i.e.

$$\mathscr{X}_e^+(A_e) \subset \operatorname{Ker} D_e. \tag{6.5}$$

To formalize the notion of a structurally stable regulator we bring in the data point

$$\mathbf{p} = (A_1, A_3, B_1) \in \mathbb{R}^N$$

as before. Then we say that a synthesis S_c is *strong at* $\mathbf{p}$ if the properties (6.4) and (6.5) are well-posed at $\mathbf{p}$: namely, hold for all data points $\tilde{\mathbf{p}}$ in a neighborhood of $\mathbf{p}$ in $\mathbb{R}^N$. Notice that the property (6.4) of internal stability is guaranteed to hold near $\mathbf{p}$ for any synthesis. The interesting feature of a strong synthesis is that it preserves the more delicate property (6.5) of output regulation.

The main result of this section is the following.

Theorem 8.5. *Subject to assumptions* (3.2) *and* (6.1), *RPIS admits a strong synthesis at* $\mathbf{p} = (A_1, A_3, B_1)$ *if and only if it is well-posed at* $\mathbf{p}$, *namely if and only if* (3.4) *is true.*

The proof is deferred to the end of the section, along with some additional remarks of interpretation.

The essential geometric property of a strong synthesis amounts to the following transversality condition.

Theorem 8.6. *A synthesis* (F_c, A_c, B_c) *is strong at* $\mathbf{p} = (A_1, A_3, B_1)$ *if and only if*

$$(A_1 \otimes 1_2' - 1_1 \otimes A_2') \operatorname{Ker}(D_1 \otimes 1_2') + (B_1 F_c \otimes 1_2') \operatorname{Ker}(A_c \otimes 1_2' - 1_c \otimes A_2') = \mathscr{X}_1 \otimes \mathscr{X}_2'. \tag{6.6}$$

PROOF. The proof is very similar to that of Theorem 8.2 and so need only be sketched. Let

$$\mathscr{X}_e^+(A_e) = \operatorname{Im} V_e = \operatorname{Im} \begin{bmatrix} V_1 \\ V_c \\ 1_2 \end{bmatrix}.$$

That the insertion map for $\mathscr{X}_e^+(A_e)$ has the form indicated follows by internal stability. Write

$$V_1 = -\check{D}_1 D_2 + V_0,$$

where $D_1 \check{D}_1 = 1_{\mathscr{Y}}$. Substitution in

$$A_e V_e = V_e A_2, \qquad D_e V_e = 0$$

yields

$$A_1 V_0 - V_0 A_2 + B_1 F_c V_c = -A_3 + A_1 \check{D}_1 D_2 - \check{D}_1 D_2 A_2 \tag{6.7a}$$

$$A_c V_c - V_c A_2 = 0 \tag{6.7b}$$

$$D_1 V_0 = 0. \tag{6.7c}$$

Now S_c is strong at $\mathbf{p}$ if and only if (6.7) is solvable for (V_0, V_c) at every $\tilde{\mathbf{p}} = (\tilde{A}_1, \tilde{A}_3, \tilde{B}_1)$ in a neighborhood of $\mathbf{p}$. Keeping $\tilde{A}_1 = A_1$, $\tilde{B}_1 = B_1$ fixed and letting $\tilde{A}_3$ vary through a neighborhood of A_3 in $\mathbb{R}^{n_1 \times n_2} \simeq \mathscr{X}_1 \otimes \mathscr{X}'_2$, we see immediately from (6.7) that (6.6) is true. Conversely, the condition (6.6) has the form

$$L_1 \mathscr{K}_1 + L_c \mathscr{K}_c = \mathscr{X}_1 \otimes \mathscr{X}'_2$$

for fixed subspaces

$$\mathscr{K}_1 \subset \mathscr{X}_1 \otimes \mathscr{X}'_2, \qquad \mathscr{K}_c \subset \mathscr{X}_c \otimes \mathscr{X}'_2,$$

and where

$$L_1 := A_1 \otimes 1'_2 - 1_1 \otimes A'_2, \qquad L_c := B_1 F_c \otimes 1'_2 \tag{6.8}$$

are continuous matrix-valued functions of $\mathbf{p}$ in the topology of $\mathbb{R}^{n_1 n_2 \times n_1 n_2}$, $\mathbb{R}^{n_1 n_2 \times n_c n_2}$, respectively. From this it is clear that (6.6) holds throughout a neighborhood of $\mathbf{p}$ in $\mathbb{R}^N$, if it is true at $\mathbf{p}$. □

The key to constructing a strong synthesis is now simply the observation that (6.6) will hold provided the subspace

$$\mathscr{K}_c := \operatorname{Ker}(A_c \otimes 1'_2 - 1_c \otimes A'_2) \tag{6.9}$$

is of suitably high dimension, and F_c is chosen so that the subspace $L_c \mathscr{K}_c$ is placed transversely with respect to $L_1 \mathscr{K}_1$. Specifically, we have

$$\mathscr{K}_1 := \operatorname{Ker}(D_1 \otimes 1'_2), \tag{6.10}$$

$$d(\mathscr{K}_1) = (n_1 - q) n_2,$$

and so we must arrange that

$$\begin{aligned} d(\mathscr{K}_c) &\geq d(\mathscr{X}_1 \otimes \mathscr{X}'_2) - d(\mathscr{K}_1) \\ &= n_1 n_2 - (n_1 - q) n_2 \\ &= q n_2. \end{aligned}$$

With this objective in mind we shall, in fact, prove the stronger

Corollary 8.3. *A synthesis* $S_c = (F_c, A_c, B_c)$ *is strong at* **p** *if and only if*

$$d(\mathscr{K}_c) = qn_2. \tag{6.11}$$

The proof depends on two lemmas, of which the first is merely a restatement of Exercise 3.10.

Lemma 8.7. *A pair* (C, A) *is observable if and only if, for every pair* $(T, \mathscr{T})$, $T: \mathscr{T} \to \mathscr{T}$, *there holds*

$$\text{Ker}(C \otimes 1') \cap \text{Ker}(A \otimes 1' - 1 \otimes T') = 0.$$

Since (D, A) is observable by assumption, so is (D_1, A_1), and it follows by Lemma 8.7 that

$$\text{Ker}(A_1 \otimes 1_2' - 1_1 \otimes A_2') \cap \text{Ker}(D_1 \otimes 1_2') = 0,$$

namely [see (6.8)–(6.10)]

$$\text{Ker } L_1 \cap \mathscr{K}_1 = 0. \tag{6.12}$$

The next lemma settles the remaining transversality questions. The notation is that of (6.8)–(6.10).

Lemma 8.8. *If* $S_c = (F_c, A_c, B_c)$ *is any synthesis (strong or not) then*

$$\text{Ker } L_c \cap \mathscr{K}_c = 0, \tag{6.13}$$

$$L_1 \mathscr{K}_1 \cap L_c \mathscr{K}_c = 0. \tag{6.14}$$

PROOF. Since (F_c, A_c) is observable by assumption and B_1 is monic, $(B_1 F_c, A_c)$ is observable, so by Lemma 8.7,

$$\text{Ker}(B_1 F_c \otimes 1_2') \cap \text{Ker}(A_c \otimes 1_2' - 1_c \otimes A_2') = 0,$$

i.e. (6.13) is true. For (6.14), suppose

$$W \in L_1 \mathscr{K}_1 \cap L_c \mathscr{K}_c.$$

Then,

$$W = A_1 V_1 - V_1 A_2 = -B_1 F_c V_c$$

for some V_1 such that $D_1 V_1 = 0$ and some V_c such that $A_c V_c - V_c A_2 = 0$. But this implies that

$$\begin{bmatrix} A_1 & B_1 F_c \\ B_c D_1 & A_c \end{bmatrix} \begin{bmatrix} V_1 \\ V_c \end{bmatrix} - \begin{bmatrix} V_1 \\ V_c \end{bmatrix} A_2 = 0. \tag{6.15}$$

By loop stability $\sigma(A_L) \cap \sigma(A_2) = \varnothing$, hence by the nonsingularity of the Sylvester map $A_L(\cdot) - (\cdot)A_2$, it follows from (6.15) that

$$\begin{bmatrix} V_1 \\ V_c \end{bmatrix} = 0,$$

i.e. $W = 0$. □

Proof (of Corollary 8.3). We know that S_c is strong if and only if

$$L_1\mathscr{K}_1 + L_c\mathscr{K}_c = \mathscr{X}_1 \otimes \mathscr{X}'_2.$$

But (6.12)–(6.14) imply

$$\begin{aligned} d(\mathscr{K}_c) &= d(L_c\mathscr{K}_c) \\ &= d(\mathscr{X}_1 \otimes \mathscr{X}'_2) - d(L_1\mathscr{K}_1) \\ &= d(\mathscr{X}_1 \otimes \mathscr{X}'_2) - d(\mathscr{K}_1) \\ &= qn_2, \end{aligned}$$

as claimed. □

To finish our characterization of a strong synthesis we show how A_c is selected to satisfy (6.11).

Theorem 8.7. *Let $S_c = (F_c, A_c, B_c)$ be a synthesis. Let $\alpha_2(\lambda)$ be the m.p. of A_2 and let $T: \mathscr{T} \to \mathscr{T}$ be cyclic with m.p. α_2. Then, S_c is strong at* **p** *if and only if there exists a monomorphism $J: \mathscr{T} \otimes \mathscr{Y}' \to \mathscr{X}_c$ such that the following diagram commutes.*

$$\begin{array}{ccc} \mathscr{T} \otimes \mathscr{Y}' & \xrightarrow{T \otimes 1'_q} & \mathscr{T} \otimes \mathscr{Y}' \\ \Big\downarrow J & & \Big\downarrow J \\ \mathscr{X}_c & \xrightarrow{A_c} & \mathscr{X}_c \end{array}$$

Our condition may be paraphrased by saying that the compensator dynamics A_c incorporates a q-fold copy of the maximal cyclic component appearing in a rational canonical decomposition of A_2. When this is true, we shall say that A_c *contains an internal model of* A_2. In systemic language, each of the q scalar error variables that make up the regulated vector $z(\cdot)$ can be thought of as driving its own single-input, single-output subcompensator modeled on the cyclic structure of the exosystem.

Proof. Suppose that S_c is strong at **p**. By Corollary 8.3, $d(\mathscr{K}_c) = qn_2$. Let $\{\delta_i\}$, $\{\epsilon_j\}$ be the lists of invariant factors of A_c, A_2, respectively. By the result (0.13.9) on the kernel of the Sylvester map,

$$d(\mathscr{K}_c) = \sum_i \sum_j \deg(\delta_i \wedge \epsilon_j), \tag{6.16}$$

where $\delta \wedge \epsilon$ is the GCD of δ and ϵ. Letting γ_c, γ_2 denote the cyclic index of A_c, A_2, respectively, we have

$$\sum_{i=1}^{\gamma_c} \sum_{j=1}^{\gamma_2} \deg(\delta_i \wedge \epsilon_j) = qn_2. \tag{6.17}$$

If Δ denotes the left side of (6.17), then

$$\Delta \le \sum_{i=1}^{\gamma_c} \sum_{j=1}^{\gamma_2} \deg \epsilon_j = \gamma_c n_2.$$

As (A_c, B_c) is controllable we know by Theorem 1.2 that $\gamma_c \le d(\mathscr{Z}) = q$, i.e. $\gamma_c n_2 \le q n_2$. Thus, $\Delta < q n_2$ if $\gamma_c < q$ or if $\deg(\delta_i \wedge \epsilon_j) < \deg \epsilon_j$ for any i, j. Therefore, (6.17) holds only if $\gamma_c = q$ and

$$\deg(\delta_i \wedge \epsilon_1) = \deg \epsilon_1, \qquad i \in \mathbf{q}.$$

This means that each of the q invariant factors of A_c is divisible by the m.p. of A_2. But this property is equivalent to the existence of a monomorphism J as described.

Conversely, let A_c satisfy the condition of the theorem. By use of the equivalence just stated, a simple calculation from (6.16) yields $d(\mathscr{K}_c) = q n_2$, as claimed. □

The following result will be needed in the actual construction of a strong synthesis.

Lemma 8.9. *Let $A\colon \mathscr{X} \to \mathscr{X}$, let (D, A, B) be complete (with D epic and B monic) and suppose that for some $S\colon \mathscr{S} \to \mathscr{S}$,*

$$(A \otimes 1' - 1 \otimes S') \operatorname{Ker}(D \otimes 1') + \operatorname{Im}(B \otimes 1') = \mathscr{X} \otimes \mathscr{S}'. \tag{6.18}$$

Let $M\colon \mathscr{M} \to \mathscr{M}$ be any map with $\sigma(M) \subset \sigma(S)$ and having cyclic index $\gamma(M) \le \gamma(A)$. Then, (6.18) holds with $(S, \mathscr{S})$ replaced by $(M, \mathscr{M})$. Furthermore, there exist maps (G, H) such that the triples

$$(G, M, H), \tag{6.19}$$

and

$$[D \quad 0], \qquad \begin{bmatrix} A & BG \\ HD & M \end{bmatrix}, \qquad \begin{bmatrix} B \\ 0 \end{bmatrix} \tag{6.20}$$

are complete.

Proof. By Theorem 8.3, (6.18) is equivalent to

$$(A - \lambda 1) \operatorname{Ker} D + \operatorname{Im} B = \mathscr{X}, \qquad \lambda \in \sigma(S). \tag{6.21}$$

From this, the first assertion is clear. Now let rank $B = m$, rank $D = q$. Then by (6.18), $m \ge q$ and by completeness, $q \ge \gamma(A)$. We shall prove the existence of G, such that the triples (6.19), (6.20) are observable for all H; namely,

$$\operatorname{Ker} G \cap \operatorname{Ker}(M - \lambda 1) = 0, \qquad \lambda \in \sigma(M) \tag{6.22}$$

and

$$\operatorname{Ker}[D \quad 0] \cap \operatorname{Ker} \begin{bmatrix} A - \lambda 1 & BG \\ HD & M - \lambda 1 \end{bmatrix} = 0, \qquad \lambda \in \sigma(A) \cup \sigma(M). \tag{6.23}$$

It is clear that in (6.23) H plays no role and may be set to zero. Now,

$$\operatorname{Ker} D \cap \operatorname{Ker}(A - \lambda 1) = 0, \qquad \lambda \in \sigma(A) \tag{6.24}$$

since (D, A) is observable. Suppose (6.22) is true. By (6.24), (6.23) will then hold, provided

$$(A - \lambda 1) \operatorname{Ker} D \cap BG \operatorname{Ker}(M - \lambda 1) = 0, \qquad \lambda \in \sigma(M). \tag{6.25}$$

But from property (0.10.4) of cyclic index,

$$d[\operatorname{Ker}(M - \lambda 1)] \leq \gamma(M) \leq q, \qquad \lambda \in \mathbb{C}; \tag{6.26}$$

and it follows from (6.21) and (6.26) that (6.25) is true for almost all $G \in \mathbb{R}^{m \times d(\mathcal{M})}$. For almost all G, $d(\operatorname{Ker} G) = 0$ if $d(\mathcal{M}) \leq m$, and

$$d(\operatorname{Ker} G) = d(\mathcal{M}) - m \leq d(\mathcal{M}) - q \tag{6.27}$$

if $d(\mathcal{M}) \geq m$. From (6.26), (6.27) it is now clear that (6.22) is true for almost all G. Therefore, both (6.22) and (6.23) hold for almost all G and all H, namely, for these (G, H) the triples (6.19), (6.20) are observable.

The result that the triples (6.19), (6.20) are controllable for all G and almost all H is immediate by duality. □

We are finally in a position to prove our main result.

PROOF (of Theorem 8.5). (Only if) Since

$$B_1 F_c \otimes 1_2' = (B_1 \otimes 1_2')(F_c \otimes 1_2'),$$

the property (6.6) of a strong synthesis clearly implies the condition of well-posedness (3.4).

(If) Referring to (6.3) and Theorem 8.7, we must construct a complete triple (F_c, A_c, B_c) such that A_L is stable and A_c contains an internal model. To this end, set

$$F_c = [F_{c1} \quad F_{c2}], \qquad A_c = \begin{bmatrix} A_{c1} & 0 \\ 0 & A_{c2} \end{bmatrix}, \qquad B_c = \begin{bmatrix} B_{c1} \\ B_{c2} \end{bmatrix}. \tag{6.28}$$

Fix $A_{c1} \simeq T \otimes 1_q'$ as described in Theorem 8.7; i.e. A_{c1} supplies the internal model. To (F_{c2}, A_{c2}, B_{c2}) we assign the role of loop stabilization, as follows. By the well-posedness condition (3.4), and an application of Lemma 8.9 to (D_1, A_1, B_1), first choose (F_{c1}, B_{c1}) such that the triples

$$(F_{c1}, A_{c1}, B_{c1}), \tag{6.29}$$

$$[D_1 \quad 0], \qquad \begin{bmatrix} A_1 & B_1 F_{c1} \\ B_{c1} D_1 & A_{c1} \end{bmatrix}, \qquad \begin{bmatrix} B_1 \\ 0 \end{bmatrix} \tag{6.30}$$

are complete; almost any pair (F_{c1}, B_{c1}) will do. Exploiting the results of Section 3.5, we stabilize the complete triple (6.30) by means of a full-order observer with dynamics A_{c2}. That is, we pick (F_{c2}, A_{c2}, B_{c2}) such that the composite system

$$\begin{bmatrix} \dot{x}_1 \\ \dot{x}_{c1} \end{bmatrix} = \begin{bmatrix} A_1 & B_1 F_{c1} \\ B_{c1} D_1 & A_{c1} \end{bmatrix} \begin{bmatrix} x_1 \\ x_{c1} \end{bmatrix} + \begin{bmatrix} B_1 \\ 0 \end{bmatrix} F_{c2} x_2$$

$$\dot{x}_{c2} = A_{c2} x_2 + B_{c2} D_1 x_1$$

is stable; equivalently, the loop map

$$A_L = \begin{bmatrix} A_1 & B_1 F_{c1} & B_1 F_{c2} \\ B_{c1} D_1 & A_{c1} & 0 \\ B_{c2} D_1 & 0 & A_{c2} \end{bmatrix} \qquad (6.31)$$

is stable. To satisfy fully our technical requirements we may arrange (by a small perturbation of A_{c2}, if necessary) that

$$\sigma(A_{c2}) \cap \sigma(A_{c1}) = \varnothing;$$

and then (by a perturbation of (F_{c2}, B_{c2}) or a state-space reduction) that (F_{c2}, A_{c2}, B_{c2}) is complete. This and the completeness of (6.29) ensure that (F_c, A_c, B_c) is complete, and we are done. □

Several features of the strong synthesis that has been constructed are worth dwelling on. The first two embody structural principles of fundamental importance.

1. The compensator input is the regulated variable z: in other words a *bona fide* error feedback structure is utilized. It will be shown in Section 8.8 that error feedback is actually necessary if the regulation property is to be well-posed.
2. The fixed part (A_{c1}) of the compensator, where we do not and cannot permit parameter variations, will be in general a highly redundant or "reduplicated" model of the actual dynamics of the exogenous signals which the designer envisages *a priori*; the precise degree of redundancy is clear from the proof of Theorem 8.7. This redundancy is the price that is paid for insisting on regulation in the presence of a very wide variety of parameter variations both internal to the feedback loop (A_1, B_1) and also in the binding map A_3 which couples the loop directly to the exosystem. A rough estimate of the total order of dynamic compensation in a strong synthesis can be obtained as follows. Let the degree of the m.p. of A_2 be μ. The internal model is then of order μq, and the system map in (6.30), for plant together with internal model, is of order $\mu q + n_1$. The controllability index of the triple (6.30), and therefore, approximately, of the stabilizing dynamics (if the latter are designed by the relatively efficient procedure of Section 3.8: see comment 4, below) would be about $(\mu q + n_1)/m$; thus, the total order of the compensator is about

$$\mu q + (\mu q + n_1)/m.$$

   This number can get large rapidly: a fifth-order plant with two inputs and two outputs, designed to track a single sinusoid of fixed frequency and reject a ramp disturbance ($\mu = 4$), would require a compensator of order about 15.
3. The foregoing objection admitted, it should be clear nevertheless that a greater degree of structural stability has been gained than was originally

sought: namely, regulation and loop stability are preserved in a neighborhood of $(F_{c1}, F_{c2}, A_{c2}, B_{c1}, B_{c2})$ as well. In fact, with (A_{c1}, D_1) fixed, all the remaining loop parameters may be varied within the limits of loop stability, and A_3 varied arbitrarily, without affecting the property of output regulation.

4. In actual implementation there is, of course, no need to stick to the procedure of the proof of Theorem 8.5. For instance, instead of a full order observer the designer could employ a minimal order observer, or a dynamic compensator in the style of Section 3.8. At this stage, optimization techniques (cf. Chapters 12 and 13) might be used to improve the quantitative features of transient response.
5. The assumption that (D, A) is observable can be relaxed to detectability. The arguments in the proofs need only slight and obvious modification if "observability" and "controllability" are replaced by "detectability" and "stabilizability" throughout.
6. If $y = Cx$ with, more generally, Ker $C \subset$ Ker D, then $z = Dx = Qy$ for some Q which could be constructed by the designer. That is, $z(\cdot)$ could be regarded as an "available" output and the procedure would be the same as before. In particular, additional outputs might be used in an "inner loop" to modify plant dynamics in some appropriate way, before the "outer loop" design is carried out. Numerous variations on the basic theme will doubtless suggest themselves to the reader.

8.7 Example 3. Well-Posed RPIS: Strong Synthesis

Let

$$A = \left[\begin{array}{cc|cc|c} -1 & 0 & & A_3 & \\ 0 & -1 & & & \\ \hline & & 0 & 1 & 0 \\ & \mathbf{0} & 0 & 0 & 0 \\ \hline 0 & 0 & 0 & 0 & 0 \end{array}\right], \qquad B = \left[\begin{array}{cc} 1 & 0 \\ 0 & 1 \\ \hline & \\ & \mathbf{0} \\ & \end{array}\right],$$

$$D = \begin{bmatrix} -1 & 0 & 1 & 0 & 0 \\ 0 & -1 & 0 & 0 & 1 \end{bmatrix},$$

and suppose first that $A_3 = 0$. The system represents two identical, decoupled first-order lags (state variables x_1, x_2) which are required to track, respectively, a ramp (x_3, x_4) or step (x_5). The pair (D, A) is observable and we assume $C = D$.

If no account is taken of possible perturbations of A_3, "naive" implementation of a feedback controller with a minimal-order observer could take the form shown in solid lines in Fig. 8.5. Here, the compensators

$$T_1(s) = \frac{6s+1}{16s^2}, \qquad T_2(s) = \frac{1}{s}$$

guarantee internal stability and output regulation for each decoupled subsystem.

It is clear, however, that if $A_3 = 0$ is perturbed to

$$A_{3\epsilon} := \begin{bmatrix} 0 & 0 & 0 \\ \epsilon & 0 & 0 \end{bmatrix}, \qquad \epsilon \neq 0,$$

as shown by the dotted branch in Fig. 8.5, then the second loop will fail to reject the ramp disturbance $x_3(\cdot)$, giving an offset (asymptotic) error

$$z_2(\infty) = -\epsilon x_4(0+).$$

A simple cure, obvious by inspection, is to replace $T_2(s)$ by $\tilde{T}_2(s) = T_1(s)$. The system then maintains internal stability and output regulation in the face of arbitrary small perturbations in (A_1, A_3, B_1), provided the "internal models" represented by the double integrators $(1/s^2)$ remain fixed in each loop. Observe that the same is true if the lower right block in A_2 (scalar 0) is extended to $\left[\begin{smallmatrix} 0 & 1 \\ 0 & 0 \end{smallmatrix}\right]$, as the effect is only to modify the flow diagram branch

$$\circ \xrightarrow{s^{-1}} \circ\, x_5 \quad \text{to} \quad \circ \xrightarrow{s^{-1}} \underset{x_6}{\circ} \xrightarrow{s^{-1}} \circ\, x_5$$

A more elaborate example, illustrating the construction in the proof of Theorem 8.5, is outlined in Exercise 8.3.

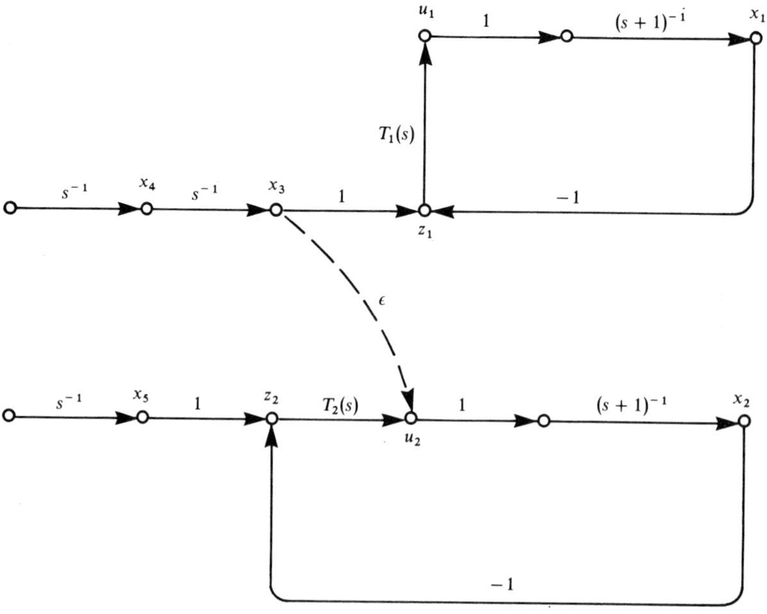

Figure 8.5 Structural Instability in a Multivariable System.

8.8 The Internal Model Principle

In Section 8.6 a strong synthesis was developed on the assumption that $y = z$, namely, the controller operated on the basis of closed-loop error feedback. In a slightly generalized version, this control structure will now be justified: in fact, shown to be itself a consequence of the requirement of structural stability of the regulator, namely, well-posedness of the properties of internal stability and output regulation.

To this end we assume no *a priori* relation between the measured output y and the regulated output z; that is, between C and D. Let the system equations take a form similar to (6.1), but now with $y = Cx$ playing the role of input to the compensator or dynamic controller. Setting $y = C_1 x_1 + C_2 x_2$, we have, in analogy to (6.2),

$$\begin{bmatrix} \dot{x}_1 \\ \dot{x}_c \\ \dot{x}_2 \end{bmatrix} = \begin{bmatrix} A_1 + B_1 K C_1 & B_1 F_c & A_3 + B_1 K C_2 \\ B_c C_1 & A_c & B_c C_2 \\ 0 & 0 & A_2 \end{bmatrix} \begin{bmatrix} x_1 \\ x_c \\ x_2 \end{bmatrix}.$$

As before, we shall set $K = 0$, and define

$$\mathscr{X}_L := \mathscr{X}_1 \oplus \mathscr{X}_c, \qquad x_L := x_1 \oplus x_c$$

$$A_L := \begin{bmatrix} A_1 & B_1 F_c \\ B_c C_1 & A_c \end{bmatrix}, \qquad B_L := \begin{bmatrix} A_3 \\ B_c C_2 \end{bmatrix}$$

$$C_L := [C_1 \quad 0], \qquad D_L := [D_1 \quad 0]$$

$$\mathscr{X}_e := \mathscr{X}_L \oplus \mathscr{X}_2, \qquad x_e := x_L \oplus x_2$$

$$A_e := \begin{bmatrix} A_L & B_L \\ 0 & A_2 \end{bmatrix}, \qquad C_e := [C_L \quad C_2], \qquad D_e := [D_L \quad D_2].$$

We assume that the triple $S_c = (F_c, A_c, B_c)$ is a synthesis in the same sense as before; namely, it is complete, A_L is stable, and output regulation holds:

$$\mathscr{X}_e^+(A_e) \subset \operatorname{Ker} D_e. \tag{8.2}$$

To simplify the development later, we impose the technical condition

$$\mathscr{X}_e^+(A_e) \not\subset \operatorname{Ker} C_e. \tag{8.3}$$

While not strictly required, (8.3) is convenient and not unreasonable. Indeed, suppose we had $\mathscr{X}_e^+(A_e) \subset \operatorname{Ker} C_e$ at every data point. Now we shall prove, without relying on (8.3), that $\operatorname{Ker} C_e \subset \operatorname{Ker} D_e$. It would follow that the regulation condition would hold and be well-posed with respect to the observed output $y = Cx$. But then the roles of y and z would scarcely be worth distinguishing. In that case, the remainder of the discussion could be based on the simpler hypothesis that $C = D$.

In defining a strong synthesis, we consider a somewhat different data point than before, namely,

$$\mathbf{p} = (A_3, B_c).$$

Again, A_3 will be allowed to vary freely through a neighborhood in $\mathbb{R}^{n_1 \times n_2}$; but variations of B_c will be confined to a subspace $\hat{\mathscr{B}}_c \subset \mathbb{R}^{n_c \times p}$ that will be specified differently (though consistently) at various stages of the discussion. In each such context we say that our synthesis is *strong at* $\mathbf{p}$ if the property (8.2) holds for all data points $\tilde{\mathbf{p}} = (\tilde{A}_3, \tilde{B}_c)$, where

$$\tilde{\mathbf{p}} \in [\mathbb{R}^{n_1 \times n_2} \times (B_c + \hat{\mathscr{B}}_c)] \cap \mathfrak{N}.$$

Here, $\mathfrak{N}$ is some neighborhood of $\mathbf{p}$ in $\mathbb{R}^{n_1 \times n_2} \times \mathbb{R}^{n_c \times p}$, and

$$B_c + \hat{\mathscr{B}}_c := \{\tilde{B}_c : \tilde{B}_c - B_c \in \hat{\mathscr{B}}_c\}.$$

Finally, we shall impose a mild detectability condition, as specified in (8.25) below.

We shall look for certain necessary conditions that our synthesis be strong. To this end it will be enough to examine the first-order variational equations derived from (8.2). Setting

$$\mathscr{X}_e^+(A_e) = \operatorname{Im} V_e = \operatorname{Im} \begin{bmatrix} V_1 \\ V_c \\ 1_2 \end{bmatrix} \tag{8.4}$$

and substituting in

$$A_e V_e = V_e A_2, \qquad D_e V_e = 0, \tag{8.5}$$

we get

$$A_1 V_1 - V_1 A_2 + B_1 F_c V_c = -A_3 \tag{8.6a}$$

$$B_c C_1 V_1 + A_c V_c - V_c A_2 = -B_c C_2 \tag{8.6b}$$

$$D_1 V_1 + D_2 = 0. \tag{8.6c}$$

Now (8.6) is to be solvable not only at $\mathbf{p}$ but also at all $\tilde{\mathbf{p}} = \mathbf{p} + t\hat{\mathbf{p}}$, where

$$\hat{\mathbf{p}} = (\hat{A}_3, \hat{B}_c) \in \mathbb{R}^{n_1 \times n_2} \times \hat{\mathscr{B}}_c,$$

and ($\hat{\mathbf{p}}$ being fixed) $0 \le t < \delta$ with $\delta > 0$ and small. Substitution in (8.6) and differentiation with respect to t at $t = 0$ yield for the induced variations $(\hat{V}_1, \hat{V}_c)$ the equations

$$A_1 \hat{V}_1 - \hat{V}_1 A_2 + B_1 F_c \hat{V}_c = -\hat{A}_3 \tag{8.7a}$$

$$B_c C_1 \hat{V}_1 + A_c \hat{V}_c - \hat{V}_c A_2 = -\hat{B}_c(C_1 V_1 + C_2) \tag{8.7b}$$

$$D_1 \hat{V}_1 = 0. \tag{8.7c}$$

Since (8.7) must have a solution $(\hat{V}_1, \hat{V}_c)$ for each admissible pair $(\hat{A}_3, \hat{B}_c)$, the variational equations take their final form:

$$\begin{bmatrix} A_1 \otimes 1_2' - 1_1 \otimes A_2' \\ B_c C_1 \otimes 1_2' \end{bmatrix} \operatorname{Ker}(D_1 \otimes 1_2') + \operatorname{Im} \begin{bmatrix} B_1 F_c \otimes 1_2' \\ A_c \otimes 1_2' - 1_c \otimes A_2' \end{bmatrix}$$
$$\supset \mathscr{X}_1 \otimes \mathscr{X}_2' \oplus \hat{\mathscr{B}}_c(C_1 V_1 + C_2). \tag{8.8}$$

We regard $\hat{B}_c: \mathscr{Y} \to \mathscr{X}_c$ in (8.7b) as an element of $\mathscr{X}_c \otimes \mathscr{Y}'$ and $C_1 V_1 + C_2: \mathscr{X}_2 \to \mathscr{Y}$ as an element of $\mathscr{Y} \otimes \mathscr{X}'_2$, so that in (8.8)

$$\hat{\mathscr{B}}_c(C_1 V_1 + C_2) := \{\hat{B}_c(C_1 V_1 + C_2): \hat{B}_c \in \hat{\mathscr{B}}_c\}$$

is a subspace of $\mathscr{X}_c \otimes \mathscr{X}'_2$. Notice that (8.8) reduces to (6.6) when $C = D$.

Our initial objective is to show that (8.8) implies

$$\operatorname{Ker} C \subset \operatorname{Ker} D. \tag{8.9}$$

Systemically, (8.9) says that the regulated vector $z = Dx$ can be computed directly from the measured vector $y = Cx$; namely, $D = QC$ for some (unique) Q that could be constructed by the designer: on our assumptions, this could be done without error, and thus $z(\cdot)$ could be made available to drive the controller.

We show first that

$$\operatorname{Ker} C_1 \subset \operatorname{Ker} D_1. \tag{8.10}$$

For this, write

$$\begin{aligned} S &:= \begin{bmatrix} A_1 \otimes 1'_2 - 1_1 \otimes A'_2 & B_1 F_c \otimes 1'_2 \\ B_c C_1 \otimes 1'_2 & A_c \otimes 1'_2 - 1_c \otimes A'_2 \end{bmatrix} \\ &= A_L \otimes 1'_2 - 1_L \otimes A'_2. \end{aligned} \tag{8.11}$$

Because $\sigma(A_L) \cap \sigma(A_2) = \varnothing$, the Sylvester map S is an automorphism (i.e. is nonsingular) on its domain $\mathscr{X}_1 \otimes \mathscr{X}'_2 \oplus \mathscr{X}_c \otimes \mathscr{X}'_2$. Let

$$P: \mathscr{X}_1 \otimes \mathscr{X}'_2 \oplus \mathscr{X}_c \otimes \mathscr{X}'_2 \to \mathscr{X}_c \otimes \mathscr{X}'_2 \tag{8.12}$$

be the natural projection. By (8.8), (8.11) and (8.12), we find in turn

$$\begin{aligned} S \operatorname{Ker}[D_1 \otimes 1'_2 \quad 0] &\supset \operatorname{Ker} P, \\ \operatorname{Ker}[D_1 \otimes 1'_2 \quad 0] &\supset S^{-1} \operatorname{Ker} P \\ &= \operatorname{Ker}(PS) \\ &= \operatorname{Ker}[B_c C_1 \otimes 1'_2 \quad A_c \otimes 1'_2 - 1_c \otimes A'_2], \\ \operatorname{Ker}(D_1 \otimes 1'_2) &\supset \operatorname{Ker}(B_c C_1 \otimes 1'_2) \\ &= \operatorname{Ker}[(B_c \otimes 1'_2)(C_1 \otimes 1'_2)] \\ &\supset \operatorname{Ker}(C_1 \otimes 1'_2), \end{aligned}$$

from which (8.10) results, as claimed. In this discussion $\hat{\mathscr{B}}_c$ played no role and might have been set to 0.

To complete the proof that $\operatorname{Ker} C \subset \operatorname{Ker} D$, we need a rather special result on the Sylvester map.

Lemma 8.10 (Variational lemma). *Let $M: \mathscr{M} \to \mathscr{M}$, $N: \mathscr{N} \to \mathscr{N}$, $\sigma(M) \cap \sigma(N) = \varnothing$. Let $\mathscr{M} = \mathscr{M}_1 \oplus \mathscr{M}_2$, $\mathscr{N}'_0 \subset \mathscr{N}'$ and*

$$\mathscr{R} := (M \otimes 1' - 1 \otimes N')^{-1}(\mathscr{M}_1 \otimes \mathscr{N}' \oplus \mathscr{M}_2 \otimes \mathscr{N}'_0).$$

[*Note that* $\mathscr{R} \subset \mathscr{M} \otimes \mathscr{N}' \simeq \mathbf{L}(\mathscr{N}, \mathscr{M})$]. *If* $\mathscr{M}_2 \otimes \mathscr{N}'_0 \neq 0$, *then*

$$\bigcup_{R \in \mathscr{R}} \operatorname{Im} R = \mathscr{M}.$$

Note that the result involves set-theoretic union, not vector sum, of the subspaces $\operatorname{Im} R \subset \mathscr{M}$.

The lemma says that if $\mathscr{M}_2 \otimes \mathscr{N}'_0 \neq 0$ and if $m \in \mathscr{M}$ is arbitrary, then there exist $T_1: \mathscr{N} \to \mathscr{M}_1$, and $T_2: \mathscr{N} \to \mathscr{M}_2$ with $\operatorname{Im} T'_2 \subset \mathscr{N}'_0$, having the following property: If R is determined (uniquely) by

$$MR - RN = \begin{bmatrix} T_1 \\ T_2 \end{bmatrix}, \tag{8.13}$$

then $m \in \operatorname{Im} R$.

Proof. Let $N'_0: \mathscr{N}'_0 \to \mathscr{N}'$ be the insertion, so in (8.13) $\operatorname{Ker} T_2 \supset \operatorname{Ker} N_0$ ($N_0: \mathscr{N} \to \mathscr{N}_0$ is the dual of N'_0), i.e.

$$T_2 = S_2 N_0 \tag{8.14}$$

for some $S_2: \mathscr{N}_0 \to \mathscr{M}_2$. By assumption $N'_0 \neq 0$, hence $N_0 \neq 0$. Consider any prime Jordan subspace $\mathscr{E}$ of $\mathscr{N}$ which contains a vector $n \in \mathscr{N}$ with $N_0 n \neq 0$. Assume that the corresponding eigenvalue ν of N is real. Let $\{e_1, e_2, \ldots\}$ be a canonical basis for $\mathscr{E}$, namely

$$(N - \nu 1)e_1 = 0, \qquad (N - \nu 1)e_{i+1} = e_i. \tag{8.15}$$

Writing

$$R \,|\, \mathscr{E} = [r_1 r_2 \cdots], \qquad \hat{T}_1 = T_1 \,|\, \mathscr{E}, \qquad \hat{N}_0 = N_0 \,|\, \mathscr{E}, \tag{8.16}$$

we have by (8.13)–(8.16) that

$$M[r_1 r_2 \cdots] - [r_1 r_2 \cdots] \begin{bmatrix} \nu & 1 & 0 & \cdot & \cdot & \cdot \\ 0 & \nu & 1 & \cdot & \cdot & \cdot \\ \cdot & \cdot & \cdot & \cdot & \cdot & \cdot \end{bmatrix} = \begin{bmatrix} \hat{T}_1 \\ S_2 \hat{N}_0 \end{bmatrix}.$$

Let e_k be the first $e.$ for which $\hat{N}_0 e_k \neq 0$. Suppose $m \in \mathscr{M}$ is given. Then we need only define $\hat{T}_1$ and S_2 to satisfy

$$\hat{T}_1 e_j = 0 \text{ for } j < k; \quad \begin{bmatrix} \hat{T}_1 \\ S_2 \hat{N}_0 \end{bmatrix} e_k = (M - \nu 1)m.$$

Since $\nu \notin \sigma(M)$ this gives $r_j = 0$ $(j < k)$, $r_k = m$, so that $m \in \operatorname{Im} R$, as required. If ν is complex, a similar argument based on the real Jordan form (0.11.10) establishes the result in that case too. □

Since $\operatorname{Ker} C_1 \subset \operatorname{Ker} D_1$, we may coordinatize

$$\mathscr{Y} \simeq \mathscr{Y}_0 \oplus \mathscr{Z} \simeq \mathbb{R}^{p-q} \oplus \mathbb{R}^q$$

in such a way that C takes the form

$$C = [C_1 \quad C_2] = \begin{bmatrix} C_{01} & C_{02} \\ D_1 & \tilde{D}_2 \end{bmatrix}. \tag{8.17}$$

To prove that $\operatorname{Ker} C \subset \operatorname{Ker} D$ it is therefore enough to show that $\tilde{D}_2 = D_2$ or, since $D_1 V_1 + D_2 = 0$, that

$$D_1 V_1 + \tilde{D}_2 = 0. \tag{8.18}$$

For this, write

$$B_c = [B_{c0} \quad B_{cz}] \in \mathbb{R}^{n_c \times (p-q)} \oplus \mathbb{R}^{n_c \times q}$$

conformably with (8.17). We now regard B_{cz} as subject to free variation, i.e. $\hat{\mathscr{B}}_c \simeq \mathbb{R}^{n_c \times q}$ is the second direct summand above, so in (8.8)

$$\hat{\mathscr{B}}_c(C_1 V_1 + C_2) = \mathscr{X}_c \otimes \operatorname{Im}(D_1 V_1 + \tilde{D}_2)'.$$

Now apply the variational Lemma 8.10, with the identifications

$$M = A_L, \qquad N = A_2, \qquad M \otimes 1' - 1 \otimes N' = S,$$

$$\mathscr{M}_1 = \mathscr{X}_1, \qquad \mathscr{M}_2 = \mathscr{X}_c, \qquad \mathscr{N}_0' = \operatorname{Im}(D_1 V_1 + \tilde{D}_2)'.$$

According to the lemma, if $D_1 V_1 + \tilde{D}_2 \neq 0$ then for every vector $x_L = x_1 \oplus x_c$ there is an element

$$\hat{W} \in \mathscr{X}_1 \otimes \mathscr{X}_2' \oplus \mathscr{X}_c \otimes \operatorname{Im}(D_1 V_1 + \tilde{D}_2)'$$

such that $S\hat{V} = \hat{W}$ implies $x_L \in \operatorname{Im} \hat{V}$. But by (8.8) any $\hat{V} = S^{-1}\hat{W}$ must satisfy the condition

$$[D_1 \otimes 1_2' \quad 0]\hat{V} = 0, \qquad \text{i.e. } D_1 x_1 = 0.$$

As x_1 was arbitrary, we conclude that $D_1 V_1 + \tilde{D}_2 = 0$ after all, and the proof of (8.18) is complete.

In the foregoing argument the role of Lemma 8.10 (under the hypothesis $D_1 V_1 + \tilde{D}_2 \neq 0$) was just to supply an admissible variation of (A_3, B_c) that caused output regulation to fail. This contradiction established the result.

We can now write

$$B_c = [B_{c0} \quad B_{cz}], \qquad C = \begin{bmatrix} C_0 \\ D \end{bmatrix}. \tag{8.19}$$

The compensator equation takes the form

$$\dot{x}_c = A_c x_c + B_{c0} y_0 + B_{cz} z,$$

where $y_0 = C_0 x$. The task remaining is to show that the compensator incorporates an internal model of A_2. For this it will be enough to establish the transversality condition (6.6) that was studied in Section 8.6. To this end, we first split the compensator dynamics in a special way, by considering just those inputs

$$y = \begin{bmatrix} y_0 \\ 0 \end{bmatrix}$$

that excite the compensator when the regulated vector $z(\cdot)$ is identically 0, namely under the condition of perfect regulation. Such inputs control the subspace

$$\mathscr{X}_{c1} := \langle A_c | \mathscr{B}_{c0} \rangle.$$

Selecting $\mathscr{X}_{c2}$ arbitrarily, such that

$$\mathscr{X}_c = \mathscr{X}_{c1} \oplus \mathscr{X}_{c2} \tag{8.20}$$

we may coordinatize (A_c, B_c) as

$$A_c = \begin{bmatrix} A_{c1} & A_{c3} \\ 0 & A_{c2} \end{bmatrix}, \qquad B_c = \begin{bmatrix} B_{c01} & B_{cz1} \\ 0 & B_{cz2} \end{bmatrix}. \tag{8.21}$$

Of course, the structure of B_c displayed in (8.21) is a refinement of that in (8.19). From (8.21) we obtain the dynamic equation

$$\dot{x}_{c2} = A_{c2} x_{c2} + B_{cz2} z$$

for that subsystem (strictly, factor system) of the compensator that is driven solely by the "error" z.

It will be shown that in fact $\mathscr{X}_{c2} \neq 0$, namely, the error-driven subcompensator is nontrivial. For this, we now take B_{c01} and B_{cz1} to be freely variable and impose, as before, the requirement of structural stability in the form of the variational condition (8.8). Suppose $\mathscr{X}_{c2} = 0$. Then, $\hat{\mathscr{B}}_c = \mathscr{X}_c \otimes \mathscr{Y}'$ and

$$\hat{\mathscr{B}}_c(C_1 V_1 + C_2) = \mathscr{X}_c \otimes \operatorname{Im}(C_1 V_1 + C_2)'.$$

On applying the variational Lemma 8.10 just as was done in the proof of (8.18), we deduce that $C_1 V_1 + C_2 = 0$, i.e. $C_e V_e = 0$. But this contradicts our assumption (8.3).

As the final step of our analysis, we shall reduce the remaining aspect of structural stability (i.e. well-posedness of regulation with respect to A_3) to the transversality condition already considered in Section 8.6. Setting $F_c = [F_{c1} \quad F_{c2}]$ in agreement with (8.20) we may write (8.1) in more detail as

$$A_L = \begin{bmatrix} A_1 & B_1 F_{c1} & B_1 F_{c2} \\ B_{c01} C_{01} + B_{cz1} D_1 & A_{c1} & A_{c3} \\ B_{cz2} D_1 & 0 & A_{c2} \end{bmatrix}$$

$$B_L = \begin{bmatrix} A_3 \\ B_{c01} C_{02} + B_{cz1} D_2 \\ B_{cz2} D_2 \end{bmatrix}.$$

Substitution of

$$V_c = \begin{bmatrix} V_{c1} \\ V_{c2} \end{bmatrix}$$

in (8.4) and (8.5) yields

$$\begin{aligned} A_1 V_1 - V_1 A_2 + B_1 F_{c1} V_{c1} + B_1 F_{c2} V_{c2} &= -A_3 \\ B_{c01} C_{01} V_1 + A_{c1} V_{c1} - V_{c1} A_2 + A_{c3} V_{c2} &= -B_{c01} C_{02} \\ A_{c2} V_{c2} - V_{c2} A_2 &= 0 \\ D_1 V_1 + D_2 &= 0. \end{aligned}$$

Considering variations in A_3 alone, we obtain the first-order variational equations

$$A_1 \hat{V}_1 - \hat{V}_1 A_2 + B_1 F_{c1} \hat{V}_{c1} + B_1 F_{c2} \hat{V}_{c2} = -\hat{A}_3 \tag{8.22a}$$

$$B_{c01} C_{01} \hat{V}_1 + A_{c1} \hat{V}_{c1} - \hat{V}_{c1} A_2 + A_{c3} \hat{V}_{c2} = 0 \tag{8.22b}$$

$$A_{c2} \hat{V}_{c2} - \hat{V}_{c2} A_2 = 0 \tag{8.22c}$$

$$D_1 \hat{V}_1 = 0. \tag{8.22d}$$

Here, $\hat{A}_3 \in \mathbb{R}^{n_1 \times n_2} \simeq \mathscr{X}_1 \otimes \mathscr{X}'_2$ is arbitrary.

It will be shown finally that (8.22) implies the desired transversality condition [cf. (6.6)], namely

$$\begin{aligned} &(A_1 \otimes 1'_2 - 1_1 \otimes A'_2)\,\mathrm{Ker}(D_1 \otimes 1'_2) \\ &\quad + (B_1 F_{c2} \otimes 1'_2)\,\mathrm{Ker}(A_{c2} \otimes 1'_2 - 1_{c2} \otimes A'_2) = \mathscr{X}_1 \otimes \mathscr{X}'_2. \end{aligned} \tag{8.23}$$

Now (8.23) is not obvious by inspection, because of the term $B_1 F_{c1} \hat{V}_{c1}$ in (8.22a). What we must show is that

$$\{\hat{A}_3 + B_1 F_{c1} \hat{V}_{c1} : \hat{A}_3 \in \mathscr{X}_1 \otimes \mathscr{X}'_2\} = \mathscr{X}_1 \otimes \mathscr{X}'_2.$$

For this, it is clearly enough to verify that if $\hat{A}_3 + B_1 F_{c1} \hat{V}_{c1} = 0$ then $\hat{A}_3 = 0$, because then the linear map

$$\hat{A}_3 \mapsto \hat{A}_3 + B_1 F_{c1} \hat{V}_{c1}$$

determined by (8.22) is nonsingular. To this end, we consider the equations

$$\begin{bmatrix} A_1 & 0 & B_1 F_{c2} \\ B_{c01} C_{01} & A_{c1} & A_{c3} \\ 0 & 0 & A_{c2} \end{bmatrix} \begin{bmatrix} \hat{V}_1 \\ \hat{V}_{c1} \\ \hat{V}_{c2} \end{bmatrix} = \begin{bmatrix} \hat{V}_1 \\ \hat{V}_{c1} \\ \hat{V}_{c2} \end{bmatrix} A_2 \tag{8.24a}$$

$$D_1 \hat{V}_1 = 0. \tag{8.24b}$$

Now (8.24) will have only the trivial solution provided the pair (D_L, A_L^0) (say) appearing on the left side of (8.24) is detectable, at least with respect to $\sigma(A_2)$: that is,

$$\mathrm{Ker}(A_L^0 - \lambda 1) \cap \mathrm{Ker}\, D_L = 0, \qquad \lambda \in \sigma(A_2). \tag{8.25}$$

But (8.25) is easily seen to be equivalent to the following four conditions:

$$\sigma(A_{c1}) \cap \sigma(A_2) = \varnothing, \tag{8.26a}$$

$$\operatorname{Ker}(A_1 - \lambda 1) \cap \operatorname{Ker} D_1 = 0, \qquad \lambda \in \sigma(A_2), \tag{8.26b}$$

$$(A_1 - \lambda 1) \operatorname{Ker} D_1 \cap B_1 F_{c2} \operatorname{Ker}(A_{c2} - \lambda 1) = 0, \qquad \lambda \in \sigma(A_2), \tag{8.26c}$$

$$\operatorname{Ker}(B_1 F_{c2}) \cap \operatorname{Ker}(A_{c2} - \lambda 1) = 0, \qquad \lambda \in \sigma(A_2), \tag{8.26d}$$

We shall assume that (8.26a) is true; otherwise, a small *ad hoc* variation in A_{c1} will make it so. Of course, (8.26b) is true if (D_1, A_1) is observable, namely if (D, A) is observable. Next, controllability of (A_c, B_c) implies that of (A_{c2}, B_{cz2}), hence the cyclic index

$$\begin{aligned}\gamma(A_{c2}) &= \max_{\lambda} d[\operatorname{Ker}(A_{c2} - \lambda 1)] \\ &\le d(\mathscr{Y}) = q;\end{aligned}$$

therefore, (8.26c) is true for almost all $F_{c2} \in \mathbb{R}^{m \times n_{c2}}$. Finally, since $m \ge q$ and B_1 is monic, the pair $(B_1 F_{c2}, A_{c2})$ is observable for almost all F_{c2}, hence for such F_{c2} (8.26d) is true.

Subject, then, to the one additional (and harmless) technical assumption (8.25), we infer from (8.22) that (8.23) is true. It remains to show that A_{c2} must contain an internal model of A_2. On the basis of (8.23), it suffices to apply Corollary 8.3 and Theorem 8.7, to the triple $(F_{c2}, A_{c2}, B_{cz2})$ corresponding to the error-driven subcompensator. We have already remarked that (A_{c2}, B_{cz2}) is controllable; as for the pair (F_{c2}, A_{c2}), examination of the proof of Corollary 8.3 (from Lemma 8.8) reveals that condition (8.26d) can serve in place of observability. The desired result now follows just as before.

The major conclusion of this section will be summarized in the following general statement, which we entitle *The Internal Model Principle: A regulator is structurally stable only if the controller utilizes feedback of the regulated variable, and incorporates in the feedback loop a suitably reduplicated model of the dynamic structure of the exogenous signals which the regulator is required to process.*

In plain terms, every good regulator must incorporate a model of the outside world.

8.9 Exercises

8.1. For Example 2, Section 8.5, find the complete class of perturbations in (A_1, A_3, B_1) for which solvability of RPIS fails.

8.2. For the ill-posed but solvable RPIS of Example 2, Section 8.5, design a standard feedback controller (e.g. just as illustrated in Section 7.7) which implements a solution of RPIS. Now introduce the perturbation specified by (5.1). What goes wrong?

8.3. Illustrate the construction in the proof of Theorem 8.5 with a numerical example. Hint: The following is transparent from its signal flow graph (Fig. 8.6). Let

$$A_1 = \begin{bmatrix} 1 & 1 & 0 & 0 \\ 0 & 1 & 1 & 0 \\ 0 & 0 & 1 & 0 \\ 0 & 0 & 0 & 1 \end{bmatrix}, \qquad B_1 = \begin{bmatrix} 0 & 0 \\ 0 & 1 \\ 1 & 0 \\ 0 & 1 \end{bmatrix}.$$

To keep the dynamic order of the example small, we shall assume that the outputs x_1 and x_4 are to track reference step inputs x_5 and x_6, respectively; and that the measured variables are the tracking errors:

$$A_2 = \begin{bmatrix} 0 & 0 \\ 0 & 0 \end{bmatrix}, \qquad C = D = \begin{bmatrix} -1 & 0 & 0 & 0 & 1 & 0 \\ 0 & 0 & 0 & -1 & 0 & 1 \end{bmatrix}.$$

The synthesis procedure is as follows.

Step 1. Establish the internal model structure: Since there are two scalar error variables, and the m.p. of the exosystem (A_2) is λ, the internal model dynamics (A_{c1}) is just a two-fold copy of the map ($0^{1 \times 1}$) having m.p. λ: i.e.

$$A_{c1} = \begin{bmatrix} 0 & 0 \\ 0 & 0 \end{bmatrix}.$$

Step 2. Select F_{c1}, B_{c1}, such that the triples (F_{c1}, A_{c1}, B_{c1}) and

$$[D_1 \quad 0], \qquad \begin{bmatrix} A_1 & B_1 F_{c1} \\ B_{c1} D_1 & A_{c1} \end{bmatrix}, \qquad \begin{bmatrix} B_1 \\ 0 \end{bmatrix} \tag{9.1}$$

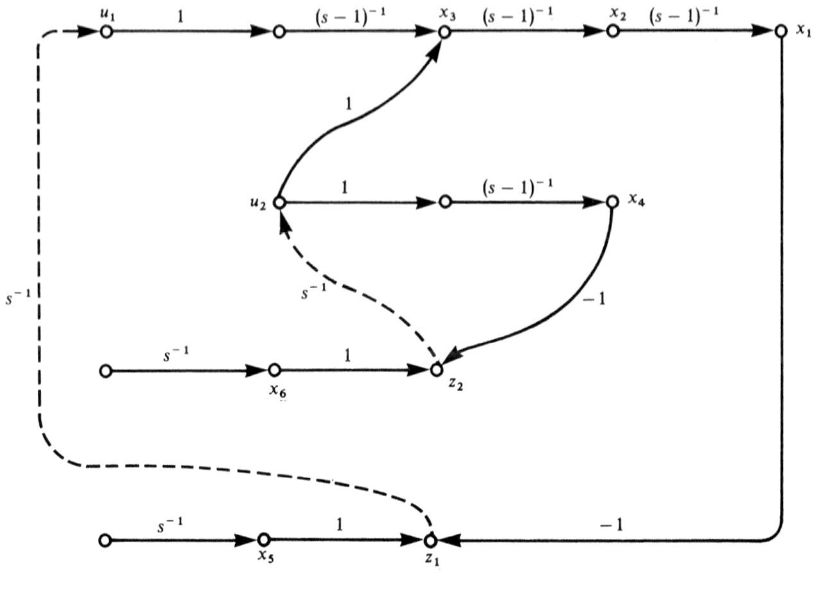

Figure 8.6 Signal Flow for Regulator Synthesis. Internal model shown dotted; stabilizing dynamics omitted.

are complete: the simplest choice is

$$F_{c1} = B_{c1} = \begin{bmatrix} 1 & 0 \\ 0 & 1 \end{bmatrix},$$

corresponding to the connections shown (dotted) in Fig. 8.6.

Step 3. Select (F_{c2}, A_{c2}, B_{c2}), such that the loop map (6.31) is stable or, more generally, has its eigenvalues in desired locations. A dynamic observer for this purpose would have minimal order 4. Alternatively, the controllability index of the triple (9.1) is 3, so that by Theorem 3.5 a dynamic compensator of order 2 would suffice, and could be designed by the procedure of Exercise 3.12. The reader is invited to complete the details.

8.4. Investigate how the compensation technique of Section 3.8 might be exploited in the construction of a strong synthesis.

8.5. By use of the results of Section 0.10, verify the equivalence property used in the proof of Theorem 8.7.

8.6. Complete the proof of Lemma 8.10 for the case where v is complex. Hint: Corresponding to the 2×2 real Jordan blocks of $N \mid \mathscr{E}$, write

$$R \mid \mathscr{E} = [r_{11} \quad r_{12} \mid r_{21} \quad r_{22} \mid \cdots];$$

let k be the first index such that $\hat{N}_0[r_{k1} \quad r_{k2}] \neq 0$; and obtain an expression for $c_1 r_{k1} + c_2 r_{k2}$, c_1 and c_2 being assignable real scalars.

8.7. Verify that the conditions (8.25) and (8.26) are equivalent.

8.8. *Direct control feedthrough.* The case where $z = Dx + Eu$ can easily be treated by the technique of Exercise 4.6. Retain the assumptions of Section 8.1 (including $D\langle A \mid \mathscr{B}\rangle = \mathscr{Z}$) and prove the following.

Theorem 8.1′. *RPIS is solvable if and only if there exists $\mathscr{V} \in \mathfrak{I}(A, B; D, E)$ such that $\mathscr{V} \cap \langle A \mid \mathscr{B}\rangle = 0$ and $\langle A \mid \mathscr{B}\rangle + \mathscr{V} = \mathscr{X}$.*

Theorem 8.2′. *RPIS is well-posed if and only if*

$$[A_1 \otimes 1_2' - 1_1 \otimes A_2' \quad B_1 \otimes 1_2'] \operatorname{Ker}[D_1 \otimes 1_2' \quad E \otimes 1_2'] = \mathscr{X}_1 \otimes \mathscr{X}_2'.$$

Corollary 8.2 remains unchanged.

Theorem 8.3(i)′. *RPIS is well-posed if and only if*

$$[A_1 - \lambda 1_1 \quad B_1] \operatorname{Ker}[D_1 \quad E] = \mathscr{X}_1, \qquad \lambda \in \sigma(A_2). \tag{9.2}$$

Check that (9.2) is equivalent to the matrix condition

$$\operatorname{Rank}\begin{bmatrix} A_1 - \lambda 1_1 & B_1 \\ D_1 & E \end{bmatrix} = n_1 + q, \qquad \lambda \in \sigma(A_2).$$

Obtain the counterpart Theorem 8.3(ii)′ using the results of Exercise 7.14. The definition of a synthesis in Section 8.6 is modified in the obvious way, using the relations

$$u = F_c x_c, \qquad z = Dx + EF_c x_c. \tag{9.3}$$

The main result, Theorem 8.5, remains unchanged; in the course of proving it one needs

Theorem 8.6′. *A synthesis* (F_c, A_c, B_c) *is strong at* $\mathbf{p} = (A_1, A_3, B_1)$ *if and only if*

$$[A_1 \otimes 1_2' - 1_1 \otimes A_2' \quad B_1 F_c \otimes 1_2'] \times \operatorname{Ker} \begin{bmatrix} D_1 \otimes 1_2' & EF_c \otimes 1_2' \\ 0 & A_c \otimes 1_2' - 1_c \otimes A_2' \end{bmatrix} = \mathscr{X}_1 \otimes \mathscr{X}_2'.$$

Corollary 8.3 and Theorem 8.7 remain the same, while the simple change required in Lemma 8.9 will be identified at once from (9.3).

8.10 Notes and References

The exposition in this chapter is based on Francis [1] and Francis and Wonham [2]. Alternative synthesis procedures leading to effectively the same results are described by Davison [1], Sebakhy and Wonham [1] and Staats and Pearson [1]. In essence, these syntheses represent a straightforward extension to multivariable systems of the classical technique of integral feedback control: the "integrator" being replaced by a single-input, single-output model of the exosystem that is inserted in each of the scalar error feedback loops. In classical linear control, offset error and sensitivity to parameter variations were reduced by means of dynamic feedback and high loop gain. The major issue was the tradeoff of gain versus stability and noise immunity: for a thorough exposition see Horowitz [1]. In this approach, the presence of an "internal model" as the central feature of control logic was generally obscured by two factors: the "exosystem" was seldom explicitly represented as such, and the limitations of analog technology usually rendered the implementation of an "unstable" dynamic element like a pure integrator infeasible. By the use of high gain, the action of such an internal model could be approximated by a stable network like a lag. In contrast, the "internal model" structure was explicitly developed by designers of nonlinear predictive controllers: for instance, O. J. M. Smith [1] proposed a design "philosophy ... that feedback systems contain both linear and nonlinear predictors of several types, and that the construction of these will include models of the mechanisms of generation of the various signals being predicted." In the context of predictive man-machine controls a detailed exposition of the "internal model" approach has been presented by Kelley [1].

The reduplication of the internal model in each error channel may be thought of as analogous to redundancy in reliability theory; on this topic von Neumann's paper [1] is classic. Intuitively, if a regulator is to operate reliably in response to external stimuli, then its internal model of the outside world must be rich enough for it to distinguish these stimuli from the otherwise confusing effects of minor internal disruptions.

Some progress has been made in extending the Internal Model Principle to nonlinear systems: the reader may consult Francis and Wonham [3], Wonham [9], [10] and (for an interesting connection with fuzzy system theory) Negoita and Keleman [1]. In the latter context a suggestive discussion of the "robust" control of uncertain ("fuzzy") systems was given by Chang and Zadeh [1].

Structural stability is a well-established concept in the qualitative theory of differential equations: an early seminal reference is the paper of Andronov and Pontryagin

[1]; for recent accounts see Hirsch and Smale [1] and Thom [1]. A differential equation (i.e. vector field) is said to be structurally stable if the qualitative (topological) behavior of its trajectories is preserved under small perturbations of the vector field. Also current in the control literature is the term "robustness" (cf. Chang and Zadeh [1], Davison [1], Staats and Pearson [1]), borrowed from statistics, where it signifies the insensitivity of the performance of an estimate to the underlying population distribution (Box and Andersen [1]).

Noninteracting Control I: Basic Principles

9

Consider a multivariable system whose scalar outputs z_{ij} have been grouped in disjoint subsets, each having a physical significance to distinguish it from the remaining subsets. Represent the output subsets by vectors

$$z_i = \text{col}(z_{i1}, \ldots, z_{ip_i}), \qquad i \in \mathbf{k}.$$

For instance, with $k = 3$ and each $p_i = 2$, z_i could represent angular position and velocity of a rigid body relative to the ith axis of rotation. Next, suppose the system is controlled by scalar inputs $u_1, \ldots, u_m$, where $m \geq k$. In many applications it is desirable to partition the input set into k disjoint subsets $U_1, \ldots, U_k$, such that for each $i \in \mathbf{k}$ the inputs of U_i control the output vector z_i completely, without affecting the behavior of the remaining $z_j, j \neq i$. Such a control action is *noninteracting*, and the system is *decoupled*. From an input-output viewpoint decoupling splits the system into k independent subsystems. Considerable advantages may result of simplicity and reliability, especially if control is partially to be executed by a human operator.

In general, decoupling in the manner described is impossible: sometimes, however, noninteraction is achievable by introducing state feedback, possibly with auxiliary integrating elements, and by regrouping the input control variables in suitable functional combinations. The objective is to cancel or compensate for inherent cross-couplings, and also to achieve satisfactory dynamic response.

For our linear multivariable system there exists an extensive theory of noninteraction based on the structural concepts of Chapters 4 and 5. In this chapter we introduce the main ideas, in Chapter 10 develop more fully the technique of dynamic compensation, and in Chapter 11 discuss generic solvability. As usual, we shall initially formulate the systems problem in terms of state equations, then extract the underlying algebraic structure.

9.1 Decoupling: Systems Formulation

Let

$$\dot{x}(t) = Ax(t) + Bu(t) \tag{1.1a}$$

$$z_i(t) = D_i x(t), \qquad i \in \mathbf{k},\ k \geq 2. \tag{1.1b}$$

Here, as usual, $A: \mathscr{X} \to \mathscr{X}$, $B: \mathscr{U} \to \mathscr{X}$ and $D_i: \mathscr{X} \to \mathscr{Z}_i$. For the space of output vectors we adopt the external direct sum

$$\mathscr{Z} = \mathscr{Z}_1 \oplus \cdots \oplus \mathscr{Z}_k.$$

We assume that state feedback is allowed with an arbitrary feedback map F, and that arbitrary "gain" maps G_i can be introduced at the input. Thus, the admissible controls are of the form

$$u(t) = Fx(t) + \sum_{i=1}^{k} G_i v_i(t),$$

where $F: \mathscr{X} \to \mathscr{U}$, $G_i: \mathscr{U} \to \mathscr{U}$ and $v_i(t) \in \mathscr{U}$. The $v_i(\cdot)$ are the new external inputs. The resulting signal flow graph is shown in Fig. 9.1.

The c.s. generated by the control $v_i(\cdot)$ is

$$\mathscr{R}_i = \langle A + BF \mid \operatorname{Im}(BG_i)\rangle. \tag{1.2}$$

Our first objective is to arrange that $v_i(\cdot)$ does not affect the outputs $z_j(\cdot)$ for $j \neq i$. Thus, we must have

$$D_j \mathscr{R}_i = 0, \qquad j \neq i,\ i \in \mathbf{k},\ j \in \mathbf{k}. \tag{1.3}$$

Secondly, if $v_i(\cdot)$ is to control the output z_i completely, we must be able to reach each vector in the image of D_i by suitable choice of $v_i(\cdot)$, which means that

$$D_i \mathscr{R}_i = \operatorname{Im} D_i, \qquad i \in \mathbf{k}. \tag{1.4}$$

Thus, our problem can be stated as follows. *Given A, B and D_i ($i \in \mathbf{k}$), find (if possible) F and G_i ($i \in \mathbf{k}$), such that* (1.3) *and* (1.4) *are true for the $\mathscr{R}_i$ defined by* (1.2).

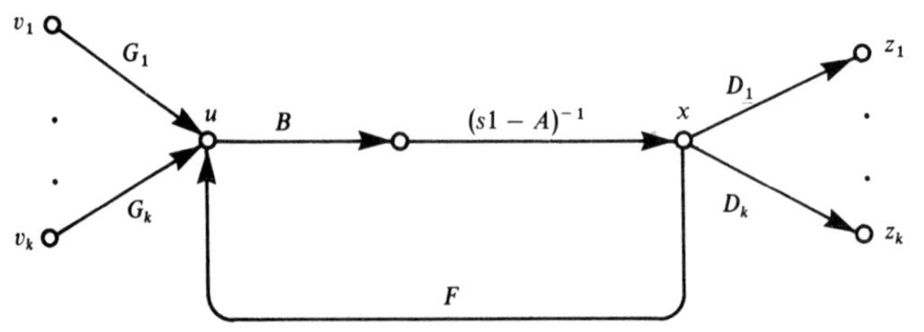

Figure 9.1 Signal Flow: Solution of RDP.
$D_j(s1 - A - BF)^{-1}BG_i = 0,\ j \neq i$; v_i completely controls z_i.

9.2 Restricted Decoupling Problem (RDP)

Recalling Proposition 5.3, we see that (1.2) can be written

$$\mathscr{R}_i = \langle A + BF \mid \mathscr{B} \cap \mathscr{R}_i \rangle, \qquad i \in \mathbf{k}. \tag{2.1}$$

Write $\mathscr{K}_i := \operatorname{Ker} D_i$ ($i \in \mathbf{k}$). Then, (1.3) becomes

$$\mathscr{R}_i \subset \bigcap_{j \neq i} \mathscr{K}_j, \qquad i \in \mathbf{k}. \tag{2.2}$$

Finally, it is easily verified that (1.4) is equivalent to

$$\mathscr{R}_i + \mathscr{K}_i = \mathscr{X}, \qquad i \in \mathbf{k}. \tag{2.3}$$

Thus, the **Restricted Decoupling Problem (RDP)** can be stated: *Given A, B and subspaces $\mathscr{K}_i \subset \mathscr{X}$ ($i \in \mathbf{k}$), find (if possible) a map $F\colon \mathscr{X} \to \mathscr{U}$ and c.s. $\mathscr{R}_i$ ($i \in \mathbf{k}$), such that* (2.1), (2.2) *and* (2.3) *are true.* The problem is "restricted" in the sense that only state feedback is to be utilized, with no augmentation of system dynamic order.

Condition (2.1) will be referred to as the *compatibility condition.* In general, if $\mathscr{V}_i \subset \mathscr{X}$ are subspaces such that

$$\mathbf{F}(\mathscr{V}_i) \neq \varnothing, \qquad i \in \mathbf{k}, \tag{2.4}$$

it by no means follows that

$$\bigcap_{i=1}^{k} \mathbf{F}(\mathscr{V}_i) \neq \varnothing. \tag{2.5}$$

That is, although there exist F_i ($i \in \mathbf{k}$) such that

$$(A + BF_i)\mathscr{V}_i \subset \mathscr{V}_i, \qquad i \in \mathbf{k},$$

it need not be true that some F exists such that

$$(A + BF)\mathscr{V}_i \subset \mathscr{V}_i$$

for all $i \in \mathbf{k}$ simultaneously. If (2.5) does hold, the family $\mathscr{V}_i$ ($i \in \mathbf{k}$) is *compatible* relative to the pair (A, B). Then all the $\mathscr{V}_i$ have at least one friend in common. A simple sufficient (but not necessary) condition for compatibility is that (2.4) hold and the $\mathscr{V}_i$ be independent.

Recalling Proposition 5.3, we see that (2.1) states that the $\mathscr{R}_i$ are compatible in the sense just defined.

Conditions (2.2) are the *noninteraction conditions* and (2.3) are the *output controllability conditions.* In these terms, RDP amounts to seeking compatible c.s. which are small enough to guarantee noninteraction, yet large enough to ensure output controllability.

No special restrictions have been placed on the subspaces $\mathscr{K}_i$. However, we may as well assume that

$$\mathscr{K}_i \neq \mathscr{X}, \qquad i \in \mathbf{k};$$

otherwise, $D_i = 0$, i.e. the ith output is identically zero, and we may take $\mathscr{R}_i = 0$ in (2). Next, if RDP is solvable, it is clearly necessary that the subspaces $\mathscr{K}_i^{\perp} \subset \mathscr{X}'$ be independent (i.e. the row spaces of arbitrary matrix representations of the D_i be independent). For if independence fails, then for some $l \in \mathbf{k}$,

$$\mathscr{K}_l^{\perp} \cap \left(\sum_{j \neq l} \mathscr{K}_j^{\perp} \right) \neq 0,$$

or

$$\mathscr{K}_l + \bigcap_{j \neq l} \mathscr{K}_j \neq \mathscr{X},$$

and (2.3) must fail at $i = l$. Intuitively, we are attempting to control a variable in the lth output "block" which also appears as a linear combination of variables in the remaining output blocks, and if the controls are noninteracting, this is clearly impossible.

Finally, we may as well assume that the pair (A, B) is controllable. Otherwise, we may structure the system as in Exercise 1.4, picking out the controllable subspace $\langle A \mid \mathscr{B} \rangle$. If the induced map $\bar{A}$ on $\mathscr{X}/\langle A \mid \mathscr{B} \rangle$ is unstable, the decoupling problem itself is unrealistic. If $\bar{A}$ is stable, then the coset $\bar{x}(t) \to 0$ as $t \to \infty$, and we shall assume that convergence is fast enough for us to neglect the corresponding transient component of $x(\cdot)$ in $\langle A \mid \mathscr{B} \rangle$.

Returning to RDP, write

$$\hat{\mathscr{K}}_i := \bigcap_{j \neq i} \mathscr{K}_j, \qquad i \in \mathbf{k}. \tag{2.6}$$

On heuristic grounds, it is plausible to attack the problem as follows: find

$$\mathscr{R}_i^* := \sup \mathfrak{C}(\hat{\mathscr{K}}_i), \qquad i \in \mathbf{k}. \tag{2.7}$$

The $\mathscr{R}_i^*$ will then satisfy (2.2). If for some i, $\mathscr{R}_i^*$ fails to satisfy (2.3), then clearly RDP is not solvable, since $\mathscr{R}_i^*$ is supremal. Suppose, then, that $\mathscr{R}_i^*$ satisfies (2.3) for each $i \in \mathbf{k}$. It remains to determine whether the $\mathscr{R}_i^*$ are compatible. If they are, we are done. If they are not, the problem remains unsettled, as there might exist a family of c.s. $\mathscr{R}_1, \ldots, \mathscr{R}_k$ for which $\mathscr{R}_i \subset \mathscr{R}_i^*$ with strict inclusion for some i, with the $\mathscr{R}_i$ all large enough to satisfy (2.3); and this set of smaller $\mathscr{R}_i$ might now be compatible. As we lack a systematic procedure for scanning over all families of compatible $\mathscr{R}_i$, necessary and sufficient conditions for the solvability of RDP, in the general case, are not yet known.

Luckily, the suggested method works in a special case, and later we shall see that it can be made to work in general, if we allow extension of the state space through dynamic compensation.

9.3 Solution of RDP: Outputs Complete

Let us make the additional assumption that

$$\bigcap_{i=1}^{k} \mathscr{K}_i = 0. \tag{3.1}$$

This means simply that if $D_i x = 0$ for all $i \in \mathbf{k}$ then $x = 0$, i.e. the map $D\colon \mathscr{X} \to \mathscr{Z}$ defined by

$$Dx := D_1 x \oplus \cdots \oplus D_k x \tag{3.2}$$

is monic. In this sense the set of outputs is "complete."

Theorem 9.1. *Subject to assumption* (3.1), *RDP is solvable if and only if*

$$\mathscr{R}_i^* + \mathscr{K}_i = \mathscr{X}, \qquad i \in \mathbf{k}, \tag{3.3}$$

where the $\mathscr{R}_i^*$ *are defined by* (2.6) *and* (2.7).

PROOF. If (3.3) holds, then (2.2) and (2.3) are true for the $\mathscr{R}_i^*$. We show next that the $\hat{\mathscr{K}}_i$ are independent. Indeed,

$$\begin{aligned}\hat{\mathscr{K}}_i \cap \sum_{j \neq i} \hat{\mathscr{K}}_j &= \left(\bigcap_{r \neq i} \mathscr{K}_r\right) \cap \sum_{j \neq i} \left(\bigcap_{s \neq j} \mathscr{K}_s\right)\\ &\subset \left(\bigcap_{r \neq i} \mathscr{K}_r\right) \cap \mathscr{K}_i = \bigcap_{r=1}^{k} \mathscr{K}_r = 0.\end{aligned}$$

Since $\mathscr{R}_i^* \subset \hat{\mathscr{K}}_i$ ($i \in \mathbf{k}$) it follows that the $\mathscr{R}_i^*$ are independent, hence compatible.

Conversely, from the fact that the $\mathscr{R}_i^*$ are supremal relative to the condition (2.2), it follows that (3.3) is necessary. □

Independence of the $\mathscr{R}_i^*$ implies not only compatibility but also that the spectrum of $A + BF$ can be assigned to a suitable region $\mathbb{C}_g$ of the complex plane. For a precise statement, write

$$\rho_i := d(\mathscr{R}_i^*), \qquad i \in \mathbf{k};\ \rho_0 := n - \sum_{i=1}^{k} \rho_i.$$

We have

Theorem 9.2. *Let* (A, B) *be controllable and assume that* (3.1) *holds. Let* Λ_i $(i = 0, 1, \ldots, k)$ *be a symmetric set of* ρ_i *complex numbers. There exists*

$$F \in \bigcap_{i=1}^{k} \mathbf{F}(\mathscr{R}_i^*),$$

such that

$$\sigma[(A + BF)|\mathscr{R}_i^*] = \Lambda_i, \qquad i \in \mathbf{k},$$

and

$$\sigma(A + BF) = \dot{\bigcup_{i=0}^{k}} \Lambda_i.$$

PROOF. By Theorem 5.1 there exist $F_i: \mathscr{X} \to \mathscr{U}$, such that

$$\sigma[(A + BF_i)|\mathscr{R}_i^*] = \Lambda_i, \qquad i \in \mathbf{k}.$$

Since (3.1) holds it follows as in the proof of Theorem 9.1 that the $\mathscr{R}_i^*$ $(i \in \mathbf{k})$ are independent, hence there exists $F_0: \mathscr{X} \to \mathscr{U}$, such that

$$F_0|\mathscr{R}_i^* = F_i|\mathscr{R}_i^*, \qquad i \in \mathbf{k}.$$

Clearly,

$$F_0 \in \bigcap_{i=1}^{k} \mathbf{F}(\mathscr{R}_i^*).$$

Write

$$\mathscr{R} := \mathscr{R}_1^* \oplus \cdots \oplus \mathscr{R}_k^*.$$

Then, $F_0 \in \mathbf{F}(\mathscr{R})$ and by Proposition 4.1 there exists $F: \mathscr{X} \to \mathscr{U}$, such that

$$F|\mathscr{R} = F_0|\mathscr{R}$$

and

$$\begin{aligned}\sigma(A + BF) &= \Lambda_0 \,\dot{\cup}\, \sigma[(A + BF)|\mathscr{R}] \\ &= \Lambda_0 \,\dot{\cup}\, \dot{\bigcup_{i=1}^{k}} \Lambda_i.\end{aligned}$$

□

Our success in solving RDP under the condition (3.1) depended strongly, of course, on the fact that (3.1) made the $\mathscr{R}_i^*$ independent. Although in general the $\mathscr{R}_i^*$ will not be independent, they can be transformed into new c.s. which are, by suitable imbedding in an extended state space, as we now establish.

9.4 Extended Decoupling Problem (EDP)

Suppose the system equations (1.1) are augmented by the equations of n_a auxiliary integrators with scalar inputs u_{ai} and outputs x_{ai}:

$$\dot{x}_{ai} = u_{ai}, \qquad i \in \mathbf{n}_a. \tag{4.1}$$

For notational convenience, rewrite (4.1) as

$$\dot{x}_a = B_a u_a,$$

where $x_a \in \mathscr{X}_a$, $u_a \in \mathscr{U}_a$ and $B_a\colon \mathscr{U}_a \simeq \mathscr{X}_a$. Thus,

$$d(\mathscr{X}_a) = d(\mathscr{U}_a) = n_a.$$

It is convenient to imbed (1.1) and (4.1) in common state and input spaces. For this, construct an extended state space as the external direct sum

$$\mathscr{X}_e = \mathscr{X} \oplus \mathscr{X}_a.$$

Similarly, define the extended input space

$$\mathscr{U}_e = \mathscr{U} \oplus \mathscr{U}_a.$$

The maps A, B, B_a have natural extensions defined as follows:

$$A_e\colon \mathscr{X}_e \to \mathscr{X}_e, \qquad x \oplus x_a \mapsto Ax$$

$$B_e\colon \mathscr{U}_e \to \mathscr{X}_e, \qquad u \oplus u_a \mapsto Bu$$

$$B_{ae}\colon \mathscr{U}_e \to \mathscr{X}_e, \qquad u \oplus u_a \mapsto B_a u_a.$$

The corresponding matrices are

$$A_e = \begin{bmatrix} A & 0 \\ 0 & 0 \end{bmatrix}, \qquad B_e = \begin{bmatrix} B & 0 \\ 0 & 0 \end{bmatrix}, \qquad B_{ae} = \begin{bmatrix} 0 & 0 \\ 0 & B_a \end{bmatrix}.$$

For simplicity we shall omit the subscript e on these maps, so that from now on

$$A\mathscr{X}_a = B\mathscr{U}_a = B_a\mathscr{U} = 0;$$

also,

$$\mathscr{B}_a = \operatorname{Im} B_a = \mathscr{X}_a.$$

Let $P\colon \mathscr{X}_e \to \mathscr{X}_e$ be the projection on $\mathscr{X}$ along $\mathscr{X}_a$:

$$P|\mathscr{X} = 1, \qquad P|\mathscr{X}_a = 0, \qquad P = \begin{bmatrix} 1 & 0 \\ 0 & 0 \end{bmatrix}.$$

Thus,

$$PA = AP = A, \qquad PB = B, \qquad PB_a = 0.$$

We now define an *extended controllability subspace* (e.c.s.) to be a c.s. for the extended pair $(A, B + B_a)$. If $\mathscr{V} \subset \mathscr{X}_e$, write $\mathbf{F}_e(\mathscr{V})$ for the family of maps $F\colon \mathscr{X}_e \to \mathscr{U}_e$, such that

$$[A + (B + B_a)F]\mathscr{V} \subset \mathscr{V}.$$

It is now natural to introduce the

Extended Decoupling Problem (EDP). *Given the original maps $A\colon \mathscr{X} \to \mathscr{X}$, $B\colon \mathscr{U} \to \mathscr{X}$ and subspaces $\mathscr{K}_i \subset \mathscr{X}$ $(i \in \mathbf{k})$, find (if possible) $\mathscr{X}_a$ (i.e. n_a) and e.c.s. $\mathscr{S}_i$ $(i \in \mathbf{k})$, such that*

$$\bigcap_{i=1}^{k} \mathbf{F}_e(\mathscr{S}_i) \neq \varnothing, \tag{4.2}$$

$$\mathscr{S}_i \subset \bigcap_{j \neq i} (\mathscr{K}_j \oplus \mathscr{X}_a), \qquad i \in \mathbf{k}, \tag{4.3}$$

and

$$\mathscr{S}_i + (\mathscr{K}_i \oplus \mathscr{X}_a) = \mathscr{X} \oplus \mathscr{X}_a, \qquad i \in \mathbf{k}. \tag{4.4}$$

Conditions (4.2)–(4.4) express the requirements, respectively, of compatibility, noninteraction and output controllability for the extended problem. Thus, EDP has the same formal appearance as RDP, but valuable flexibility is gained from the special structure of the extended pair $(A, B + B_a)$ and the extended output kernels $\mathscr{K}_i \oplus \mathscr{X}_a$.

We can easily justify EDP as the "correct" description of decoupling by dynamic compensation: the output relations $z_i = D_i x$ $(i \in \mathbf{k})$ of the original system are preserved on replacing $\mathscr{K}_i$ by $\mathscr{K}_i \oplus \mathscr{X}_a$, or equivalently by defining extensions D_{ie} of the D_i to vanish on $\mathscr{X}_a$; no additional inputs (vectors in $\mathscr{B}$) to the original system (1.1) are postulated; and, subject to the latter constraint, full linear coupling is allowed between the two systems (1.1) and (4.1). The corresponding, more elaborate signal flow is shown in Fig. 9.2, where the extended control has matrix representation

$$\begin{bmatrix} u \\ u_a \end{bmatrix} = \begin{bmatrix} F_{11} & F_{12} \\ F_{21} & F_{22} \end{bmatrix} \begin{bmatrix} x \\ x_a \end{bmatrix} + \begin{bmatrix} G_{11} & \cdots & G_{1k} \\ G_{21} & \cdots & G_{2k} \end{bmatrix} \begin{bmatrix} v_1 \\ \vdots \\ v_k \end{bmatrix}.$$

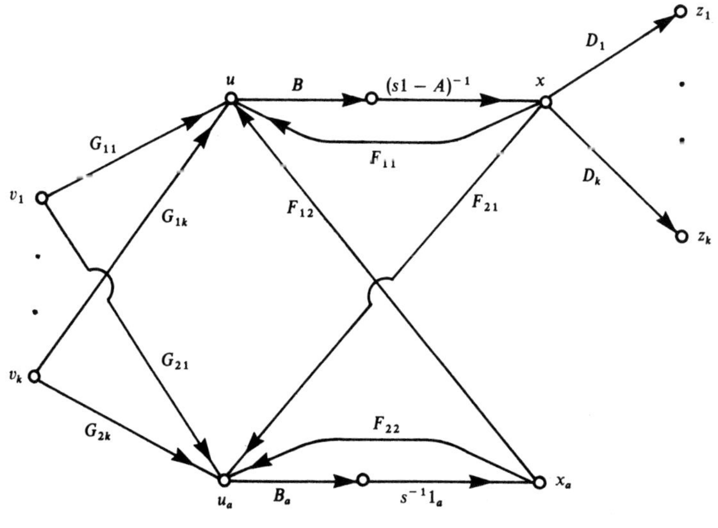

Figure 9.2 Signal Flow: Solution of EDP.

9.5 Solution of EDP

The fundamental result of decoupling theory is the following.

Theorem 9.3. *For the RDP defined in Section 9.2 let*

$$\mathscr{R}_i^* := \sup \mathfrak{C}\left(\bigcap_{j \neq i} \mathscr{K}_j\right), \qquad i \in \mathbf{k}.$$

Then the corresponding EDP of Section 9.4 is solvable if and only if

$$\mathscr{R}_i^* + \mathscr{K}_i = \mathscr{X}, \qquad i \in \mathbf{k}. \tag{5.1}$$

Informally, the theorem states that decoupling can be achieved by dynamic compensation if and only if the largest noninteracting c.s. of RDP satisfy merely the output controllability conditions of RDP. The crucial fact is that dynamic compensation makes it possible to satisfy the compatibility condition of EDP. As will be seen from the proof of Theorem 9.3, below, a geometric interpretation of dynamic compensation is that the additional components of state space provide room for the $\mathscr{R}_i^*$ to be disentangled from one another: in particular, to be made independent and, therefore, compatible. An alternative, systemic interpretation is displayed in Fig. 9.3. Here, k identical "models" of the original system, with state vectors $x_1, \ldots, x_k$, are wired up with feedback and gain maps (F_i, G_i) that synthesize the $\mathscr{R}_i^*$. Each pair $(F_i x_i, G_i v_i)$ of feedback and external input signals is fed to the control node of the original system. A simple computation shows that the response $x(\cdot)$ to $v_i(\cdot)$ alone is given by

$$\dot{x} - \dot{x}_i = A(x - x_i).$$

Thus, if $x_i(0) = x(0) = 0$, say, then in principle $x(t) = x_i(t)$ $(t \geq 0)$ just as desired. Of course, this style of open-loop compensation is crude, and impractical if A is unstable, but it does suggest that some form of dynamic compensation ought to work. The actual techniques discussed below are much more economical and never lead to instability.

To prove Theorem 9.3 we need two preliminary results which relate c.s. to their extensions.

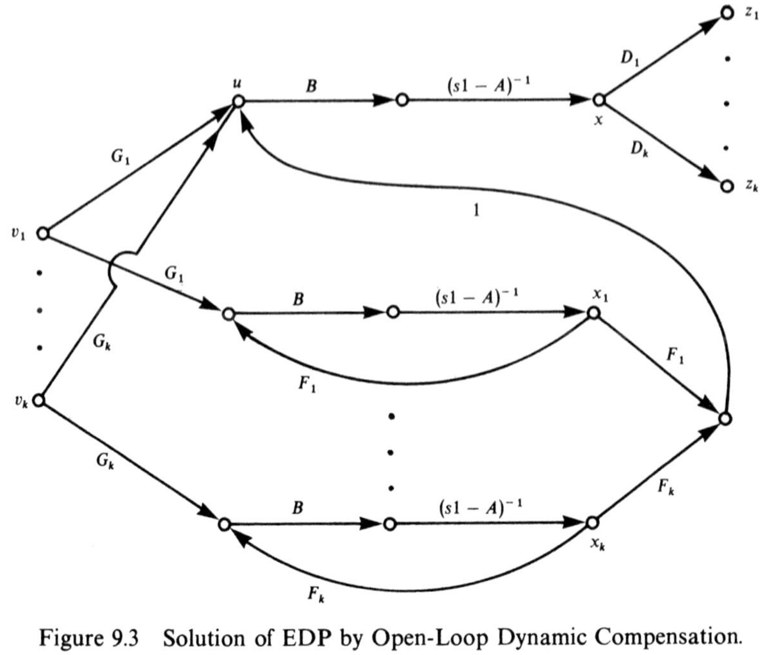

Figure 9.3 Solution of EDP by Open-Loop Dynamic Compensation.

Lemma 9.1. *If $\mathscr{S}$ is an e.c.s. then $P\mathscr{S}$ is a c.s.*

PROOF. We shall apply to $\mathscr{S}$ the results of Chapter 5.
Since $\mathscr{S}$ is an e.c.s., $A\mathscr{S} \subset \mathscr{S} + \mathscr{B} + \mathscr{B}_a$. Therefore,

$$A(P\mathscr{S}) = PA\mathscr{S} \subset P\mathscr{S} + \mathscr{B},$$

so that $\mathbf{F}(P\mathscr{S}) \neq \emptyset$. Also, by Theorem 5.3, $\mathscr{S}^\mu \uparrow \mathscr{S}$, where

$$\mathscr{S}^0 = 0;\ \mathscr{S}^\mu = \mathscr{S} \cap (A\mathscr{S}^{\mu-1} + \mathscr{B} + \mathscr{B}_a), \qquad \mu = 1, 2, \ldots \tag{5.2}$$

Since Ker $P = \mathscr{X}_a = \mathscr{B}_a$, we have

$$\begin{aligned}(\mathscr{S} + A\mathscr{S}^{\mu-1} + \mathscr{B} + \mathscr{B}_a) \cap \operatorname{Ker} P &= \mathscr{X}_a \\ &= \mathscr{S} \cap \operatorname{Ker} P \\ &\quad + (A\mathscr{S}^{\mu-1} + \mathscr{B} + \mathscr{B}_a) \cap \operatorname{Ker} P.\end{aligned}$$

Hence, we can apply (0.4.3) to (5.2) to obtain

$$\begin{aligned}P\mathscr{S}^\mu &= P\mathscr{S} \cap P(A\mathscr{S}^{\mu-1} + \mathscr{B} + \mathscr{B}_a) \\ &= P\mathscr{S} \cap (AP\mathscr{S}^{\mu-1} + \mathscr{B}).\end{aligned}$$

Since $P\mathscr{S}^\mu \uparrow P\mathscr{S}$, Theorem 5.3 implies that $P\mathscr{S}$ is a c.s. □

Lemma 9.2. *Let $\mathscr{R}$ be a c.s. and let $E: \mathscr{X} \oplus \mathscr{X}_a \to \mathscr{X} \oplus \mathscr{X}_a$ be an arbitrary map with* Im $E \subset \mathscr{X}_a$. *Then, $\mathscr{S} := (P + E)\mathscr{R}$ is an e.c.s.*

PROOF. Since

$$\begin{aligned}A\mathscr{S} - A\mathscr{R} &\subset \mathscr{R} + \mathscr{B} \\ &= P\mathscr{S} + \mathscr{B} \subset \mathscr{S} + \mathscr{B} + \operatorname{Ker} P \\ &= \mathscr{S} + \mathscr{B} + \mathscr{B}_a,\end{aligned}$$

we have $\mathbf{F}_e(\mathscr{S}) \neq \emptyset$. Also, $\mathscr{R}^\mu \uparrow \mathscr{R}$, where

$$\mathscr{R}^0 = 0;\ \mathscr{R}^\mu = \mathscr{R} \cap (A\mathscr{R}^{\mu-1} + \mathscr{B}), \qquad \mu = 1, 2, \ldots.$$

Define

$$\mathscr{S}^0 = 0;\ \mathscr{S}^\mu = \mathscr{S} \cap (A\mathscr{S}^{\mu-1} + \mathscr{B} + \mathscr{B}_a), \qquad \mu = 1, 2, \ldots. \tag{5.3}$$

Then, $\mathscr{S}^0 \supset (P + E)\mathscr{R}^0$, and if $\mathscr{S}^{\mu-1} \supset (P + E)\mathscr{R}^{\mu-1}$, (5.3) yields

$$\begin{aligned}\mathscr{S}^\mu &\supset [(P + E)\mathscr{R}] \cap [A(P + E)\mathscr{R}^{\mu-1} + \mathscr{B} + \mathscr{X}_a] \\ &= [(P + E)\mathscr{R}] \cap [A\mathscr{R}^{\mu-1} + \mathscr{B} + \mathscr{X}_a] \\ &\supset [(P + E)\mathscr{R}] \cap [(P + E)(A\mathscr{R}^{\mu-1} + \mathscr{B} + \mathscr{X}_a)] \\ &\supset (P + E)[\mathscr{R} \cap (A\mathscr{R}^{\mu-1} + \mathscr{B} + \mathscr{X}_a)] \\ &= (P + E)[\mathscr{R} \cap (A\mathscr{R}^{\mu-1} + \mathscr{B})] \\ &= (P + E)\mathscr{R}^\mu.\end{aligned}$$

Thus, for all μ,

$$\mathscr{S} \supset \mathscr{S}^{\mu} \supset (P + E)\mathscr{R}^{\mu} \uparrow (P + E)\mathscr{R} = \mathscr{S},$$

hence $\mathscr{S}^{\mu} \uparrow \mathscr{S}$, so $\mathscr{S}$ is an e.c.s., as claimed. □

PROOF (of Theorem 9.3). (Only if) Suppose $\mathscr{S}_i$ $(i \in \mathbf{k})$ is a solution of EDP. By (4.3)

$$P\mathscr{S}_i \subset P\left[\bigcap_{j \neq i} (\mathscr{K}_j \oplus \mathscr{X}_a)\right] = \bigcap_{j \neq i} \mathscr{K}_j, \tag{5.4}$$

and by (4.4)

$$P\mathscr{S}_i + \mathscr{K}_i = \mathscr{X}. \tag{5.5}$$

By Lemma 9.1 and (5.4)

$$P\mathscr{S}_i \in \mathfrak{C}\left(\bigcap_{j \neq i} \mathscr{K}_j\right).$$

Therefore, $P\mathscr{S}_i \subset \mathscr{R}_i^*$ $(i \in \mathbf{k})$, and (5.1) follows from (5.5).

(If) Suppose (5.1) is true. Let $\mathscr{X}_{ai}$ be a linear space over $\mathbb{R}$ with $\mathscr{X}_{ai} \simeq \mathscr{R}_i^*$ $(i \in \mathbf{k})$, and introduce the external direct sum

$$\mathscr{X}_a := \bigoplus_{i=1}^{k} \mathscr{X}_{ai}.$$

Let $E_i: \mathscr{X} \oplus \mathscr{X}_a \to \mathscr{X} \oplus \mathscr{X}_a$ be a map with the properties

$$\mathscr{R}_i^* \cap \operatorname{Ker} E_i = 0, \qquad i \in \mathbf{k},$$

and

$$\operatorname{Im} E_i = E_i \mathscr{R}_i^* = \mathscr{X}_{ai}, \qquad i \in \mathbf{k}.$$

Thus, E_i is monic on $\mathscr{R}_i^*$ and zero on a complement of $\mathscr{R}_i^*$ in $\mathscr{X} \oplus \mathscr{X}_a$. Define

$$\mathscr{S}_i = (P + E_i)\mathscr{R}_i^*, \qquad i \in \mathbf{k}. \tag{5.6}$$

By Lemma 9.2, $\mathscr{S}_i$ is an e.c.s. It is easily checked as well that the $\mathscr{S}_i$ $(i \in \mathbf{k})$ are independent, hence compatible. Next, from

$$\mathscr{R}_i^* \subset \bigcap_{j \neq i} \mathscr{K}_j, \qquad i \in \mathbf{k},$$

there follows

$$\begin{aligned}\mathscr{S}_i \subset (P + E_i) \bigcap_{j \neq i} \mathscr{K}_j &\subset \left(\bigcap_{j \neq i} \mathscr{K}_j\right) \oplus \mathscr{X}_a \\ &= \bigcap_{j \neq i} (\mathscr{K}_j \oplus \mathscr{X}_a), \qquad i \in \mathbf{k},\end{aligned}$$

which verifies (4.3). From (5.1) and (5.6), we have

$$\mathscr{S}_i + (P + E_i)\mathscr{K}_i = (P + E_i)\mathscr{X}. \tag{5.7}$$

Because

$$(P + E_i)\mathcal{K}_i + \mathcal{X}_a = \mathcal{K}_i \oplus \mathcal{X}_a$$

and

$$(P + E_i)\mathcal{X} + \mathcal{X}_a = \mathcal{X} \oplus \mathcal{X}_a,$$

(4.4) results on adding $\mathcal{X}_a$ to both sides of (5.7). □

Since the e.c.s. $\mathcal{S}_i$ constructed in the proof of Theorem 9.3 are independent it follows, as in the proof of Theorem 9.2, that $F \in \bigcap_{i=1}^{k} \mathbf{F}_e(\mathcal{S}_i)$ can be chosen such that, for each $i \in \mathbf{k}$,

$$\sigma[(A + (B + B_a)F) \mid \mathcal{S}_i]$$

is a given symmetric set of $d(\mathcal{S}_i)$ complex numbers.

9.6 Naive Extension

The order of dynamic compensation introduced in the proof of Theorem 9.3, namely,

$$n_a = \sum_{i=1}^{k} d(\mathcal{R}_i^*), \tag{6.1}$$

is unnecessarily large. On grounds of reliability, economy or esthetics it may be desirable to keep the dynamic order of compensation small. While determination of the strictly minimal number of auxiliary integrators required is apparently quite difficult (see Chapter 10), it is nevertheless easy to improve somewhat on the bound (6.1). We have the following.

Theorem 9.4. *If EDP is solvable, the order of dynamic compensation required is no greater than*

$$n_a := \sum_{i=1}^{k} d(\mathcal{R}_i^*) - d\left(\sum_{i=1}^{k} \mathcal{R}_i^*\right). \tag{6.2}$$

For the proof, we need

Lemma 9.3. *Let $\mathcal{R}_1, \ldots, \mathcal{R}_k$ be arbitrary subspaces of $\mathcal{X}$. Define*

$$n_a := \sum_{i=1}^{k} d(\mathcal{R}_i) - d\left(\sum_{i=1}^{k} \mathcal{R}_i\right).$$

Then if $d(\mathcal{X}_a) = n_a$ and $\mathcal{X}_e = \mathcal{X} \oplus \mathcal{X}_a$, there exist maps $E_i\colon \mathcal{X}_e \to \mathcal{X}_e$ ($i \in \mathbf{k}$), such that

$$\operatorname{Im} E_i \subset \mathcal{X}_a, \qquad i \in \mathbf{k},$$

and the subspaces $\mathscr{S}_i := (1 + E_i)\mathscr{R}_i$ $(i \in \mathbf{k})$ *are independent.*

Proof. Write $\mathscr{T}_1 = 0$ and

$$\mathscr{T}_i = \mathscr{R}_i \cap \sum_{j=1}^{i-1} \mathscr{R}_j, \qquad i = 2, \ldots, k.$$

Then,

$$\begin{aligned}\sum_{i=1}^{k} d(\mathscr{T}_i) &= \sum_{i=2}^{k} \left[d(\mathscr{R}_i) + d\left(\sum_{j=1}^{i-1} \mathscr{R}_j\right) - d\left(\sum_{j=1}^{i} \mathscr{R}_j\right)\right] \\ &= \sum_{i=1}^{k} d(\mathscr{R}_i) - d\left(\sum_{i=1}^{k} \mathscr{R}_i\right) \\ &= d(\mathscr{X}_a).\end{aligned} \tag{6.3}$$

By (6.3) there exist maps E_i: $\mathscr{X}_e \to \mathscr{X}_e$, such that

$$\mathscr{T}_i \simeq E_i \mathscr{T}_i = \text{Im } E_i \subset \mathscr{X}_a, \qquad i \in \mathbf{k},$$

and the subspaces Im E_i $(i \in \mathbf{k})$ are independent. Suppose, contrary to what is claimed, that the $\mathscr{S}_i = (1 + E_i)\mathscr{R}_i$ are not independent. Let $i \geq 2$ be the greatest integer such that

$$\mathscr{S}_i \cap \sum_{j \neq i} \mathscr{S}_j \neq 0.$$

There is $x \neq 0$, such that

$$x = (1 + E_i)r_i = \sum_{j=1}^{i-1} (1 + E_j)r_j,$$

where $r_j \in \mathscr{R}_j$ $(j \in \mathbf{i})$; so

$$r_i = \sum_{j=1}^{i-1} r_j,$$

and therefore $r_i \in \mathscr{T}_i$. By independence of the Im E_i and the fact that Im $E_i \subset \mathscr{X}_a$, there follows $E_i r_i = 0$, and as Ker $E_i \cap \mathscr{T}_i = 0$, there results $r_i = 0$, hence $x = 0$, a contradiction. □

Proof (of Theorem 9.4). With E_i and $\mathscr{S}_i$ as in Lemma 9.3 (with $\mathscr{R}_i^*$ in place of $\mathscr{R}_i$), we have that the $\mathscr{S}_i$ $(i \in \mathbf{k})$ are independent. This with Lemma 9.2 implies that the $\mathscr{S}_i$ are compatible e.c.s. That the $\mathscr{S}_i$ also satisfy (4.3) and (4.4) follows exactly as in the proof of Theorem 9.3. That is, the $\mathscr{S}_i$ $(i \in \mathbf{k})$ provide a solution of EDP, with n_a given by (6.2). □

Just as before, independence of the $\mathscr{S}_i$ implies that the spectra of the maps

$$[A + (B + B_a)F] \mid \mathscr{S}_i$$

can be assigned arbitrarily and independently by suitable choice of F.

9.7 Example

Let $n = 5$ and

$$A = \begin{bmatrix} 0 & 1 & 0 & 0 & 0 \\ 0 & 0 & 0 & 0 & 0 \\ 0 & 0 & 1 & 1 & 0 \\ 1 & 0 & 0 & 0 & 1 \\ 0 & 0 & 0 & 0 & 0 \end{bmatrix}, \qquad B = \begin{bmatrix} 0 & 1 \\ 1 & 0 \\ 0 & 0 \\ 0 & 0 \\ 0 & 1 \end{bmatrix},$$

$$D_1 = [1 \;\; 0 \;\; 0 \;\; 0 \;\; 0], \qquad D_2 = [0 \;\; 0 \;\; 0 \;\; 0 \;\; 1].$$

It is easily checked that

$$\mathscr{R}_1^* = \mathscr{K}_2, \qquad \mathscr{R}_2^* = \mathscr{K}_1.$$

Since

$$\mathscr{R}_1^* \cap \mathscr{R}_2^* = \operatorname{Im} \begin{bmatrix} 0 & 0 & 0 \\ 1 & 0 & 0 \\ 0 & 1 & 0 \\ 0 & 0 & 1 \\ 0 & 0 & 0 \end{bmatrix}$$

is not (A, B)-invariant, the subspaces $\mathscr{R}_1^*$ and $\mathscr{R}_2^*$ are certainly not compatible. Nevertheless, by Theorem 9.3 EDP is solvable, and by Theorem 9.4, we may take

$$\begin{aligned} n_a &= d(\mathscr{R}_1^*) + d(\mathscr{R}_2^*) - d(\mathscr{R}_1^* + \mathscr{R}_2^*) \\ &= d(\mathscr{R}_1^* \cap \mathscr{R}_2^*) \\ &= 3. \end{aligned}$$

In the notation of the proof of Lemma 9.3, we have $\mathscr{T}_1 = 0$ and $\mathscr{T}_2 = \mathscr{R}_1^* \cap \mathscr{R}_2^*$. Thus, we may define

$$\mathscr{S}_1 = \operatorname{Im} \begin{bmatrix} 1 & 0 & 0 & 0 \\ 0 & 1 & 0 & 0 \\ 0 & 0 & 1 & 0 \\ 0 & 0 & 0 & 1 \\ 0 & 0 & 0 & 0 \\ 0 & 0 & 0 & 0 \\ 0 & 0 & 0 & 0 \\ 0 & 0 & 0 & 0 \end{bmatrix}, \qquad \mathscr{S}_2 = \operatorname{Im} \begin{bmatrix} 0 & 0 & 0 & 0 \\ 1 & 0 & 0 & 0 \\ 0 & 1 & 0 & 0 \\ 0 & 0 & 1 & 0 \\ 0 & 0 & 0 & 1 \\ 1 & 0 & 0 & 0 \\ 0 & 1 & 0 & 0 \\ 0 & 0 & 1 & 0 \end{bmatrix}.$$

The design can now be completed by choosing $F \in \mathbf{F}_e(\mathscr{S}_1) \cap \mathbf{F}_e(\mathscr{S}_2)$, such that $A + (B + B_a)F$ has a suitable spectrum.

9.8 Partial Decoupling

In certain applications it may be appropriate to replace the stringent requirement of complete dynamic noninteraction by a weaker constraint under which the outputs z_i are only partially decoupled. Suppose, for instance, that the z_i ($i \in \mathbf{k}$) are to be controlled sequentially rather than simultaneously. First, v_1 controls z_1, possibly changing the values of $z_2, \ldots, z_k$; then, v_2 controls z_2, with the requirement that z_1 be left unaffected, but possibly changing the values of $z_3, \ldots, z_k$; and so forth, with v_k controlling z_k without influencing $z_1, \ldots, z_{k-1}$. It is easy to see that this situation can be formalized, in our previous notation, as the

Triangular Decoupling Problem (TDP). *Given A, B and $D_1, \ldots, D_k$, find F and c.s. $\mathscr{R}_1, \ldots, \mathscr{R}_k$, such that*

$$\mathscr{R}_i = \langle A + BF \mid \mathscr{B} \cap \mathscr{R}_i \rangle, \qquad i \in \mathbf{k},$$

$$\mathscr{R}_i \subset \bigcap_{j=1}^{i-1} \operatorname{Ker} D_j, \qquad i \in \mathbf{k}, \tag{8.1}$$

and

$$\mathscr{R}_i + \operatorname{Ker} D_i = \mathscr{X}, \qquad i \in \mathbf{k}.$$

In (8.1) the vacuous condition at $i = 1$ just says $\mathscr{R}_1 \subset \mathscr{X}$. Proof of the following easy result is left to the reader.

Theorem 9.5. *TDP is solvable if and only if*

$$\mathscr{R}_i^* + \operatorname{Ker} D_i = \mathscr{X}, \qquad i \in \mathbf{k},$$

where $\mathscr{R}_i^*$ is the supremal c.s. subject to (8.1). *Furthermore, if $\Lambda_i \subset \mathbb{C}$ ($i \in \mathbf{k}$) is symmetric, with $|\Lambda_i| = d(\mathscr{R}_i^*/\mathscr{R}_{i+1}^*)$ ($i \in \mathbf{k-1}$), $|\Lambda_k| = d(\mathscr{R}_k^*)$, there exists*

$$F \in \bigcap_{i=1}^{k} \mathbf{F}(\mathscr{R}_i^*),$$

such that

$$\sigma\left[(A + BF) \,\Big|\, \sum_{i=1}^{k} \mathscr{R}_i^*\right] = \dot{\bigcup_{i=1}^{k}} \Lambda_i.$$

Note that, by (8.1), $\mathscr{R}_1^* \supset \cdots \supset \mathscr{R}_k^*$, so that $\sum_{i=1}^{k} \mathscr{R}_i^* = \mathscr{R}_1^*$.

Noninteraction constraints other than the "triangular" are also readily treated; in general, one must exploit extension. In this direction, we are led finally to the

General Extended Decoupling Problem (GEDP).[1] *Given* A, B *and subspaces* $\mathscr{K}_i \subset \mathscr{X}$, $\hat{\mathscr{K}}_i \subset \mathscr{X}$ $(i \in \mathbf{k})$, *find* $n_a = d(\mathscr{X}_a) = d(\mathscr{U}_a)$, *a map*

$$F\colon \mathscr{X} \oplus \mathscr{X}_a \to \mathscr{U} \oplus \mathscr{U}_a,$$

and e.c.s. $\mathscr{S}_i$ $(i \in \mathbf{k})$, *such that*

$$\mathscr{S}_i = \langle A + (B + B_a)F \mid (\mathscr{B} \oplus \mathscr{B}_a) \cap \mathscr{S}_i \rangle, \qquad i \in \mathbf{k},$$

$$\mathscr{S}_i \subset \hat{\mathscr{K}}_i \oplus \mathscr{X}_a, \qquad i \in \mathbf{k},$$

and

$$\mathscr{S}_i + (\mathscr{K}_i \oplus \mathscr{X}_a) = \mathscr{X} \oplus \mathscr{X}_a, \qquad i \in \mathbf{k}.$$

Here, no special relation is postulated among the $\mathscr{K}_i$ and $\hat{\mathscr{K}}_i$, although it is clear that GEDP is solvable only if $\mathscr{K}_i + \hat{\mathscr{K}}_i = \mathscr{X}$ $(i \in \mathbf{k})$. We have

Theorem 9.6. *GEDP is solvable if and only if*

$$\mathscr{R}_i^* + \mathscr{K}_i = \mathscr{X}, \qquad i \in \mathbf{k}, \tag{8.2}$$

where $\mathscr{R}_i^* := \sup \mathfrak{C}(\hat{\mathscr{K}}_i)$. *Furthermore, if* (8.2) *holds one can take*

$$n_a \leq \sum_i d(\mathscr{R}_i^*) - d\left(\sum_i \mathscr{R}_i^*\right).$$

The proof is straightforward mimicry of that of Theorem 9.3, combined with the result of Theorem 9.4, and is left to the reader.

Theorem 9.6 can be used to study a wide variety of control structures including hierarchical and decentralized configurations. Each choice of $\hat{\mathscr{K}}_i$ bounds the subspace reachable by the ith control input $v_i(\cdot)$ and thus the $\hat{\mathscr{K}}_i$ parametrize the allowed interactions among the v_i. Of course, this control over signal flow is only achieved at the expense of implementing suitable internal feedback paths, and the theory specifies precisely what these must be.

9.9 Exercises

9.1. *Compatibility of a family of subspaces*

i. Let $\mathscr{V}_1$, $\mathscr{V}_2$ be (A, B)-invariant. Show that $\mathscr{V}_1$, $\mathscr{V}_2$ are compatible if and only if $\mathscr{V}_1 \cap \mathscr{V}_2$ is (A, B)-invariant.

ii. Show that the family $\{\mathscr{V}_i, i \in \mathbf{k}\}$ is compatible if and only if the set of linear matrix equations

$$W_i BFV_i = -W_i AV_i, \qquad i \in \mathbf{k},$$

has an $m \times n$ solution matrix F, for suitably chosen matrices W_i, V_i $(i \in \mathbf{k})$.

[1] The notation is again that of Section 9.4.

iii. Let $\mathfrak{B} = \{\mathscr{V}_i, i \in \mathbf{k}\}$ be an arbitrary family of subspaces of $\mathscr{X}$. Let $\mathfrak{L}(\mathfrak{B}) = \mathfrak{L}$ be the smallest family of subspaces of $\mathscr{X}$ which contains each $\mathscr{V}_i$ $(i \in \mathbf{k})$ and is closed under subspace addition and intersection (it should be verified that $\mathfrak{L}$ exists and is unique!). $\mathfrak{L}$ is the *enveloping lattice* of the $\mathscr{V}_i$. Show that if $\mathfrak{L}$ is a distributive lattice (i.e., the intersection operation distributes over *arbitrary* sums), then $\mathfrak{B}$ is decomposable in the following sense: there exist an independent family $\mathscr{W}_j \subset \mathscr{X}$ $(j \in \mathbf{l})$ and index subsets $J_i \subset \mathbf{l}$ $(i \in \mathbf{k})$, such that

$$\mathscr{V}_i = \bigoplus_{j \in J_i} \mathscr{W}_j, \qquad i \in \mathbf{k}.$$

Note: It is not claimed that the $\mathscr{W}_j$ are unique or that they all belong to $\mathfrak{L}$. Hint: Recall the usual Boolean decomposition of an arbitrary union of k subsets into a union of $2^k - 1$ disjoint subsets; first do the problem for $k = 3$, then generalize.

iv. Show that a family $\{\mathscr{V}_i, i \in \mathbf{k}\}$ of (A, B)-invariant subspaces is compatible if each $\mathscr{V} \in \mathfrak{L}(\mathfrak{B})$ is (A, B)-invariant and $\mathfrak{L}(\mathfrak{B})$ is distributive. Show that the first of these conditions is necessary for compatibility, but not sufficient if $k \geq 3$. Show, however, that distributivity of $\mathfrak{L}(\mathfrak{B})$ is not necessary.

9.2. Show that conditions (1.4) and (2.3) are equivalent.

9.3. Show that the following decoupling problem is solvable without dynamic compensation. Design the decoupled system so that its poles all lie in the region

$$-3 \leq \Re e\, \lambda \leq -1, \qquad |\Im m\, \lambda| \leq 1$$

and give the signal flow diagram for the final result.

$$A = \begin{bmatrix} 0 & 0 & 0 & 0 & 0 & 0 & 1 & 0 \\ 0 & 0 & 0 & 1 & 0 & 0 & 0 & 0 \\ 0 & 1 & 0 & -1 & 1 & -1 & 3 & 2 \\ 0 & 0 & 1 & 0 & 0 & 0 & 0 & 0 \\ 1 & 0 & 0 & 0 & 0 & 0 & 0 & 0 \\ 1 & -1 & 1 & 2 & 2 & 2 & 0 & -1 \\ -1 & 2 & -3 & 1 & 3 & 1 & 0 & -2 \\ 0 & 0 & 0 & 0 & 0 & 1 & 0 & 0 \end{bmatrix},$$

$$B = \begin{bmatrix} 0 & 0 & 0 \\ 0 & 0 & 0 \\ 1 & 2 & -1 \\ 0 & 0 & 0 \\ 0 & 0 & 0 \\ 2 & 3 & 1 \\ -1 & 1 & 1 \\ 0 & 0 & 0 \end{bmatrix},$$

$$D_1 = \begin{bmatrix} 1 & 0 & -1 & 0 & 0 & 0 & 0 & 0 \\ 1 & 0 & 0 & 1 & -1 & 0 & 0 & 0 \\ 0 & 1 & 0 & 0 & 0 & 0 & -1 & 0 \end{bmatrix},$$

$$D_2 = \begin{bmatrix} 1 & 0 & 0 & 0 & -1 & 1 & 0 & -1 \\ 1 & 0 & 0 & 0 & 2 & -2 & 3 & -1 \end{bmatrix}.$$

9.4. Prove the following converse to Lemma 9.3: Let $\mathscr{R}_i \subset \mathscr{X}$ $(i \in \mathbf{k})$ and suppose there exist $\mathscr{S}_i \subset \mathscr{X} \oplus \mathscr{X}_a$ $(i \in \mathbf{k})$, such that $P\mathscr{S}_i = \mathscr{R}_i$ $(i \in \mathbf{k})$ and the $\mathscr{S}_i$ are independent. Then,

$$d(\mathscr{X}_a) \geq \sum_i d(\mathscr{R}_i) - d\left(\sum_i \mathscr{R}_i\right).$$

9.5. Show that the following decoupling problem is solvable with dynamic compensation of order 3. Design the decoupled system with spectrum as in Exercise 9.3 and give the signal flow graph.

$$A = \begin{bmatrix} 5 & -2 & 2 & 6 & 9 & 4 & -5 & 6 \\ -7 & -9 & -1 & -9 & -7 & 9 & 6 & 5 \\ -2 & -9 & 2 & 1 & -9 & 1 & -3 & -6 \\ -3 & 3 & 4 & 5 & 1 & -5 & 7 & -3 \\ 4 & -4 & -9 & -8 & 1 & 6 & -6 & -9 \\ -9 & -7 & 5 & -5 & -9 & -1 & -1 & 1 \\ -2 & -7 & -5 & 7 & 4 & 2 & 4 & -4 \\ -9 & 9 & 1 & -9 & 8 & -9 & -3 & -1 \end{bmatrix},$$

$$B = \begin{bmatrix} -8 & 9 & -9 & -9 \\ -1 & 6 & 9 & -5 \\ -3 & 6 & -1 & 5 \\ 4 & 4 & 6 & -9 \\ -6 & -6 & 2 & 4 \\ 3 & -3 & -6 & -6 \\ 4 & -7 & 5 & -7 \\ -9 & 4 & -8 & 9 \end{bmatrix},$$

$$D_1 = \begin{bmatrix} 2 & 5 & 4 & 1 & -9 & -8 & 9 & -7 \\ -3 & 3 & 3 & -8 & 2 & 9 & -3 & 7 \end{bmatrix}$$

$$D_2 = \begin{bmatrix} 9 & -5 & -5 & -2 & 7 & 8 & -6 & 6 \\ 4 & -1 & -5 & 4 & -8 & -6 & -8 & 1 \\ -9 & -7 & 2 & -2 & -9 & 4 & -3 & 7 \end{bmatrix}.$$

9.6. Prove Theorem 9.5.

9.7. *Decoupling by output feedback.* With D defined as in (3.2), assume (D, A) is observable. Consider the constraint on RDP that Ker $F \supset$ Ker D, i.e. $F = \tilde{F}D$ for some $\tilde{F}$. Show that if $\{\mathscr{R}_i, i \in \mathbf{k}\}$ are c.s. which satisfy (2.2) and (2.3), they furnish a solution of the constrained RDP if and only if (i) the $\mathscr{R}_i$ $(i \in \mathbf{k})$ are independent, and (ii) $A(\mathscr{R}_i \cap \operatorname{Ker} D) \subset \mathscr{R}_i$ $(i \in \mathbf{k})$. Hint: For (i), show that dependence and (constrained) compatibility contradict observability.

9.8. Call the triple (D, A, B) *output controllable* if $D: \mathscr{X} \to \mathscr{Z}$ is epic and $D\langle A \mid \mathscr{B}\rangle = \mathscr{Z}$. Consider the system

$$\dot{x} = Ax + Bu, \qquad z = Dx, \qquad \dot{w} = z.$$

Show that (D, A, B) is output controllable if and only if

$$\left([0 \quad 1], \quad \begin{bmatrix} A & 0 \\ D & 0 \end{bmatrix}, \quad \begin{bmatrix} B \\ 0 \end{bmatrix} \right)$$

is output controllable. Interpret, and generalize to the case $z = Dx + Eu$.

9.9. *Decoupling with direct control feedthrough.* Discuss the decoupling problem in the case $z_i = D_i x + E_i u$ ($i \in \mathbf{k}$). Hint: Introduce additional state variables w_i as the integrals of the z_i:

$$\dot{w}_i = D_i x + E_i u, \qquad i \in \mathbf{k},$$

or alternatively (with stability as a constraint)

$$\dot{w}_i = -w_i + D_i x + E_i u, \qquad i \in \mathbf{k}.$$

With the help of Exercise 9.8, prove that noninteraction and output controllability can be achieved with respect to the z_i if and only if the same is true for the w_i. In this way the problem is reduced to the one treated in the text.

9.10 Notes and References

Noninteraction is a long-established topic in control theory, dating back at least to Voznesenskii [1]. For reviews of early work with transfer matrices, see Tsien [1] and Kavanagh [1]. The state space approach to decoupling was initiated by Morgan [1] and Rekasius [1], and developed further by Falb and Wolovich [1], Gilbert [2] and Gilbert and Pivnichny [1]; these authors confined their investigations to the case of scalar output blocks, with an equal number of scalar inputs. The more general problems discussed in this chapter were formulated and solved by Wonham and Morse [1] and Morse and Wonham [1]; see also Wonham [6], and Morse and Wonham [2], [3]. A significant alternative approach has been developed by Silverman and Payne [1]. For complementary details see in addition Silverman [1], Cremer [1] and Mufti [1], [2]. The result of Exercise 9.7 is due to Denham [2].

Applications of decoupling theory to the lateral and longitudinal control systems of aircraft are described by Cliff and Lutze [1], [2]; for an additional, adaptive feature to compensate for parameter variations, see Yuan [1] and Yuan and Wonham [1]. Parameter variations can also be accommodated by combining the decoupling problem with that of disturbance decoupling (DDP); this approach is developed by Fabian and Wonham [2], [3] and Chang and Rhodes [1]. Finally, in some applications it may happen that decoupling cannot be exactly achieved or maintained but that coupling effects can be minimized; for a quantitative approach to this problem using optimization and model-following techniques, see Yore [1].

10 Noninteracting Control II: Efficient Compensation

In this chapter we continue the discussion in Chapter 9 on solution of EDP by dynamic compensation. A refinement of the construction used to prove Theorem 9.4 permits a further reduction of the bound (9.6.2) on dynamic order. The reduced bound turns out to be strictly minimal if the number of independent control inputs is equal to the number of output blocks to be decoupled. As these results are somewhat specialized and their proofs are intricate, the reader interested only in the main features of the theory is advised to skip to Chapter 11.

10.1 The Radical

We have seen that the geometric role of dynamic compensation is to supply an auxiliary component of state space. This allows untangling of the $\mathscr{R}_i^*$ in the sense that their extensions can be made independent, hence compatible. To achieve compatibility more efficiently, we first introduce a construction which "localizes" the mutual dependence of an arbitrary collection of subspaces.

Let $\mathscr{V}_1, \ldots, \mathscr{V}_k$ be a family of subspaces of $\mathscr{X}$. The *radical* of the family, written $\check{\mathscr{V}}$ or $(\mathscr{V}_\bullet)^\vee$ ($\bullet$ stands for dummy index), is

$$\check{\mathscr{V}} := \sum_{i=1}^{k} \left[\mathscr{V}_i \cap \left(\sum_{\substack{j=1 \\ j \neq i}}^{k} \mathscr{V}_j \right) \right]. \tag{1.1}$$

From the definition, $\check{\mathscr{V}} = 0$ if and only if the $\mathscr{V}_i$ ($i \in \mathbf{k}$) are independent. As a quantitative measure of mutual dependence among the $\mathscr{V}_i$ we introduce also the function

$$\underset{1\le i\le k}{\Delta}\ \mathscr{V}_i := \sum_{i=1}^{k} d(\mathscr{V}_i) - d\left(\sum_{i=1}^{k} \mathscr{V}_i\right).$$

Thus, $\Delta_i \mathscr{V}_i \ge 0$, with equality if and only if the $\mathscr{V}_i$ are independent.

The computation of dimensional relations is often rendered more efficient by use of the following identities.[1]

Lemma 10.1.

i. *Write*

$$\check{\mathscr{V}}_i := \sum_{j\ne i} \mathscr{V}_j, \qquad i\in\mathbf{k}. \tag{1.2}$$

The radical $\check{\mathscr{V}}$ *has the following properties.*

$$\check{\mathscr{V}} = \bigcap_i \check{\mathscr{V}}_i \tag{1.3}$$

$$= \sum_{j\ne i} (\mathscr{V}_j \cap \check{\mathscr{V}}_j), \qquad i\in\mathbf{k} \tag{1.4}$$

$$= \sum_{j\ne i} (\mathscr{V}_j \cap \check{\mathscr{V}}), \qquad i\in\mathbf{k} \tag{1.5}$$

$$= \sum_{i} (\mathscr{V}_i \cap \check{\mathscr{V}}) \tag{1.6}$$

$$= (\mathscr{V}_\bullet \cap \check{\mathscr{V}})^\vee \tag{1.7}$$

$$= (\mathscr{V}_\bullet \cap \mathscr{V})^\vee, \qquad \text{for all } \mathscr{V}\supset\check{\mathscr{V}} \tag{1.8}$$

$$= (\mathscr{V}_\bullet + \check{\mathscr{V}})^\vee. \tag{1.9}$$

ii. *If* $\mathscr{W}\subset\mathscr{X}$ *and* $\mathscr{V} := (\mathscr{V}_\bullet \cap \mathscr{W})^\vee$, *then*

$$\mathscr{V} = (\mathscr{V}_\bullet \cap \mathscr{V})^\vee.$$

iii.[2] $\check{\mathscr{V}}$ *is the smallest subspace* $\mathscr{V}_0\subset\mathscr{X}$ *with the property: the factor spaces* $(\mathscr{V}_i + \mathscr{V}_0)/\mathscr{V}_0$ $(i\in\mathbf{k})$ *are independent subspaces of* $\mathscr{X}/\mathscr{V}_0$.

iv.
$$\Delta_i \mathscr{V}_i = \Delta_i(\mathscr{V}_i \cap \mathscr{V}) \text{ for all } \mathscr{V}\supset\check{\mathscr{V}}. \tag{1.10}$$

v. *For all* $\mathscr{V}\supset\check{\mathscr{V}}$,

$$\underset{i}{\Delta}\left(\frac{\mathscr{V}_i + \mathscr{V}}{\mathscr{V}}\right) = d\left[\frac{\mathscr{V}\cap\sum_i \mathscr{V}_i}{\sum_i (\mathscr{V}\cap\mathscr{V}_i)}\right].$$

PROOF.

i. Write $\mathscr{W} := \bigcap_i \check{\mathscr{V}}_i$. By an easy induction,

$$\mathscr{W} = (\mathscr{V}_2\cap\check{\mathscr{V}}_2 + \cdots + \mathscr{V}_r\cap\check{\mathscr{V}}_r + \mathscr{V}_{r+1} + \cdots + \mathscr{V}_k)\cap\check{\mathscr{V}}_{r+1}\cap\cdots\cap\check{\mathscr{V}}_k$$

[1] In the spirit of high-school trigonometry.

[2] It is this property which suggested the designation "radical."

for $r = 2, 3, \ldots$. Setting $r = k$ yields

$$\mathscr{W} = \sum_{j=2}^{k} (\mathscr{V}_j \cap \check{\mathscr{V}}_j),$$

hence, by symmetry

$$\mathscr{W} = \sum_{j \neq i} (\mathscr{V}_j \cap \check{\mathscr{V}}_j) \subset \check{\mathscr{V}}_i, \qquad i \in \mathbf{k}. \tag{1.11}$$

By (1.11), $\mathscr{W} \supset \mathscr{V}_i \cap \check{\mathscr{V}}_i$ $(i \in \mathbf{k})$, hence

$$\mathscr{W} \supset \sum_i (\mathscr{V}_i \cap \check{\mathscr{V}}_i) = \check{\mathscr{V}} \qquad \text{(by 1.1) and (1.2))},$$

and this proves (1.3) and (1.4). Also,

$$\begin{aligned}
\mathscr{V}_j \cap \check{\mathscr{V}}_j &= \left(\mathscr{V}_j \cap \bigcap_{l \neq j} \check{\mathscr{V}}_l \right) \cap \check{\mathscr{V}}_j \\
&= \mathscr{V}_j \cap \bigcap_l \check{\mathscr{V}}_l \\
&= \mathscr{V}_j \cap \check{\mathscr{V}} \qquad \text{(by (1.3))},
\end{aligned}$$

and summing over $j \neq i$ yields (1.5). By (1.5)

$$\check{\mathscr{V}} = \sum_i (\mathscr{V}_i \cap \check{\mathscr{V}}) \tag{1.6 bis}$$

$$= \sum_i \left[\mathscr{V}_i \cap \check{\mathscr{V}} \cap \sum_{j \neq i} (\mathscr{V}_j \cap \check{\mathscr{V}}) \right] \qquad \text{(by (1.5))}$$

$$= (\mathscr{V}_\bullet \cap \check{\mathscr{V}})^\vee. \tag{1.7 bis}$$

If $\check{\mathscr{V}} \subset \mathscr{V}$,

$$\begin{aligned}
\check{\mathscr{V}} &\subset (\mathscr{V}_\bullet \cap \mathscr{V})^\vee \qquad \text{(by (1.7))} \\
&\subset (\mathscr{V}_\bullet)^\vee = \check{\mathscr{V}},
\end{aligned}$$

proving (1.8). For (1.9), write $\mathscr{W}_j := \mathscr{V}_j + \check{\mathscr{V}}$ $(j \in \mathbf{k})$. By application of (1.6) to the $\mathscr{W}_j$,

$$\begin{aligned}
\check{\mathscr{W}} &= \sum_j (\mathscr{W}_j \cap \check{\mathscr{W}}) = \sum_j \left[(\mathscr{V}_j + \check{\mathscr{V}}) \cap \sum_{l \neq j} (\mathscr{V}_l + \check{\mathscr{V}}) \right] \\
&= \sum_j [(\mathscr{V}_j + \check{\mathscr{V}}) \cap (\check{\mathscr{V}}_j + \check{\mathscr{V}})] \\
&= \sum_j (\mathscr{V}_j + \check{\mathscr{V}}) \cap \check{\mathscr{V}}_j \qquad (\text{since } \check{\mathscr{V}}_j \supset \check{\mathscr{V}}) \\
&= \sum_j (\mathscr{V}_j \cap \check{\mathscr{V}}_j + \check{\mathscr{V}}) = \check{\mathscr{V}}.
\end{aligned}$$

ii. By application of (1.6) to the family $\mathscr{V}_j \cap \mathscr{W}$ $(j \in \mathbf{k})$,

$$\begin{aligned}
\mathscr{V} &= (\mathscr{V}_\bullet \cap \mathscr{W})^\vee = [(\mathscr{V}_\bullet \cap \mathscr{W}) \cap (\mathscr{V}_\bullet \cap \mathscr{W})^\vee]^\vee = (\mathscr{V}_\bullet \cap \mathscr{W} \cap \mathscr{V})^\vee \\
&= (\mathscr{V} \cap \mathscr{V})^\vee \qquad (\text{since } \mathscr{V} \subset \mathscr{W}).
\end{aligned}$$

iii. The subspaces $(\mathscr{V}_i + \mathscr{V}_0)/\mathscr{V}_0 \subset \mathscr{X}/\mathscr{V}_0$ are independent if and only if

$$(\mathscr{V}_i + \mathscr{V}_0) \cap \sum_{j \neq i} (\mathscr{V}_j + \mathscr{V}_0) = \mathscr{V}_0, \qquad i \in \mathbf{k},$$

or equivalently,

$$(\mathscr{V}_i + \mathscr{V}_0) \cap (\check{\mathscr{V}}_i + \mathscr{V}_0) = \mathscr{V}_0, \qquad i \in \mathbf{k},$$

or

$$\sum_i [(\mathscr{V}_i + \mathscr{V}_0) \cap (\check{\mathscr{V}}_i + \mathscr{V}_0)] = \mathscr{V}_0. \tag{1.12}$$

Let $\hat{\mathscr{V}} = \lim \mathscr{V}^\mu$, where

$$\mathscr{V}^{\mu+1} = \sum_i [(\mathscr{V}_i + \mathscr{V}^\mu) \cap (\check{\mathscr{V}}_i + \mathscr{V}^\mu)], \qquad \mu = 0, 1, 2, \ldots;$$

$$\mathscr{V}^0 = 0.$$

Then, $\mathscr{V}^\mu\uparrow$ as $\mu\uparrow$, hence $\hat{\mathscr{V}}$ exists and is easily seen to be the infimal solution of (1.12); furthermore,

$$\check{\mathscr{V}} = \sum_i (\mathscr{V}_i \cap \check{\mathscr{V}}_i) = \mathscr{V}^1 \subset \hat{\mathscr{V}}. \tag{1.13}$$

Finally, $\check{\mathscr{V}}$ is a solution of (1.12); indeed,

$$\begin{aligned}\sum_i [(\mathscr{V}_i + \check{\mathscr{V}}) \cap (\check{\mathscr{V}}_i + \check{\mathscr{V}})] &= \sum_i [(\mathscr{V}_i + \check{\mathscr{V}}) \cap \check{\mathscr{V}}_i] \\ &= \sum_i (\mathscr{V}_i \cap \check{\mathscr{V}}_i + \check{\mathscr{V}}) = \check{\mathscr{V}}.\end{aligned}$$

Hence, $\check{\mathscr{V}} \supset \hat{\mathscr{V}}$ and this with (1.13) proves $\check{\mathscr{V}} = \hat{\mathscr{V}}$.

iv. By the independence proved in (iii),

$$\sum_i d\left[\frac{\mathscr{V}_i + \check{\mathscr{V}}}{\check{\mathscr{V}}}\right] = d\left[\sum_i \frac{\mathscr{V}_i + \check{\mathscr{V}}}{\check{\mathscr{V}}}\right],$$

so that

$$\sum_i [d(\mathscr{V}_i) - d(\mathscr{V}_i \cap \check{\mathscr{V}})] = d\left(\sum_i \mathscr{V}_i\right) - d(\check{\mathscr{V}}). \tag{1.14}$$

Since

$$d(\check{\mathscr{V}}) = d\left[\sum_i (\mathscr{V}_i \cap \check{\mathscr{V}})\right] \qquad \text{(by (1.6))},$$

(1.14) yields (1.10) for the case $\mathscr{V} = \check{\mathscr{V}}$. Applying this result to the family $\{\mathscr{V}_i \cap \mathscr{V}, i \in \mathbf{k}\}$,

$$\begin{aligned}\Delta_i (\mathscr{V}_i \cap \mathscr{V}) &= \Delta_i [\mathscr{V}_i \cap \mathscr{V} \cap (\mathscr{V}_\bullet \cap \mathscr{V})^\vee] \\ &= \Delta_i (\mathscr{V}_i \cap \mathscr{V} \cap \check{\mathscr{V}}) \qquad \text{(by (1.8))} \\ &= \Delta_i (\mathscr{V}_i \cap \check{\mathscr{V}}) = \Delta_i \mathscr{V}_i,\end{aligned}$$

as claimed.

v. Write $\mathscr{V}_\sigma := \sum_i \mathscr{V}_i$. We have

$$\underset{i}{\Delta}\left(\frac{\mathscr{V}_i + \mathscr{V}}{\mathscr{V}}\right) = \sum_i d\left(\frac{\mathscr{V}_i + \mathscr{V}}{\mathscr{V}}\right) - d\left(\frac{\mathscr{V}_\sigma + \mathscr{V}}{\mathscr{V}}\right)$$

$$= \sum_i d\left(\frac{\mathscr{V}_i}{\mathscr{V}_i \cap \mathscr{V}}\right) - d\left(\frac{\mathscr{V}_\sigma}{\mathscr{V}_\sigma \cap \mathscr{V}}\right)$$

$$= \sum_i d(\mathscr{V}_i) - \sum_i d(\mathscr{V}_i \cap \mathscr{V}) - d(\mathscr{V}_\sigma) + d(\mathscr{V}_\sigma \cap \mathscr{V}).$$

Using (1.10) to evaluate $\sum_i d(\mathscr{V}_i \cap \mathscr{V})$, we get

$$\underset{i}{\Delta}\left(\frac{\mathscr{V}_i + \mathscr{V}}{\mathscr{V}}\right) = -d\left[\sum_i (\mathscr{V}_i \cap \mathscr{V})\right] + d(\mathscr{V}_\sigma \cap \mathscr{V})$$

$$= d\left[\frac{\mathscr{V}_\sigma \cap \mathscr{V}}{\sum_i (\mathscr{V}_i \cap \mathscr{V})}\right],$$

the required result. □

10.2 Efficient Extension

The key to efficient extension (contrast Section 9.6) lies in the property (iii) of the radical stated in Lemma 10.1. The idea will be to reduce the radical of a given family to a smaller subspace having better properties, by means of the following construction.

Lemma 10.2. *Let $\mathscr{R}_1, \ldots, \mathscr{R}_k$ be a family of subspaces of $\mathscr{X}$, let $\mathscr{V} \subset \check{\mathscr{R}}$, and*

$$\mathscr{R}_0 := (\mathscr{R}_\bullet \cap \mathscr{V})^\vee.$$

Let

$$n_a := \underset{1 \le i \le k}{\Delta}\left[\frac{\mathscr{R}_i + \mathscr{R}_0}{\mathscr{R}_0}\right]$$

and take the extended space $\mathscr{X}_e = \mathscr{X} \oplus \mathscr{X}_a$, with $d(\mathscr{X}_a) = n_a$. Then there exist maps $E_i: \mathscr{X}_e \to \mathscr{X}_e$ ($i \in \mathbf{k}$), such that

$$\operatorname{Im} E_i \subset \mathscr{X}_a, \qquad \operatorname{Ker} E_i \supset \mathscr{R}_0, \qquad i \in \mathbf{k},$$

and if

$$\mathscr{V}_i := (1 + E_i)\mathscr{R}_i, \qquad i \in \mathbf{k},$$

then

$$\check{\mathscr{V}} = \mathscr{R}_0.$$

Proof. Let $\bar{\mathscr{X}} := \mathscr{X}/\mathscr{R}_0$ and $P: \mathscr{X} \to \bar{\mathscr{X}}$ be canonical. By Lemma 9.3 there exist maps

$$\bar{E}_i: \bar{\mathscr{X}} \oplus \mathscr{X}_a \to \bar{\mathscr{X}} \oplus \mathscr{X}_a, \qquad i \in \mathbf{k},$$

(where the direct sum is external), such that $\operatorname{Im} \bar{E}_i \subset \mathscr{X}_a$ and the subspaces $(\bar{1} + \bar{E}_i)\bar{\mathscr{R}}_i$ $(i \in \mathbf{k})$ are independent in $\bar{\mathscr{X}} \oplus \mathscr{X}_a$. Let $E_i := \bar{E}_i P$ and $\mathscr{V}_i := (1 + E_i)\mathscr{R}_i$. With $\bar{\mathscr{R}}_i := P\mathscr{R}_i$, we have that

$$\bar{\mathscr{V}}_i = (\bar{1} + \bar{E}_i)\bar{\mathscr{R}}_i = \frac{\mathscr{V}_i + \mathscr{R}_0}{\mathscr{R}_0}$$

are independent in $(\mathscr{X} \oplus \mathscr{X}_a)/\mathscr{R}_0$, and by Lemma 10.1 (iii) there follows $\mathscr{R}_0 \supset \check{\mathscr{V}}$. For the reverse inclusion, note

$$\begin{aligned} \mathscr{R}_0 &= (\mathscr{R}_\bullet \cap \mathscr{R}_0)^\vee \qquad \text{(by Lemma 10.1(ii))} \\ &= \sum_i \left[\mathscr{R}_i \cap \mathscr{R}_0 \cap \sum_{j \neq i} (\mathscr{R}_j \cap \mathscr{R}_0)\right]. \end{aligned}$$

Thus, $x \in \mathscr{R}_0$ implies

$$x = \sum_i x_i$$

with $x_i \in \mathscr{R}_i \cap \mathscr{R}_0$, and

$$x_i = \sum_{j \neq i} x_{ij}, \qquad i \in \mathbf{k},$$

with $x_{ij} \in \mathscr{R}_j \cap \mathscr{R}_0$. Since $\operatorname{Ker} E_i \supset \mathscr{R}_0$ $(i \in \mathbf{k})$,

$$x_i = (1 + E_i)x_i \in \mathscr{V}_i$$

and

$$x_{ij} = (1 + E_j)x_{ij} \in \mathscr{V}_j.$$

Therefore,

$$x_i \in \mathscr{V}_i \cap \hat{\mathscr{V}}_i, \qquad i \in \mathbf{k},$$

hence, $x \in \check{\mathscr{V}}$. □

Suppose now that the $\mathscr{R}_i$ are (A, B)-invariant. In general, it is not true that $\check{\mathscr{R}}$ is (A, B)-invariant. Nevertheless, in the case of interest one can generate a useful class of (A, B)-invariant subspaces contained in $\check{\mathscr{R}}$.

Lemma 10.3. *As in RDP, let $\mathscr{R}_i^*$ $(i \in \mathbf{k})$ be the supremal c.s. contained in*

$$\hat{\mathscr{K}}_i := \bigcap_{j \neq i} \mathscr{K}_j,$$

and let $\check{\mathscr{R}}^*$ denote their radical. Then,

$$\check{\mathscr{R}}^* \subset \bigcap_i \mathscr{K}_i, \tag{2.1}$$

and if $\mathscr{V} \subset \check{\mathscr{R}}^*$ is (A, B)-invariant, so is

$$\check{\mathscr{R}}_0(\mathscr{V}) := (\mathscr{R}_\bullet^* \cap \mathscr{V})^\vee. \tag{2.2}$$

Proof. Dropping the superscript (*), we have

$$\check{\mathscr{R}} = \bigcap_i \sum_{j \neq i} \mathscr{R}_j \qquad \text{(by (1.3))}$$
$$= \bigcap_i \sum_{j \neq i} \bigcap_{l \neq j} \mathscr{K}_l = \bigcap_i \mathscr{K}_i,$$

proving (2.1). Since $\check{\mathscr{R}} \subset \hat{\mathscr{K}}_i$ ($i \in \mathbf{k}$), there results $\mathscr{V} \subset \hat{\mathscr{K}}_i$, hence $\mathscr{V} \subset \mathscr{V}_i$ ($i \in \mathbf{k}$), where

$$\mathscr{V}_i := \sup \mathfrak{I}(A, B; \hat{\mathscr{K}}_i).$$

There follows

$$\varnothing \neq \mathbf{F}(\mathscr{V}_i) \cap \mathbf{F}(\mathscr{V}) \subset \mathbf{F}(\mathscr{R}_i) \cap \mathbf{F}(\mathscr{V}) \subset \mathbf{F}(\mathscr{R}_i \cap \mathscr{V}),$$

i.e. $\mathscr{R}_i \cap \mathscr{V}$ is (A, B)-invariant, hence so is

$$\tilde{\mathscr{V}}_i := \sum_{j \neq i} (\mathscr{R}_j \cap \mathscr{V}).$$

Now,

$$\tilde{\mathscr{V}}_i \subset \sum_{j \neq i} (\mathscr{R}_j \cap \check{\mathscr{R}}) = \check{\mathscr{R}} \qquad \text{(by (1.5))}$$
$$\subset \hat{\mathscr{K}}_i;$$

so $\tilde{\mathscr{V}}_i \subset \mathscr{V}_i$, and applying the same argument as before we get that $\mathscr{R}_i \cap \tilde{\mathscr{V}}_i$ is (A, B)-invariant. Finally,

$$\check{\mathscr{R}}_0(\mathscr{V}) = \sum_i \left[\mathscr{R}_i \cap \mathscr{V} \cap \sum_{j \neq i} (\mathscr{R}_j \cap \mathscr{V})\right]$$
$$= \sum_i \left[\mathscr{R}_i \cap \sum_{j \neq i} (\mathscr{R}_j \cap \mathscr{V})\right] = \sum_i (\mathscr{R}_i \cap \tilde{\mathscr{V}}_i)$$

must be (A, B)-invariant, as claimed. □

Next, we relate the radical to the concept of compatibility. Recall that a family $\{\mathscr{T}_i \subset \mathscr{X}, i \in \mathbf{k}\}$ is *compatible* relative to (A, B) if

$$\bigcap_{i=1}^{k} \mathbf{F}(\mathscr{T}_i) \neq \varnothing.$$

Lemma 10.4. *Let $\mathscr{T}_i$ ($i \in \mathbf{k}$) be (A, B)-invariant. If $\check{\mathscr{T}}$ is (A, B)-invariant, then the family*

$$\check{\mathscr{T}}, \mathscr{T}_1 + \check{\mathscr{T}}, \ldots, \mathscr{T}_k + \check{\mathscr{T}}$$

is compatible.

Proof. Let $P: \mathscr{X} \to \bar{\mathscr{X}} := \mathscr{X}/\check{\mathscr{T}}$ be canonical, let $F_0 \in \mathbf{F}(\check{\mathscr{T}})$ and $A_0 := A + BF_0$. By Lemma 10.1 (iii) the subspaces $\bar{\mathscr{T}}_i := P\mathscr{T}_i$ ($i \in \mathbf{k}$) are independent, and are clearly $(\bar{A}_0, \bar{B})$-invariant relative to the maps $\bar{A}_0: \bar{\mathscr{X}} \to \bar{\mathscr{X}}$ and $\bar{B}: \mathscr{U} \to \bar{\mathscr{X}}$ induced in $\bar{\mathscr{X}}$. Hence, there exists $\bar{F}_1: \bar{\mathscr{X}} \to \mathscr{U}$, such that $(\bar{A}_0 + \bar{B}\bar{F}_1)\bar{\mathcal{T}}_i \subset \bar{\mathcal{T}}_i$ $(i \in \mathbf{k})$. With $F := F_0 + F_1 P$, we have $\overline{A + BF}\,\bar{\mathcal{T}}_i \subset \bar{\mathcal{T}}_i$ $(i \in \mathbf{k})$ and so

$$(A + BF)\mathcal{T}_i \subset \mathcal{T}_i + \check{\mathcal{T}}, \qquad i \in \mathbf{k}.$$

Since also

$$(A + BF)\check{\mathcal{T}} = A_0\check{\mathcal{T}} \subset \check{\mathcal{T}},$$

the lemma follows. □

Combining results we now show how to exploit extension to construct a compatible family with compatible radical. In the following the notation is that of Section 9.4 for the extended spaces $\mathcal{X}_a$, $\mathcal{X}_e$, and extended maps A, B, B_a, introduced in EDP.

Lemma 10.5. *Under the assumptions of Lemma 10.3, take*

$$d(\mathcal{X}_a) \geq n_0 := \Delta_i \left(\frac{\mathcal{R}_i^* + \mathcal{R}_0}{\mathcal{R}_0} \right), \tag{2.3}$$

where $\mathcal{R}_0 := \check{\mathcal{R}}_0(\mathcal{V})$ *is defined by* (2.2). *Then there exist maps* $E_i: \mathcal{X}_e \to \mathcal{X}_e$ $(i \in \mathbf{k})$ *with the properties:*

$$\operatorname{Im} E_i \subset \mathcal{X}_a, \qquad i \in \mathbf{k}, \tag{2.4}$$

$$\operatorname{Ker} E_i \supset \mathcal{R}_0, \qquad i \in \mathbf{k}; \tag{2.5}$$

the subspaces

$$\mathcal{V}_i := (1 + E_i)\mathcal{R}_i^*, \qquad i \in \mathbf{k}, \tag{2.6}$$

are such that

$$\check{\mathcal{V}} = \mathcal{R}_0 \tag{2.7}$$

$$\subset \bigcap_i \mathcal{K}_i; \tag{2.8}$$

and the family

$$\check{\mathcal{V}},\ \mathcal{V}_1 + \check{\mathcal{V}}, \ldots, \mathcal{V}_k + \check{\mathcal{V}}$$

is compatible relative to $(A, B + B_a)$.

PROOF. Lemma 10.2 provides E_i and $\mathcal{V}_i$ with the properties (2.4)–(2.7), and (2.8) follows by Lemma 10.3 and the fact that $\mathcal{R}_0 \subset \check{\mathcal{R}}^*$. Again by Lemma 10.3, $\mathcal{R}_0$ is (A, B)-invariant, hence $(A, B + B_a)$-invariant. Thus, $\check{\mathcal{V}}$ and the $\mathcal{V}_i$ $(i \in \mathbf{k})$ are $(A, B + B_a)$-invariant, and the result follows by application of Lemma 10.4 with $\mathcal{V}_i$ in place of $\mathcal{T}_i$ and $(A, B + B_a)$ in place of (A, B). □

Remark. By Lemma 9.2, the $\mathcal{V}_i$ defined by (2.6) are extended controllability subspaces (e.c.s.) contained in $\hat{\mathcal{K}}_i \oplus \mathcal{X}_a$. They need not, however, be $(A, B + B_a)$-compatible. This difficulty will be treated next.

10.3 Efficient Decoupling

Assume EDP is solvable, i.e.

$$\mathscr{R}_i^* + \mathscr{K}_i = \mathscr{X}, \qquad i \in \mathbf{k}. \tag{3.1}$$

With $d(\mathscr{X}_a)$ subject to (2.3) it will be shown how to construct in $\mathscr{X}_e$ a compatible family of e.c.s. which solves EDP, and also permits assignment of closed-loop eigenvalues to a "good" subset $\mathbb{C}_g \subset \mathbb{C}$. Let $\mathscr{V} \subset \mathscr{X}_e := \mathscr{X} \oplus \mathscr{X}_a$. Then, $\mathbf{F}(\mathscr{V})$ (resp. $\mathbf{F}_e(\mathscr{V})$) will denote the set of maps $F: \mathscr{X}_e \to \mathscr{U}_e$, such that $(A + BF)\mathscr{V} \subset \mathscr{V}$ [resp. $(A + (B + B_a)F)\mathscr{V} \subset \mathscr{V}$]. Now according to Lemma 10.5, where the $\mathscr{V}_i$ ($i \in \mathbf{k}$) are defined, there exists

$$F \in \mathbf{F}_e(\check{\mathscr{V}}) \cap \bigcap_{i=1}^{k} \mathbf{F}_e(\mathscr{V}_i + \check{\mathscr{V}}). \tag{3.2}$$

We define e.c.s. $\mathscr{S}_i$ ($i \in \mathbf{k}$) by means of

$$\mathscr{S}_i = \langle A + (B + B_a)F \,|\, (\mathscr{B} + \mathscr{B}_a) \cap (\mathscr{V}_i + \check{\mathscr{V}})\rangle, \tag{3.3}$$

i.e. $\mathscr{S}_i$ is the supremal e.c.s. in $\mathscr{V}_i + \check{\mathscr{V}}$.

It will be shown that the $\mathscr{S}_i$ solve EDP, namely

$$\mathscr{S}_i + \mathscr{K}_i + \mathscr{X}_a = \mathscr{X} \oplus \mathscr{X}_a, \qquad i \in \mathbf{k}, \tag{3.4}$$

and

$$\mathscr{S}_i \subset \hat{\mathscr{K}}_i \oplus \mathscr{X}_a, \qquad i \in \mathbf{k}. \tag{3.5}$$

By the remark after Lemma 10.5, the $\mathscr{V}_i$ are themselves e.c.s., and clearly satisfy

$$\begin{aligned} \mathscr{V}_i + \mathscr{K}_i + \mathscr{X}_a &= \mathscr{R}_i^* + \mathscr{K}_i + \mathscr{X}_a && \text{(by (2.6))} \\ &= \mathscr{X} \oplus \mathscr{X}_a && \text{(by (3.1))}. \end{aligned} \tag{3.6}$$

Since $\mathscr{V}_i \subset \mathscr{V}_i + \check{\mathscr{V}}$ and $\mathscr{S}_i$ is supremal, (3.4) now follows from (3.6). Finally, as

$$\mathscr{V}_i \subset \mathscr{R}_i^* + \mathscr{X}_a \subset \hat{\mathscr{K}}_i + \mathscr{X}_a, \tag{3.7}$$

we have

$$\begin{aligned} \mathscr{S}_i &\subset \mathscr{V}_i + \check{\mathscr{V}} && \text{(by (3.3))} \\ &\subset \hat{\mathscr{K}}_i + \mathscr{X}_a + \bigcap_j \mathscr{K}_j && \text{(by (2.8), (3.7))} \\ &= \hat{\mathscr{K}}_i \oplus \mathscr{X}_a, \qquad i \in \mathbf{k}, \end{aligned}$$

proving (3.5).

It remains to describe our freedom to assign $\sigma[A + (B + B_a)F]$. This is controlled by the choice of $\mathscr{V} \subset \check{\mathscr{R}}^*$, which serves to fix the subspace

$$\mathscr{R}_0 := \check{\mathscr{R}}_0(\mathscr{V}) := (\mathscr{R}_\bullet^* \cap \mathscr{V})^\vee. \tag{2.2 bis}$$

Starting with a symmetric partition $\mathbb{C} = \mathbb{C}_g \cup \mathbb{C}_b$, take $\mathscr{V} = \mathscr{V}_g$ to be the supremal (A, B)-invariant subspace in $\check{\mathscr{R}}^*$ with the property: there is $\tilde{F} \in \mathbf{F}(\mathscr{V})$, such that

$$\sigma[(A + B\tilde{F}) | \mathscr{V}] \subset \mathbb{C}_g.$$

That $\mathscr{V}_g$ exists as just defined was proved in Lemma 5.7. Indeed, let

$$\mathscr{W} := \sup \mathfrak{J}(A, B; \check{\mathscr{R}}^*). \tag{3.8}$$

If $\mathscr{R} := \sup \mathfrak{C}(A, B; \check{\mathscr{R}}^*)$, then

$$\mathscr{R} \subset \mathscr{V}_g \subset \mathscr{W} \subset \check{\mathscr{R}}^*;$$

taking arbitrary $\hat{F} \in \mathbf{F}(\mathscr{W})$ and with $P: \mathscr{X} \to \mathscr{X}/\mathscr{R}$ canonical, we have explicitly

$$\mathscr{V}_g = P^{-1}[P\mathscr{X}_g(A + B\hat{F}) \cap P\mathscr{W}]. \tag{3.9}$$

Now setting $\mathscr{V} = \mathscr{V}_g$ in (2.2), we obtain

$$\mathscr{R}_0 = (\mathscr{R}_\bullet^* \cap \mathscr{V}_g)^\vee. \tag{3.10}$$

We claim there is $F_0 \in \mathbf{F}(\mathscr{R}_0)$, such that

$$\sigma[(A + BF_0) | \mathscr{R}_0] \subset \mathbb{C}_g. \tag{3.11}$$

As $\mathscr{V}_g$ is (A, B)-invariant so, by Lemma 10.3, is $\mathscr{R}_0$. Furthermore, $\mathscr{R}_0 \supset \mathscr{R}$; indeed, $\mathscr{R}$ is a c.s. such that

$$\mathscr{R} \subset \check{\mathscr{R}}^* \subset \hat{\mathscr{K}}_i, \qquad i \in \mathbf{k} \text{ (by (2.1))},$$

and as the $\mathscr{R}_i^*$ are supremal in $\hat{\mathscr{K}}_i$, we have $\mathscr{R} \subset \mathscr{R}_i^*$ $(i \in \mathbf{k})$, hence, $\mathscr{R} \subset \mathscr{R}_i^* \cap \mathscr{V}_g$ $(i \in \mathbf{k})$, and by (3.10), $\mathscr{R} \subset \mathscr{R}_0$. Choose $F_1 \in \mathbf{F}(\mathscr{R}) \cap \mathbf{F}(\mathscr{R}_0)$ with the property

$$\sigma[(A + BF_1) | \mathscr{R}] \subset \mathbb{C}_g.$$

Such F_1 certainly exists, and $F_1 | \mathscr{R}_0$ clearly has an extension $F_0 \in \mathbf{F}(\mathscr{W})$. Now,

$$\mathbf{F}(\mathscr{W}) \subset \mathbf{F}(\mathscr{R}) \cap \mathbf{F}(\mathscr{V}_g)$$

so that

$$F_0 \in \mathbf{F}(\mathscr{R}) \cap \mathbf{F}(\mathscr{R}_0) \cap \mathbf{F}(\mathscr{V}_g) \cap \mathbf{F}(\mathscr{W}).$$

For the induced map $\overline{A + BF_0}$ on $\mathscr{X}/\mathscr{R}$,

$$\sigma\left[\overline{A + BF_0} \,\middle|\, \frac{\mathscr{V}_g}{\mathscr{R}}\right] \subset \mathbb{C}_g;$$

and finally,

$$\sigma[(A + BF_0) | \mathscr{R}_0] \subset \sigma[(A + BF_0) | \mathscr{R}] \cup \sigma\left[\overline{(A + BF_0)} \,\middle|\, \frac{\mathscr{V}_g}{\mathscr{R}}\right] \subset \mathbb{C}_g,$$

as claimed in (3.11).

The next step is to construct $F_1 \in \mathbf{F}_e(\check{\mathscr{V}})$, such that

$$\sigma[(A + (B + B_a)F_1)|\check{\mathscr{V}}] \subset \mathbb{C}_g. \tag{3.12}$$

As $\check{\mathscr{V}} = \mathscr{R}_0$ (by (2.7)), we arrange that

$$F_1|\check{\mathscr{V}} = F_0|\mathscr{R}_0, \qquad F_1\check{\mathscr{V}} \subset \mathscr{U}, \tag{3.13a,b}$$

with F_0 as in (3.11); here, (3.13b) ensures that $(B + B_a)F_1 = BF_1$, hence $F_1 \in \mathbf{F}_e(\check{\mathscr{V}})$, and (3.12) is true.

Set $A_1 := A + (B + B_a)F_1$. To complete the definition of F, recall that the $(\mathscr{V}_i + \check{\mathscr{V}})/\check{\mathscr{V}}$ are independent c.s. for the pair induced by $(A_1, B + B_a)$ in $(\mathscr{X} \oplus \mathscr{X}_a)/\check{\mathscr{V}}$, and so there exists $F_2 \in \mathbf{F}_e(\mathscr{V}_i + \check{\mathscr{V}})$ $(i \in \mathbf{k})$, such that

$$\operatorname{Ker} F_2 \supset \check{\mathscr{V}} \tag{3.14a}$$

and

$$\sigma[(A_1 + (B + B_a)F_2)|\,\mathscr{V}_i + \check{\mathscr{V}}] \subset \mathbb{C}_g, \qquad i \in \mathbf{k}. \tag{3.14b}$$

Setting $F := F_1 + F_2$, we obtain that (3.2) is true and

$$\sigma\left[(A + (B + B_a)F)\,\middle|\,\check{\mathscr{V}} + \sum_i \mathscr{V}_i\right] \subset \mathbb{C}_g,$$

so

$$\sigma\left[(A + (B + B_a)F)\,\middle|\,\sum_i \mathscr{S}_i\right] \subset \mathbb{C}_g.$$

Finally, we shall assume that $\langle A|\mathscr{B}\rangle = \mathscr{X}$, hence $(A, B + B_a)$ is controllable. Projecting modulo $\sum_i \mathscr{S}_i$ we proceed, in the standard way, to modify F, if necessary, to get

$$\sigma[A + (B + B_a)F] \subset \mathbb{C}_g,$$

the desired result.

Summarizing, we have

Theorem 10.1. *For the RDP of Section 9.2, let (A, B) be controllable, and $\mathscr{R}_i^*$ be the supremal c.s. in $\mathscr{K}_i$, with radical $\check{\mathscr{R}}^*$. Assume EDP is solvable, i.e. $\mathscr{R}_i^* + \mathscr{K}_i = \mathscr{X}$ $(i \in \mathbf{k})$. Let $\mathbb{C} = \mathbb{C}_g \cup \mathbb{C}_b$ be a symmetric partition. Define*

$$\mathscr{V}_g := \sup\{\mathscr{V} : \mathscr{V} \subset \check{\mathscr{R}}^* \;\&\; \exists F \in \mathbf{F}(\mathscr{V}),\ \sigma[(A + BF)|\mathscr{V}] \subset \mathbb{C}_g\}, \tag{3.15a}$$

and

$$\mathscr{R}_0 := (\mathscr{R}_\bullet^* \cap \mathscr{V}_g)^\vee. \tag{3.15b}$$

Then EDP is solvable with extension bound

$$d(\mathscr{X}_a) \le \Delta_i \left(\frac{\mathscr{R}_i^* + \mathscr{R}_0}{\mathscr{R}_0}\right). \tag{3.19}$$

Furthermore, the extended feedback map $F\colon \mathscr{X} \oplus \mathscr{X}_a \to \mathscr{U} \oplus \mathscr{U}_a$ *can be chosen so that*

$$\sigma[A + (B + B_a)F] \subset \mathbb{C}_g.$$

We remark that the bound (3.19) is, in general, lower than the bound

$$d(\mathscr{X}_a) \leq \underset{i}{\Delta}\, \mathscr{R}_i^* \tag{3.20}$$

obtained in Section 9.6. As illustration consider the example of Section 9.7. In the present notation, we have

$$\check{\mathscr{R}}^* = \mathscr{R}_1^* \cap \mathscr{R}_2^* = \operatorname{Im} \begin{bmatrix} 0 & 0 & 0 \\ 1 & 0 & 0 \\ 0 & 1 & 0 \\ 0 & 0 & 1 \\ 0 & 0 & 0 \end{bmatrix}.$$

From (3.8),

$$\mathscr{W} = \operatorname{Im} \begin{bmatrix} 0 & 0 \\ 0 & 0 \\ 1 & 0 \\ 0 & 1 \\ 0 & 0 \end{bmatrix},$$

which yields $A\mathscr{W} \subset \mathscr{W}$ and $\mathscr{R} = 0$. Now,

$$\sigma(A \mid \mathscr{W}) = \{0, 1\}.$$

Suppose $0 \in \mathbb{C}_g$ and $1 \in \mathbb{C}_b$. Then,

$$\mathscr{V}_g = \mathscr{W} \cap \operatorname{Ker} A = \operatorname{Im} \begin{bmatrix} 0 \\ 0 \\ 1 \\ -1 \\ 0 \end{bmatrix}$$

and

$$\mathscr{R}_0 = \mathscr{R}_1^* \cap \mathscr{R}_2^* \cap \mathscr{V}_g = \operatorname{Im} \begin{bmatrix} 0 \\ 0 \\ 1 \\ -1 \\ 0 \end{bmatrix}.$$

This gives

$$d\left[\frac{\mathscr{R}_i^* + \mathscr{R}_0}{\mathscr{R}_0}\right] = 3, \qquad i \in \mathbf{2},$$

and

$$d\left[\frac{\mathscr{R}_1^* + \mathscr{R}_2^*}{\mathscr{R}_0}\right] = 4.$$

Thus (3.19) yields $d(\mathscr{X}_a) \le 2$ in contrast to the bound $d(\mathscr{X}_a) \le 3$ obtained from (3.20).

Computation of efficient decoupling e.c.s. for this example is completed in Exercise 10.2.

10.4 Minimal Order Compensation: $d(\mathscr{B}) = 2$

The solution of EDP provided by Theorem 10.1, while "efficient," is not generally "minimal," in the sense of requiring least possible order of dynamic compensation subject to the constraint that the closed loop spectrum be "good." However, if the number of output blocks to be decoupled happens to equal the number of independent scalar controls, i.e.

$$d(\mathscr{B}) = k, \tag{4.1}$$

this is actually so: the bound (3.19) on $d(\mathscr{X}_a)$ cannot be improved. Quickly stated, the reason is the following: (4.1) means that $d(\mathscr{B})$ has the least value required if EDP is to be solvable at all; then the *only* nontrivial c.s. in $\hat{\mathscr{K}}_i$ is the supremal c.s. $\mathscr{R}_i^*$; and for the $\mathscr{R}_i^*$ the extension described in Theorem 10.1 is always minimal.

While of marginal practical interest, this result has a modest esthetic appeal. In this section we shall prove it in the simplest case

$$d(\mathscr{B}) = k = 2, \tag{4.2}$$

deferring the generalization to Section 10.5. Actually, the central fact required is the following "projective" property of extensions which in no way depends on decoupling, but is interesting in its own right.

Lemma 10.6. *Let $\mathscr{W} \subset \mathscr{X} \oplus \mathscr{X}_a$ be $(A, B + B_a)$-invariant, $\mathscr{S}$ the supremal e.c.s. in $\mathscr{W}$, $P: \mathscr{X} \oplus \mathscr{X}_a \to \mathscr{X} \oplus \mathscr{X}_a$ the projection on $\mathscr{X}$ along $\mathscr{X}_a$, and write*

$$\mathscr{V} := P\mathscr{W}, \qquad \mathscr{R} := P\mathscr{S}.$$

Then, (i) *$\mathscr{V}$ is (A, B)-invariant and $\mathscr{R}$ is the supremal c.s. in $\mathscr{V}$;* (ii) *$\mathscr{V}/\mathscr{R} \simeq \mathscr{W}/\mathscr{S}$; and* (iii) *for all $F \in \mathbf{F}_e(\mathscr{W})$ and $F_0 \in \mathbf{F}(\mathscr{V})$, the induced map $\overline{A + BF_0}$ in $\mathscr{V}/\mathscr{R}$ is similar to the induced map $\overline{A + (B + B_a)F}$ in $\mathscr{W}/\mathscr{S}$.*

PROOF.

i. Recall that $PA = AP$ and $\operatorname{Im}(B + B_a) = \mathscr{B} \oplus \mathscr{B}_a$, so

$$A\mathscr{W} \subset \mathscr{W} + \mathscr{B} + \mathscr{B}_a$$

implies $A\mathscr{V} \subset \mathscr{V} + \mathscr{B}$. By Theorem 5.6, $\mathscr{S} = \lim \mathscr{S}^\mu(\mu\uparrow)$, where $\mathscr{S}^0 = 0$ and

$$\mathscr{S}^\mu := \mathscr{W} \cap (A\mathscr{S}^{\mu-1} + \mathscr{B} + \mathscr{B}_a), \qquad \mu = 1, 2, \ldots.$$

Since Ker $P = \mathscr{X}_a = \mathscr{B}_a$ there follows

$$P\mathscr{S}^\mu = \mathscr{V} \cap (AP\mathscr{S}^{\mu-1} + \mathscr{B}).$$

Again, by Theorem 5.6, lim $P\mathscr{S}^\mu$ is the supremal c.s. in $\mathscr{V}$, and lim $P\mathscr{S}^\mu = P$ lim $\mathscr{S}^\mu = P\mathscr{S} = \mathscr{R}$.

ii. Let $F \in \mathbf{F}_e(\mathscr{W})$, and write

$$\mathscr{W} = \mathscr{W} \cap (\mathscr{S} + \mathscr{X}_a) \oplus \mathscr{T}.$$

Since $\mathscr{W} \cap \mathscr{X}_a \subset \mathscr{S} \subset \mathscr{W}$, we have

$$\mathscr{W} \cap (\mathscr{S} + \mathscr{X}_a) = \mathscr{S}, \tag{4.3}$$

hence

$$\mathscr{W} = \mathscr{S} \oplus \mathscr{T} \tag{4.4}$$

and

$$\mathscr{W} + \mathscr{X}_a = (\mathscr{S} + \mathscr{X}_a) \oplus \mathscr{T}.$$

As $\mathscr{B}_a = \mathscr{X}_a$, $\mathscr{S} + \mathscr{X}_a$ is (A, B)-invariant, so we take $F_0: \mathscr{X} \oplus \mathscr{X}_a \to \mathscr{U} \oplus \mathscr{U}_a$, such that $F_0 \in \mathbf{F}(\mathscr{S} + \mathscr{X}_a)$ and $F_0 | \mathscr{T} = F | \mathscr{T}$. Then, by (4.4)

$$\begin{aligned} &[(A + BF_0) - (A + (B + B_a)F)]\mathscr{W} \\ &\quad \subset (A + BF_0)\mathscr{S} + (A + (B + B_a)F)\mathscr{S} + B_a F\mathscr{T} \subset \mathscr{S} + \mathscr{X}_a. \end{aligned} \tag{4.5}$$

Also

$$\begin{aligned} (A + BF_0)(\mathscr{W} + \mathscr{X}_a) &= (A + BF_0)(\mathscr{S} + \mathscr{T} + \mathscr{X}_a) \\ &\subset (A + BF_0)\mathscr{T} + \mathscr{S} + \mathscr{X}_a \\ &= [A + (B + B_a)F_0]\mathscr{T} + \mathscr{S} + \mathscr{X}_a \\ &= [A + (B + B_a)F]\mathscr{T} + \mathscr{S} + \mathscr{X}_a \\ &\subset \mathscr{W} + \mathscr{X}_a; \end{aligned}$$

and as $\mathscr{V} \subset \mathscr{W} + \mathscr{X}_a$,

$$(A + BF_0)\mathscr{V} = P(A + BF_0)\mathscr{V} \subset P(\mathscr{W} + \mathscr{X}_a) = \mathscr{V};$$

so that finally,

$$F_0 \in \mathbf{F}(\mathscr{S} + \mathscr{X}_a) \cap \mathbf{F}(\mathscr{W} + \mathscr{X}_a) \cap \mathbf{F}(\mathscr{V}). \tag{4.6}$$

By (4.6), together with the standard inclusions $\mathbf{F}_e(\mathscr{W}) \subset \mathbf{F}_e(\mathscr{S})$ and $\mathbf{F}(\mathscr{V}) \subset \mathbf{F}(\mathscr{R})$, the first, third, and fifth squares of the displayed diagrams commute (here, the Q_i are canonical projections and bars as usual denote the induced maps). We claim that isomorphisms J_1, J_2 exist as shown.

For J_1, let $W: \mathscr{W} \to \mathscr{W} + \mathscr{X}_a$ be the insertion map and define J_1 according to $J_1 Q_1 = Q_2 W$; as Ker $Q_1 = \mathscr{S} = (\mathscr{S} + \mathscr{X}_a) \cap \mathscr{W}$ (by (4.3)) $=$ Ker $Q_2 \cap \mathscr{W} =$ Ker$(Q_2 W)$, J_1 exists and is unique.

$$
\begin{array}{ccc}
\mathscr{W} & \xrightarrow{A+(B+B_a)F} & \mathscr{W} \\
{\scriptstyle Q_1}\big\downarrow & & \big\downarrow{\scriptstyle Q_1} \\
\mathscr{W}/\mathscr{S} & \xrightarrow{\overline{A+(B+B_a)F}} & \mathscr{W}/\mathscr{S} \\
{\scriptstyle J_1}\big\downarrow\approx & & \approx\big\downarrow{\scriptstyle J_1} \\
\dfrac{\mathscr{W}+\mathscr{X}_a}{\mathscr{S}+\mathscr{X}_a} & \xrightarrow{\overline{A+BF_0}} & \dfrac{\mathscr{W}+\mathscr{X}_a}{\mathscr{S}+\mathscr{X}_a}
\end{array}
$$

$$
\begin{array}{ccc}
\mathscr{W}+\mathscr{X}_a & \xrightarrow{A+BF_0} & \mathscr{W}+\mathscr{X}_a \\
{\scriptstyle Q_2}\big\downarrow & & \big\downarrow{\scriptstyle Q_2} \\
\dfrac{\mathscr{W}+\mathscr{X}_a}{\mathscr{S}+\mathscr{X}_a} & \xrightarrow{\overline{A+BF_0}} & \dfrac{\mathscr{W}+\mathscr{X}_a}{\mathscr{S}+\mathscr{X}_a} \\
{\scriptstyle J_2}\big\downarrow\approx & & \approx\big\downarrow{\scriptstyle J_2} \\
\mathscr{V}/\mathscr{R} & \xrightarrow{\overline{\overline{A+BF_0}}} & \mathscr{V}/\mathscr{R}
\end{array}
$$

$$
\begin{array}{ccc}
\mathscr{V} & \xrightarrow{A+BF_0} & \mathscr{V} \\
{\scriptstyle Q_3}\big\downarrow & & \big\downarrow{\scriptstyle Q_3} \\
\mathscr{V}/\mathscr{R} & \xrightarrow{\overline{\overline{A+BF_0}}} & \mathscr{V}/\mathscr{R}
\end{array}
$$

With $x \in \mathscr{W}$, $J_1(Q_1 x) = 0$ implies $x \in \operatorname{Ker} Q_2 W = \mathscr{S}$, so $Q_1 x = 0$ and J_1 is monic; also

$$Q_2 W \mathscr{W} = Q_2 \mathscr{W} = Q_2(\mathscr{W} + \mathscr{X}_a) = \frac{\mathscr{W} + \mathscr{X}_a}{\mathscr{S} + \mathscr{X}_a},$$

hence, J_1 is epic. For the second square, with $x \in \mathscr{W}$,

$$
\begin{aligned}
(\overline{A + BF_0})J_1(Q_1 x) &= (\overline{A + BF_0})(Q_2 W x) \\
&= Q_2(A + BF_0)Wx \\
&= Q_2 W[A + (B + B_a)F]x \qquad \text{(by (4.5))} \\
&= J_1 Q_1[A + (B + B_a)F]x \\
&= J_1[\overline{A + (B + B_a)F}](Q_1 x),
\end{aligned}
$$

as claimed.

Define J_2 according to $J_2 Q_2 = Q_3 P$; since

$$
\begin{aligned}
\operatorname{Ker} Q_2 = \mathscr{S} + \mathscr{X}_a = \mathscr{S} + \operatorname{Ker} P &= P^{-1}(P\mathscr{S}) \\
&= P^{-1}(\mathscr{X} \cap \mathscr{R}) = P^{-1}(\operatorname{Im} P \cap \operatorname{Ker} Q_3) \\
&= \operatorname{Ker}(Q_3 P),
\end{aligned}
$$

J_2 exists and is unique; with $x \in \mathscr{W} + \mathscr{X}_a$, $J_2(Q_2 x) = 0$ implies

$$x \in \operatorname{Ker}(Q_3 P) = \mathscr{S} + \mathscr{X}_a = \operatorname{Ker} Q_2,$$

so $Q_2 x = 0$, and J_2 is monic; as $Q_3 P | (\mathscr{W} + \mathscr{X}_a)$ is epic, J_2 is epic too. For the fourth square, with $x \in \mathscr{W} + \mathscr{X}_a$,

$$\begin{aligned}
(\overline{\overline{A + BF_0}})J_2(Q_2 x) &= (\overline{\overline{A + BF_0}})Q_3 P x \\
&= Q_3(A + BF_0)Px \\
&= Q_3 P(A + BF_0)Px \quad [\text{since } (A + BF_0)Px \in \mathscr{V}] \\
&= J_2 Q_2(A + BF_0)Px \quad [\text{since } \mathscr{V} \subset \mathscr{W} + \mathscr{X}_a] \\
&= J_2(\overline{A + BF_0})Q_2 P x \\
&= J_2(\overline{A + BF_0})(Q_2 x) \\
&\qquad [\text{since } x - Px \in \mathscr{X}_a \subset \operatorname{Ker} Q_2],
\end{aligned}$$

and the fourth square commutes, as claimed.

iii. Cut out the second and fourth squares. Attach the top edge of the fourth to the bottom edge of the second. Then appeal to Theorem 5.7.

Returning to the decoupling problem, we have on the assumption (4.2)

$$\breve{\mathscr{R}}^* = \mathscr{R}_1^* \cap \mathscr{R}_2^*,$$

$$\mathscr{R}_0 = \mathscr{V}_g, \tag{4.7}$$

and the bound (3.19) becomes

$$n_0 = d(\mathscr{R}_1^* \cap \mathscr{R}_2^*) - d(\mathscr{R}_0). \tag{4.8}$$

Our aim is to show that for any solution of EDP, such that

$$\sigma[A + (B + B_a)F] \subset \mathbb{C}_g \tag{4.9}$$

we must have

$$d(\mathscr{X}_a) \geq n_0. \tag{4.10}$$

The proof depends on Lemma 10.6 together with some easier relations which we establish next. It will be assumed throughout that (4.2) holds and $\langle A | \mathscr{B} \rangle = \mathscr{X}$.

Lemma 10.7. *Let $\mathscr{R}_1$, $\mathscr{R}_2$ be c.s., such that*

$$\begin{aligned}
\mathscr{R}_1 \subset \mathscr{K}_2, &\qquad \mathscr{R}_1 + \mathscr{K}_1 = \mathscr{X} \\
\mathscr{R}_2 \subset \mathscr{K}_1, &\qquad \mathscr{R}_2 + \mathscr{K}_2 = \mathscr{X},
\end{aligned} \tag{4.11}$$

where $0 \neq \mathscr{K}_i \neq \mathscr{X}$ ($i \in \mathbf{2}$). Then,

$$\mathscr{R}_i = \mathscr{R}_i^* \qquad (i \in \mathbf{2}).$$

PROOF. By (4.2), we must have

$$d(\mathscr{B} \cap \mathscr{R}_i) = 0,\ 1,\ \text{or}\ 2, \qquad i \in \mathbf{2},$$

and the first and third possibilities are ruled out by (4.11). Then, $\mathscr{B} \cap \mathscr{R}_i = \mathscr{B} \cap \mathscr{R}_i^*$, and choosing $F_i \in \mathbf{F}(\mathscr{R}_i) \cap \mathbf{F}(\mathscr{R}_i^*)$, we get

$$\mathscr{R}_i = \langle A + BF_i | \mathscr{B} \cap \mathscr{R}_i \rangle = \langle A + BF_i | \mathscr{B} \cap \mathscr{R}_i^* \rangle = \mathscr{R}_i^*,$$

as claimed. □

Lemma 10.8. *Let* $\mathscr{T}_i \subset \mathscr{X}$ $(i \in \mathbf{2})$, $\mathscr{G} \subset \mathscr{X}$, *and* $Q\colon \mathscr{X} \to \mathscr{X}/\mathscr{G}$ *the canonical projection. Then,*

$$d(\mathscr{G}) \geq d\left[\frac{Q\mathscr{T}_1 \cap Q\mathscr{T}_2}{Q(\mathscr{T}_1 \cap \mathscr{T}_2)}\right].$$

PROOF.

$$d\left[\frac{Q\mathscr{T}_1 \cap Q\mathscr{T}_2}{Q(\mathscr{T}_1 \cap \mathscr{T}_2)}\right] = d\left[\frac{(\mathscr{T}_1 + \mathscr{T}_2) \cap \mathscr{G}}{\mathscr{T}_1 \cap \mathscr{G} + \mathscr{T}_2 \cap \mathscr{G}}\right] \qquad \text{(by Exercise 0.5(ix))}$$

$$\leq d(\mathscr{G}).$$ □

As an immediate application, we obtain

Lemma 10.9. *Let P be the projection on* $\mathscr{X}$ *along* $\mathscr{X}_a$. *Let* $\mathscr{S}_i \subset \mathscr{X} \oplus \mathscr{X}_a$ $(i \in \mathbf{2})$ *and* $\mathscr{R}_i = P\mathscr{S}_i$ $(i \in \mathbf{2})$. *If* $\mathscr{R}_0$ *is such that*

$$P(\mathscr{S}_1 \cap \mathscr{S}_2) \subset \mathscr{R}_0 \subset \mathscr{R}_1 \cap \mathscr{R}_2$$

then

$$d(\mathscr{X}_a) \geq d\left[\frac{\mathscr{R}_1 \cap \mathscr{R}_2}{\mathscr{R}_0}\right].$$

PROOF. Apply Lemma 10.8 with $\mathscr{S}_i$, $\mathscr{X} \oplus \mathscr{X}_a$, $\mathscr{X}_a$ and P in place of $\mathscr{T}_i$, $\mathscr{X}$, $\mathscr{G}$, and Q. □

As our last preliminary result, we have

Lemma 10.10. *If e.c.s.* $\mathscr{S}_i$ $(i \in \mathbf{2})$ *provide a solution of EDP, such that* (4.9) *is true, then*

$$P(\mathscr{S}_1 \cap \mathscr{S}_2) \subset \mathscr{R}_0,$$

where $\mathscr{R}_0$ *is given by* (4.7).

PROOF. By (4.9) there exists $F \in \mathbf{F}_e(\mathscr{S}_1 \cap \mathscr{S}_2)$, such that

$$\sigma[(A + (B + B_a)F) | \mathscr{S}_1 \cap \mathscr{S}_2] \subset \mathbb{C}_g. \tag{4.12}$$

Writing $\mathscr{S}^*$ for the supremal e.c.s. in $\mathscr{S}_1 \cap \mathscr{S}_2$, and $\mathscr{R}^*$ for the supremal c.s. in $P(\mathscr{S}_1 \cap \mathscr{S}_2)$ we have by Lemma 10.6 that (in our usual notation)

$$\sigma\left[\overline{A + (B + B_a)F} \,\middle|\, \frac{\mathscr{S}_1 \cap \mathscr{S}_2}{\mathscr{S}^*}\right] = \sigma\left[\overline{\overline{A + BF_0}} \,\middle|\, \frac{P(\mathscr{S}_1 \cap \mathscr{S}_2)}{\mathscr{R}^*}\right]$$

for all $F_0 \in \mathbf{F}[P(\mathscr{S}_1 \cap \mathscr{S}_2)]$. By (4.12) and (4.13) there exists $F_0 \in \mathbf{F}[P(\mathscr{S}_1 \cap \mathscr{S}_2)]$, such that

$$\sigma[(A + BF_0) \mid P(\mathscr{S}_1 \cap \mathscr{S}_2)] \subset \mathbb{C}_g. \tag{4.14}$$

But as the $\mathscr{S}_i$ solve EDP, we have $P\mathscr{S}_i \subset \mathscr{R}_i^*$ ($i \in \mathbf{2}$), so that

$$P(\mathscr{S}_1 \cap \mathscr{S}_2) \subset \mathscr{R}_1^* \cap \mathscr{R}_2^*. \tag{4.15}$$

Noting again that $\mathscr{R}_1^* \cap \mathscr{R}_2^* = \check{\mathscr{R}}^*$, and recalling the supremal property (3.15a) of $\mathscr{V}_g$, we see from (4.14) and (4.15) that

$$P(\mathscr{S}_1 \cap \mathscr{S}_2) \subset \mathscr{V}_g = \mathscr{R}_0 \qquad \text{(by (4.7))},$$

as asserted. □

It is now easy to prove our main result, the inequality (4.10). Assuming $\mathscr{S}_i$ ($i \in \mathbf{2}$) solve EDP, we have that $P\mathscr{S}_i$ satisfy (4.11), hence by Lemma 10.7, $P\mathscr{S}_i = \mathscr{R}_i^*$ ($i \in \mathbf{2}$). By Lemmas 10.9 and 10.10 there results finally

$$d(\mathscr{X}_a) \geq d\left[\frac{\mathscr{R}_1^* \cap \mathscr{R}_2^*}{\mathscr{R}_0}\right] = n_0 \qquad \text{(by (4.8))}$$

as we set out to show.

10.5 Minimal Order Compensation: $d(\mathscr{B}) = k$

In this section we prove that the bound (3.19) on $d(\mathscr{X}_a)$ is best possible when $d(\mathscr{B}) = k$, $k \geq 2$ arbitrary. The argument follows the same lines as in Section 10.4, but with $\mathscr{R}_1^* \cap \mathscr{R}_2^*$ now replaced by the radical $\check{\mathscr{R}}^* := \{\mathscr{R}_i^*, i \in \mathbf{k}\}^\vee$. The first step is to generalize Lemma 10.7.

Lemma 10.11. *Let $\mathscr{R}_i$ ($i \in \mathbf{k}$) be c.s., such that*

$$\mathscr{R}_i \subset \hat{\mathscr{K}}_i, \qquad \mathscr{R}_i + \mathscr{K}_i = \mathscr{X}, \qquad i \in \mathbf{k}, \tag{5.1}$$

where $0 \neq \mathscr{K}_i \neq \mathscr{X}$, $i \in \mathbf{k}$. Then,

$$\mathscr{R}_i = \mathscr{R}_i^*, \qquad i \in \mathbf{k}.$$

PROOF. Just as in the proof of Lemma 10.7, it is enough to show that

$$d(\mathscr{B} \cap \mathscr{R}_i^*) = 1, \qquad i \in \mathbf{k}. \tag{5.2}$$

To verify (5.2) start from

$$d\left(\mathscr{B} \cap \sum_{i=1}^{j} \mathscr{R}_i^*\right) \le d\left(\mathscr{B} \cap \sum_{i=1}^{j+1} \mathscr{R}_i^*\right), \qquad j \in \mathbf{k-1}. \tag{5.3}$$

If (5.3) holds with equality for $j = l$, then

$$\mathscr{B} \cap \sum_{i=1}^{l} \mathscr{R}_i^* = \mathscr{B} \cap \sum_{i=1}^{l+1} \mathscr{R}_i^*. \tag{5.4}$$

Write $\mathscr{E} := \sum_{i=1}^{l} \mathscr{R}_i^*$. By (5.4), trivially,

$$\mathscr{B} \cap (\mathscr{E} + \mathscr{R}_{l+1}^*) = \mathscr{B} \cap \mathscr{E} + \mathscr{B} \cap \mathscr{R}_{l+1}^*$$

and so, by (0.3.2)–(0.3.3),

$$\mathscr{E} \cap (\mathscr{B} + \mathscr{R}_{l+1}^*) = \mathscr{E} \cap \mathscr{B} + \mathscr{E} \cap \mathscr{R}_{l+1}^*.$$

Then

$$A(\mathscr{E} \cap \mathscr{R}_{l+1}^*) \subset (\mathscr{B} + \mathscr{E}) \cap (\mathscr{B} + \mathscr{R}_{l+1}^*) \subset \mathscr{B} + \mathscr{E} \cap \mathscr{R}_{l+1}^*. \tag{5.5}$$

By (5.5), $\mathbf{F}(\mathscr{E} \cap \mathscr{R}_{l+1}^*) \neq \varnothing$, hence there exists $F \in \mathbf{F}(\mathscr{E}) \cap \mathbf{F}(\mathscr{R}_{l+1}^*)$, and for such F,

$$\begin{aligned} \mathscr{R}_{l+1}^* &= \langle A + BF \,|\, \mathscr{B} \cap \mathscr{R}_{l+1}^* \rangle \subset \langle A + BF \,|\, \mathscr{E} \rangle \qquad \text{(using (5.4))} \\ &\subset \mathscr{E} \subset \sum_{i=1}^{l} \bigcap_{j \neq i} \mathscr{K}_j \subset \mathscr{K}_{l+1}, \end{aligned}$$

which is ruled out by (5.1). Therefore, (5.3) holds with strict inequality at each $j \in \mathbf{k-1}$. Since

$$d\left(\mathscr{B} \cap \sum_{i=1}^{k} \mathscr{R}_i^*\right) \le d(\mathscr{B}) = k,$$

and since, again by (5.1), $d(\mathscr{B} \cap \mathscr{R}_1^*) \ge 1$, we conclude that

$$d\left(\mathscr{B} \cap \sum_{i=1}^{j} \mathscr{R}_i^*\right) = j, \qquad j \in \mathbf{k}.$$

In particular, $d(\mathscr{B} \cap \mathscr{R}_1^*) = 1$; and since the ordering of the $\mathscr{R}_i^*$ is immaterial there follows $d(\mathscr{B} \cap \mathscr{R}_i^*) = 1$ ($i \in \mathbf{k}$) as claimed. □

The remaining task is to generalize the dimensional inequality of Lemma 10.8.

Lemma 10.12. *Let $\mathscr{T}_i$ ($i \in \mathbf{k}$) and $\mathscr{G}$ be subspaces of $\mathscr{X}$, and let $\mathscr{T}_0 \subset \mathscr{X}$ be any subspace with the properties*

$$\check{\mathscr{T}} + \mathscr{G} \subset \mathscr{T}_0 = [(\mathscr{T}_\bullet + \mathscr{G}) \cap \mathscr{T}_0]^\vee. \tag{5.6}$$

Then

$$\Delta_i \left(\frac{\mathscr{T}_i + \mathscr{T}_0}{\mathscr{T}_0} \right) = d \left[\frac{\mathscr{G} + \sum_i (\mathscr{T}_i \cap \mathscr{T}_0)}{\sum_i (\mathscr{T}_i \cap \mathscr{T}_0)} \right] - d \left[\frac{\mathscr{T}_0 + \sum_i \mathscr{T}_i}{\sum_i \mathscr{T}_i} \right] \tag{5.7}$$

$$\leq d(\mathscr{G}). \tag{5.8}$$

PROOF. Write $\mathscr{T}_\sigma := \sum_i \mathscr{T}_i$. By Lemma 10.1(v),

$$\begin{aligned}
\Delta_i \left(\frac{\mathscr{T}_i + \mathscr{T}_0}{\mathscr{T}_0} \right) &= d \left[\frac{\mathscr{T}_0 \cap \mathscr{T}_\sigma}{\sum_i (\mathscr{T}_0 \cap \mathscr{T}_i)} \right] \\
&= d \left[\frac{\mathscr{T}_0}{\sum_i (\mathscr{T}_0 \cap \mathscr{T}_i)} \right] - d \left[\frac{\mathscr{T}_0}{\mathscr{T}_0 \cap \mathscr{T}_\sigma} \right] \\
&= d \left[\frac{\mathscr{T}_0}{\sum_i (\mathscr{T}_0 \cap \mathscr{T}_i)} \right] - d \left[\frac{\mathscr{T}_0 + \mathscr{T}_\sigma}{\mathscr{T}_\sigma} \right].
\end{aligned} \tag{5.9}$$

Now

$$\begin{aligned}
\mathscr{T}_0 = \mathscr{T}_0 + \mathscr{G} &= [(\mathscr{T}_\bullet + \mathscr{G}) \cap \mathscr{T}_0]^\vee + \mathscr{G} \\
&= \sum_i \{[(\mathscr{T}_i + \mathscr{G}) \cap \mathscr{T}_0] \cap [(\mathscr{T}_\bullet + \mathscr{G}) \cap \mathscr{T}_0]^\vee\} + \mathscr{G} \qquad \text{(by (1.6))} \\
&= \sum_i [(\mathscr{T}_i + \mathscr{G}) \cap \mathscr{T}_0] + \mathscr{G} \qquad \text{(by (5.6))} \\
&= \sum_i (\mathscr{T}_i \cap \mathscr{T}_0) + \mathscr{G}.
\end{aligned} \tag{5.10}$$

By (5.10), trivially,

$$d \left[\frac{\mathscr{T}_0}{\sum_i (\mathscr{T}_0 \cap \mathscr{T}_i)} \right] = d \left[\frac{\mathscr{G} + \sum_i (\mathscr{T}_i \cap \mathscr{T}_0)}{\sum_i (\mathscr{T}_0 \cap \mathscr{T}_i)} \right];$$

substitution of this in (5.9) yields (5.7), and (5.8) is then obvious. □

Applying this result in the context of EDP, we obtain

Lemma 10.13. *Let P be the projection on $\mathscr{X}$ along $\mathscr{X}_a$. Let $\mathscr{S}_i \subset \mathscr{X} \oplus \mathscr{X}_a$ $(i \in \mathbf{k})$ and $\mathscr{R}_i := P\mathscr{S}_i$ $(i \in \mathbf{k})$. Suppose $\mathscr{R}_0 \subset \mathscr{X}$ and $\mathscr{V} \subset \mathscr{X}$ are such that*

$$P\check{\mathscr{S}} \subset \mathscr{R}_0 := (\mathscr{R}_\bullet \cap \mathscr{V})^\vee.$$

Then

$$d(\mathscr{X}_a) \geq \Delta_i \left(\frac{\mathscr{R}_i + \mathscr{R}_0}{\mathscr{R}_0} \right).$$

PROOF. In Lemma 10.12 replace $(\mathscr{T}_i, \mathscr{G}, \mathscr{X})$ by $(\mathscr{S}_i, \mathscr{X}_a, \mathscr{X} \oplus \mathscr{X}_a)$ and define

$$\mathscr{S}_0 := [(\mathscr{S}_\bullet + \mathscr{X}_a) \cap (\mathscr{V} + \mathscr{X}_a)]^\vee.$$

Since $P\breve{\mathscr{S}} \subset \mathscr{V}$, we have $\breve{\mathscr{S}} \subset \mathscr{V} + \mathscr{X}_a$, so

$$\mathscr{S}_0 \supset [(\mathscr{S}_\bullet + \mathscr{X}_a) \cap \breve{\mathscr{S}}]^\vee \supset (\mathscr{S}_\bullet \cap \breve{\mathscr{S}})^\vee = \breve{\mathscr{S}} \qquad \text{(by (1.7))},$$

hence, $\mathscr{S}_0 \supset \breve{\mathscr{S}} + \mathscr{X}_a$. Also, by Lemma 10.1(ii), applied to the family $\{\mathscr{S}_i + \mathscr{X}_a, i \in \mathbf{k}\}$,

$$\mathscr{S}_0 = [(\mathscr{S}_\bullet + \mathscr{X}_a) \cap \mathscr{S}_0]^\vee .$$

Thus, $\mathscr{S}_0$ has the properties required of $\mathscr{T}_0$ in (5.6), and (5.8) yields

$$\begin{aligned} d(\mathscr{X}_a) &\geq \Delta_i \left(\frac{\mathscr{S}_i + \mathscr{S}_0}{\mathscr{S}_0}\right) = \Delta_i \left(\frac{\mathscr{S}_i + \mathscr{S}_0 + \mathscr{X}_a}{\mathscr{S}_0 + \mathscr{X}_a}\right) \\ &= \Delta_i \left[\frac{(\mathscr{S}_i + \mathscr{S}_0 + \mathscr{X}_a)/\mathscr{X}_a}{(\mathscr{S}_0 + \mathscr{X}_a)/\mathscr{X}_a}\right] \\ &= \Delta_i \left[\frac{P\mathscr{S}_i + P\mathscr{S}_0}{P\mathscr{S}_0}\right] \\ &= \Delta_i \left(\frac{\mathscr{R}_i + \mathscr{R}_0}{\mathscr{R}_0}\right). \end{aligned}$$

□

To prove our main result, that the bound (3.19) is minimal, it now suffices to check the hypotheses of Lemma 10.13 (with $\mathscr{R}_i$ replaced by $\mathscr{R}_i^*$) for any solution $\{\mathscr{S}_i, i \in \mathbf{k}\}$ of EDP which satisfies the additional requirement

$$\sigma[A + (B + B_a)F] \subset \mathbb{C}_g. \qquad \text{(4.9 bis)}$$

By exactly the same argument as in the proof of Lemma 10.10 (with $\mathscr{S}_1 \cap \mathscr{S}_2$ replaced by $\breve{\mathscr{S}}$), (4.9) implies $P\breve{\mathscr{S}} \subset \mathscr{V}_g$. But

$$\begin{aligned} \breve{\mathscr{S}} &= (\mathscr{S}_\bullet \cap \breve{\mathscr{S}})^\vee && \text{(by (1.7))} \\ &= \bigcap_i \sum_{j \neq i} (\mathscr{S}_j \cap \breve{\mathscr{S}}) && \text{(by (1.3))}; \end{aligned}$$

also, Lemma 10.11 applied to $\mathscr{R}_i = P\mathscr{S}_i$ yields $P\mathscr{S}_i = \mathscr{R}_i^*$ $(i \in \mathbf{k})$; and so

$$\begin{aligned} P\breve{\mathscr{S}} &\subset \bigcap_i \sum_{j \neq i} (\mathscr{R}_j^* \cap \mathscr{V}_g) \\ &= (\mathscr{R}_\bullet^* \cap \mathscr{V}_g)^\vee && \text{(by (1.3))} \\ &= \mathscr{R}_0, \end{aligned}$$

by the definition (3.15b). The main result is proved.

10.6 Exercises

10.1. Develop a computational procedure for decoupling with efficient compensation. Hint: Given A, B, D_i $(i \in \mathbf{k})$ and $\mathbb{C} = \mathbb{C}_g \cup \mathbb{C}_b$, compute in the following order: $\mathscr{K}_i$, $\hat{\mathscr{K}}_i$, $\mathscr{R}_i^*$ $(i \in \mathbf{k})$; solvability verification of EDP, (3.1); $\breve{\mathscr{R}}^*$ by (1.1);

$\mathscr{W} := \sup \mathfrak{I}(A, B; \check{\mathscr{R}}^*)$, $\mathscr{R} := \sup \mathfrak{C}(A, B; \mathscr{W})$; $\hat{F} \in \mathbf{F}(\mathscr{W})$; $P: \mathscr{X} \to \mathscr{X}/\mathscr{R}$; $\mathscr{X}_g(A + B\hat{F})$; $\mathscr{V}_g$ by (3.9); $\mathscr{R}_0 := \check{\mathscr{R}}_0(\mathscr{V}_g)$ by (2.2); $d(\mathscr{X}_a) = n_0$ by (2.3); E_i $(i \in \mathbf{k})$ by Lemma 10.5; $\mathscr{V}_i$ $(i \in \mathbf{k})$ by (2.6); F by (3.2); $\mathscr{S}_i$ $(i \in \mathbf{k})$ by (3.3). To assign the spectrum to $\mathbb{C}_g$, modify F by the procedure starting with (3.11).

10.2. Construct a numerical example to illustrate the procedure of Exercise 10.1. Hint: The example at the end of Section 10.3 illustrates the procedure up to the computation of $n_0 = 2$. To compute the E_i by Lemma 10.5, it is easiest first to pick a basis for $\mathscr{R}_1^* + \mathscr{R}_2^*$ which exhibits $\mathscr{R}_0$ and $\mathscr{R}_1^* \cap \mathscr{R}_2^*$; thus,

$$\mathscr{R}_1^* = \operatorname{Im} \begin{bmatrix} 1 & 0 & 0 & 0 \\ 0 & 1 & 0 & 0 \\ 0 & 0 & 1 & 1 \\ 0 & 0 & 0 & -1 \\ 0 & 0 & 0 & 0 \end{bmatrix}, \qquad \mathscr{R}_2^* = \operatorname{Im} \begin{bmatrix} 0 & 0 & 0 & 0 \\ 1 & 0 & 0 & 0 \\ 0 & 1 & 0 & 1 \\ 0 & 0 & 0 & -1 \\ 0 & 0 & 1 & 0 \end{bmatrix}.$$

According to Lemma 10.5, one may pick $E_1 = 0$, and define E_2 such that $E_2 \mathscr{R}_0 = 0$ and E_2 maps the complement of $\mathscr{R}_0$, in $\mathscr{R}_1^* \cap \mathscr{R}_2^*$, onto $\mathscr{X}_a$: e.g.

$$E_2 : \begin{bmatrix} 0 & 0 \\ 1 & 0 \\ 0 & 1 \\ 0 & 0 \\ 0 & 0 \end{bmatrix} \mapsto \left[\begin{array}{cc} 0 & 0 \\ 0 & 0 \\ 0 & 0 \\ 0 & 0 \\ 0 & 0 \\ \hline 1 & 0 \\ 0 & 1 \end{array}\right].$$

The subspaces $\mathscr{V}_i = (1 + E_i)\mathscr{R}_i^* \subset \mathscr{X} \oplus \mathscr{X}_a$ are then

$$\mathscr{V}_1 = \operatorname{Im} \left[\begin{array}{cccc} 1 & 0 & 0 & 0 \\ 0 & 1 & 0 & 0 \\ 0 & 0 & 1 & 1 \\ 0 & 0 & 0 & -1 \\ 0 & 0 & 0 & 0 \\ \hline 0 & 0 & 0 & 0 \\ 0 & 0 & 0 & 0 \end{array}\right], \qquad \mathscr{V}_2 = \operatorname{Im} \left[\begin{array}{cccc} 0 & 0 & 0 & 0 \\ 1 & 0 & 0 & 0 \\ 0 & 1 & 0 & 1 \\ 0 & 0 & 0 & -1 \\ 0 & 0 & 1 & 0 \\ \hline 1 & 0 & 0 & 0 \\ 0 & 1 & 0 & 0 \end{array}\right]$$

It is a routine matter to select B_a, and then compute an F to satisfy (3.6); one choice yields

$$B + B_a = \left[\begin{array}{cccc} 0 & 1 & 0 & 0 \\ 1 & 0 & 0 & 0 \\ 0 & 0 & 0 & 0 \\ 0 & 0 & 0 & 0 \\ 0 & 1 & 0 & 0 \\ \hline 0 & 0 & 1 & 0 \\ 0 & 0 & 0 & 1 \end{array}\right],$$

$$F = \left[\begin{array}{c|ccc} & 0 & 0 & 0 \\ 0^{4 \times 4} & 0 & -1 & 0 \\ & 0 & 0 & 0 \\ & 1 & 0 & 1 \end{array}\right].$$

From this there follows

$$A + (B + B_a)F = \left[\begin{array}{ccccc|cc} 0 & 1 & 0 & 0 & 0 & -1 & 0 \\ 0 & 0 & 0 & 0 & 0 & 0 & 0 \\ 0 & 0 & 1 & 1 & 0 & 0 & 0 \\ 1 & 0 & 0 & 0 & 1 & 0 & 0 \\ 0 & 0 & 0 & 0 & 0 & -1 & 0 \\ \hline 0 & 0 & 0 & 0 & 0 & 0 & 0 \\ 0 & 0 & 0 & 0 & 1 & 0 & 1 \end{array}\right],$$

which leaves $\mathscr{V}_1$, $\mathscr{V}_2$ invariant. Computing the $\mathscr{S}_i$ by (3.3), we find that $\mathscr{S}_i = \mathscr{V}_i$ ($i \in \mathbf{2}$) and that

$$\mathscr{S}_1 \cap \mathscr{S}_2 = \operatorname{Im} \operatorname{col}[0 \quad 0 \quad 1 \quad -1 \quad 0 \quad 0 \quad 0].$$

While in this synthesis the action of $A + (B + B_a)F$ on $\mathscr{S}_1 \cap \mathscr{S}_2$ is fixed (and is "good," since $0 \in \mathbb{C}_g$ by definition) one may, of course, assign at will the spectrum of the induced maps on $\mathscr{S}_i/(\mathscr{S}_1 \cap \mathscr{S}_2)$, and also, by controllability of $(A, B + B_a)$, the spectrum on $(\mathscr{X} \oplus \mathscr{X}_a)/(\mathscr{S}_1 + \mathscr{S}_2)$. This is achieved by pole assignment procedures with which the reader will at this stage be familiar.

10.3. Develop an example to show that efficient decoupling in the sense of Section 10.3 need not be minimal in the sense of Section 10.4.

10.4. For some concrete examples, compare the sensitivity (suitably defined) of decoupling controllers designed "naively" and "efficiently."

10.7 Notes and References

The material in this chapter is adapted from Morse and Wonham [1]. The concept of "radical" of a family of vector spaces was exploited there, but without being named as such; the term is suggested by vaguely analogous usage in ring theory.

Noninteracting Control III: Generic Solvability

11

In this chapter we discuss solvability of the noninteraction problem from the viewpoint of genericity, in the parameter space of the matrices A, B, and the D_i ($i \in \mathbf{k}$). It turns out that noninteraction is possible for almost all data sets $(A, B, D_1, \ldots, D_k)$ if and only if the array dimensions of the given matrices satisfy appropriate, rather mild, constraints. When these conditions fail decoupling is possible, if at all, only for system structures which are rather special. Finally, in the generically solvable case we determine the generic bounds on dynamic order of a decoupling compensator, corresponding to the "naive" and "efficient" extension procedures of Chapters 9 and 10, respectively.

11.1 Generic Solvability of EDP

Consider as usual the system

$$\dot{x} = Ax + Bu; \qquad z_i = D_i x, \qquad i \in \mathbf{k}.$$

To discuss genericity we regard A, B and the D_i as real matrix representations of the corresponding maps, computed relative to fixed bases in $\mathscr{U}$, $\mathscr{X}$, and $\mathscr{Z}_i$ ($i \in \mathbf{k}$). We take A: $n \times n$, B: $n \times m$, and D_i: $q_i \times n$, with $n \geq 1$, $1 \leq m \leq n$, $1 \leq q_i \leq n$, and $k \geq 2$. Listing the matrix elements in some arbitrary order we introduce the data point

$$\mathbf{p} := (A, B, D_1, \ldots, D_k)$$

in $\mathbb{R}^N$, with $N = n^2 + nm + (q_1 + \cdots + q_k)n$.

Write

$$\hat{D}_i := \begin{bmatrix} D_1 \\ \vdots \\ D_{i-1} \\ D_{i+1} \\ \vdots \\ D_k \end{bmatrix}, \qquad i \in \mathbf{k}. \tag{1.1}$$

By Theorem 9.3 the extended decoupling problem (EDP) is solvable if and only if

$$\mathscr{R}_i^* + \mathscr{K}_i = \mathscr{X}, \qquad i \in \mathbf{k}, \tag{1.2}$$

where

$$\begin{aligned} \mathscr{K}_i &:= \operatorname{Ker} D_i, & i \in \mathbf{k}, \\ \mathscr{R}_i^* &:= \sup \mathfrak{C}(\hat{\mathscr{K}}_i), & i \in \mathbf{k}, \end{aligned} \tag{1.3}$$

and

$$\begin{aligned} \hat{\mathscr{K}}_i &:= \operatorname{Ker} \hat{D}_i \\ &= \bigcap_{j \neq i} \mathscr{K}_j, \qquad i \in \mathbf{k}. \end{aligned}$$

Solvability of EDP is thus a property $\Pi: \mathbb{R}^N \to \{0, 1\}$; that is, $\Pi(\mathbf{p}) = 1$ (or 0) according as (1.2) does (or does not) hold at $\mathbf{p}$. Our first result is a criterion for generic solvability.

Theorem 11.1. *EDP is generically solvable if and only if*

$$\sum_{i=1}^{k} q_i \leq n \tag{1.4}$$

and

$$m \geq 1 + \sum_{i=1}^{k} q_i - \min_{1 \leq i \leq k} q_i. \tag{1.5}$$

It will be clear from the proof that (1.4) states simply that the row spaces of the D_i are generically independent, while (1.5) means that, generically, the number of independent controls is large enough to ensure that

$$\mathscr{B} \cap \hat{\mathscr{K}}_i \neq 0, \qquad i \in \mathbf{k}.$$

By the discussion in Chapter 9, it should be obvious that the first condition is necessary for noninteraction, while the second is necessary for output controllability.

As an example, if $n = 15$, $k = 2$, $q_1 = 3$, and $q_2 = 5$, then EDP is generically solvable if and only if $m \geq 6$. In general, since $q_i \geq 1$ we always need $m \geq k$; and if, for instance, $m = k$ we can only have $q_i = 1$ for all i.

The following notation will be used in the proof. A prime denotes matrix transpose, dual linear transformation or dual space. If n, m are integers,

$$n \vee m := \max(n, m), \qquad n \wedge m := \min(n, m).$$

If for each $\mathbf{p} \in \mathbb{R}^N$, $\mathscr{R}(\mathbf{p}) \subset \mathscr{X}$ is a linear subspace, we write

$$d(\mathscr{R}) = r(g)$$

to mean that the generic dimension of $\mathscr{R}$ is r, i.e. that $d(\mathscr{R}(\mathbf{p})) \neq r$ only for $\mathbf{p}$ in some fixed proper variety $\mathbf{V} \subset \mathbb{R}^N$ depending on the function $\mathscr{R}(\cdot)$. Subspace inclusions written $\mathscr{R} \subset \mathscr{S}(g)$ are to be interpreted in the same fashion. We observe that a finite union of proper varieties is a proper variety, hence if a finite set of propositions each holds (g), the entire set holds simultaneously (g).

Proof (of Theorem 11.1).

1. *Preliminaries.* It is clear that

$$d(\mathscr{K}_i) = n - q_i(g), \qquad i \in \mathbf{k},$$

and

$$d(\operatorname{Im} D_i') = q_i(g), \qquad i \in \mathbf{k}, \tag{1.6}$$

since the dimensional evaluations fail at $\mathbf{p}$ only if all $q_i \times q_i$ minors of the $q_i \times n$ matrix D_i vanish. Similarly,

$$d\left(\sum_{j \neq i} \operatorname{Im} D_j'\right) = n \wedge \sum_{j \neq i} q_j(g), \qquad i \in \mathbf{k},$$

and

$$\begin{aligned} d(\hat{\mathscr{K}}_i) &= n - d(\hat{\mathscr{K}}_i^{\perp}) \\ &= n - n \wedge \sum_{j \neq i} q_j(g), \qquad i \in \mathbf{k}. \end{aligned} \tag{1.7}$$

By the same reasoning

$$d(\mathscr{B}) = m(g) \tag{1.8}$$

and

$$\begin{aligned} d(\mathscr{B} \cap \hat{\mathscr{K}}_i) &= d(\mathscr{B}) + d(\hat{\mathscr{K}}_i) - d(\mathscr{B} + \hat{\mathscr{K}}_i) \\ &= m + \left(n - n \wedge \sum_{j \neq i} q_j\right) \\ &\quad - n \wedge \left[m + \left(n - n \wedge \sum_{j \neq i} q_j\right)\right](g) \\ &= 0 \vee \left(m - n \wedge \sum_{j \neq i} q_j\right)(g), \qquad i \in \mathbf{k}. \end{aligned} \tag{1.9}$$

2. *Necessity.* Suppose EDP is solvable at **p**. By (1.2) and (1.3),

$$\mathscr{K}_i + \hat{\mathscr{K}}_i = \mathscr{X}, \qquad i \in \mathbf{k},$$

or

$$\left(\sum_{j \neq i} \operatorname{Im} D_j'\right) \cap \operatorname{Im} D_i' = 0, \qquad i \in \mathbf{k},$$

that is, the subspaces $\operatorname{Im} D_i' \subset \mathscr{X}'$ are independent. It follows from this and (1.6) that EDP is generically solvable only if

$$\sum_{i=1}^{k} q_i \leq n, \tag{1.10}$$

as claimed in (1.4). By (1.7) and (1.10),

$$d(\hat{\mathscr{K}}_i) = n - \sum_{j \neq i} q_j(g), \qquad i \in \mathbf{k};$$

and by (1.9) and (1.10),

$$d(\mathscr{B} \cap \hat{\mathscr{K}}_i) = 0 \vee \left(m - \sum_{j \neq i} q_j\right)(g). \tag{1.11}$$

Now if $D_i \neq 0$, (1.2) implies $\mathscr{R}_i^* \neq 0$ and therefore, $\mathscr{B} \cap \hat{\mathscr{K}}_i \neq 0$. Thus by (1.11) generic solvability of EDP implies

$$m - \sum_{j \neq i} q_j \geq 1, \qquad i \in \mathbf{k},$$

which is equivalent to (1.5).

3. *Sufficiency.* Suppose (1.4) and (1.5) hold. Write $\hat{q}_i := \sum_{j \neq i} q_j$. By (1.4) and (1.7)

$$d(\hat{\mathscr{K}}_i) = n - \hat{q}_i(g). \tag{1.12}$$

Using (1.7), (1.8) and (1.12), we have

$$\begin{aligned} d(\hat{\mathscr{K}}_i + \mathscr{B}) &= n \wedge (n - \hat{q}_i + m) \\ &\geq n \wedge (n + 1) \qquad \text{(by (1.5))} \\ &= n. \end{aligned}$$

It follows that $\hat{\mathscr{K}}_i + \mathscr{B} = \mathscr{X}(g)$, and thus

$$A\hat{\mathscr{K}}_i \subset \hat{\mathscr{K}}_i + \mathscr{B}(g), \qquad i \in \mathbf{k}.$$

By Theorem 5.6, we have

$$\mathscr{R}_i^* = \mathscr{R}_i^n(g),$$

where

$$\mathscr{R}_i^{\mu+1} = \hat{\mathscr{K}}_i \cap (A\mathscr{R}_i^\mu + \mathscr{B}), \qquad \mu = 0, 1, \ldots, n \tag{1.13a}$$

and

$$\mathscr{R}_i^0 = 0. \tag{1.13b}$$

It will be shown that $\mathscr{R}_i^n = \hat{\mathscr{K}}_i(g)$. Because of the multiple appearances of A, B and $\hat{D}_i$ in the full expression for $\mathscr{R}_i^\mu$ determined by (1.13), the generic dimensions of the $\mathscr{R}_i^\mu$ are not obvious by inspection. For this reason it is convenient to use a more refined method than heretofore. Replace $\mathbf{p} \in \mathbb{R}^N$ by the indeterminate $\boldsymbol{\lambda} = (\lambda_1, \ldots, \lambda_N)$; i.e. $\boldsymbol{\lambda}$ is simply a list representing the N entries of the matrices $A, B, D_1, \ldots, D_k$ regarded as literal variables. We shall consider $A, B, D_1, \ldots, D_k$ as matrices over the integral domain $\mathbb{R}[\lambda]$ or, more properly, its fraction field $\mathbb{R}(\boldsymbol{\lambda})$ (cf. Section 0.9). We then regard the $\mathscr{R}_i^\mu$ defined by (1.13) as subspaces of the vector space $\mathbb{R}(\boldsymbol{\lambda})$. Let $r_{i\mu}$ (resp. $s_{i\mu}$) be the dimension of $\mathscr{R}_i^\mu$ (resp. $\mathscr{S}_i^\mu := A\mathscr{R}_i^\mu + \mathscr{B}$) over $\mathbb{R}(\boldsymbol{\lambda})$. We now compute the $r_{i\mu}$ and $s_{i\mu}$, dropping the subscript i for convenience.

Lemma 11.1. *In the vector space $\mathbb{R}^n(\boldsymbol{\lambda})$ just described, let*

$$\begin{aligned} \mathscr{S}^\mu &= A\mathscr{R}^\mu + \mathscr{B}, \\ \mathscr{R}^{\mu+1} &= \hat{\mathscr{K}} \cap \mathscr{S}^\mu, \qquad \mu = 0, 1, \ldots, n; \\ \mathscr{R}^0 &= 0. \end{aligned} \tag{1.14}$$

Then

$$\begin{aligned} s_\mu &= n \wedge (r_\mu + m), \\ r_{\mu+1} &= 0 \vee (s_\mu - \hat{q}), \qquad \mu = 0, 1, \ldots, n; \\ r_0 &= 0. \end{aligned} \tag{1.15}$$

Furthermore, if the subspaces $\mathscr{R}^\mu(\mathbf{p})$ are computed by (1.13) with $\boldsymbol{\lambda} = \mathbf{p}$, then

$$d[\mathscr{R}^\mu(\mathbf{p})] = r_\mu(g). \tag{1.16}$$

PROOF. Let R_μ be an $n \times r_\mu$ matrix over $\mathbb{R}(\boldsymbol{\lambda})$ whose columns form a basis of $\mathscr{R}^\mu$, introduce the $n \times (r_\mu + m)$ matrix

$$\hat{S}_\mu = [AR_\mu \quad B], \tag{1.17}$$

let S_μ be an $n \times s_\mu$ matrix such that $\operatorname{Im} S_\mu = \operatorname{Im} \hat{S}_\mu$, and let $S_\mu^\perp$ be an $(n - s_\mu) \times n$ matrix such that $\operatorname{Ker} S_\mu^\perp = \operatorname{Im} S_\mu$. With $\hat{D}$ defined as in (1.1), write

$$T_\mu := \begin{bmatrix} S_\mu^\perp \\ \hat{D} \end{bmatrix}. \tag{1.18}$$

Then, by (1.14)

$$\mathscr{R}^{\mu+1} = \hat{\mathscr{K}} \cap \mathscr{S}^\mu = \operatorname{Ker} \hat{D} \cap \operatorname{Ker} S_\mu^\perp = \operatorname{Ker} T_\mu. \tag{1.19}$$

It will be shown that at each stage (μ) a specific choice can be made of the $\mathbb{R}(\boldsymbol{\lambda})$-matrices R_μ, S_μ and $S_\mu^\perp$, that exhibits the dimensions r_μ, s_μ to be just as computed by (1.15). To this end, we first prescribe exactly how the R_μ etc. are to be calculated, then construct a suitable numerical example over $\mathbb{R}$.

i. *Matrix calculations.* Let T be an $r \times n$ matrix over a field $\mathbb{K}$ and let Rank $T = r \leq n$. Let $\check{T}$ be an $r \times r$ nonsingular submatrix of T. Suppose first that T is in standard form $T = [T_1 \quad T_2]$ with $T_1 = \check{T}$. Then, Ker $T =$ Im R, where

$$R := \begin{bmatrix} -T_1^{-1}T_2 \\ 1_{n-r} \end{bmatrix}.$$

More generally, if $\check{T}$ is some other submatrix of T then $\tilde{T} := TP$ is in standard form for a suitable $n \times n$ permutation matrix P; so Ker $\tilde{T} =$ Im $\tilde{R}$, say, and we set $R = P\tilde{R}$.

Next, suppose $\hat{S}$: $n \times \hat{s}$ with Rank $\hat{S} = s \geq 1$. Define S: $n \times s$, such that Im $S =$ Im $\hat{S}$ by deleting each column in $\hat{S}$ that is contained in the span of those to the left.

Finally, let $\check{S}$ be an $s \times s$ nonsingular submatrix of S and suppose S is in standard form $S = \left[\begin{smallmatrix} S_1 \\ S_2 \end{smallmatrix}\right]$ with $S_1 = \check{S}$. Then, Im $S =$ Ker $S^{\perp}$, where

$$S^{\perp} := [-S_2 S_1^{-1} \quad 1_{n-s}].$$

More generally, if $\tilde{S} = PS$ is in standard form, set $S^{\perp} = \tilde{S}^{\perp} P$.

ii. *Numerical example.* Let $e_1, \ldots, e_n$ be the standard unit vectors in $\mathbb{R}^n$, put $e_v = 0$ if $v > n$, and define the specific data point $\mathbf{p} \in \mathbb{R}^N$ according to

$$Ae_j = 0, \qquad j = 1, \ldots, \hat{q} \tag{1.20a}$$

$$Ae_{\hat{q}+r} = e_{m+r}, \qquad r = 1, \ldots, n - \hat{q} \tag{1.20b}$$

$$\dot{D} = [1_{\hat{q}} \quad 0], \qquad B = \begin{bmatrix} 1_m \\ 0 \end{bmatrix}. \tag{1.20c}$$

Easy computations establish that the subspaces $\mathscr{R}^{\mu}(\mathbf{p})$ and $\mathscr{S}^{\mu}(\mathbf{p})$ in $\mathscr{X}$ generated by the algorithm (1.14) from the data (1.20) indeed have the dimensions r_{μ} and s_{μ} given by (1.15). Furthermore, by inspection the $\mathbb{R}$-matrices $S_{\mu}(\mathbf{p})$, $T_{\mu}(\mathbf{p})$ of our example each admit a specific square submatrix, say $\check{S}_{\mu}(\mathbf{p})$, $\check{T}_{\mu}(\mathbf{p})$ respectively, of maximal rank compatible with the array size of $S_{\mu}(\mathbf{p})$, $T_{\mu}(\mathbf{p})$; and inversion of $\check{S}_{\mu}(\mathbf{p})$, $\check{T}_{\mu}(\mathbf{p})$ will determine, as in (i), suitable matrices $S_{\mu}^{\perp}(\mathbf{p})$ and $R_{\mu+1}(\mathbf{p})$.

Reverting now to the setting $\mathbb{R}(\boldsymbol{\lambda})$ we compute specific $\mathbb{R}(\boldsymbol{\lambda})$-matrices with the properties (1.17)–(1.19). Arguing inductively with respect to μ, suppose that $0 = R_0(\boldsymbol{\lambda}), \ldots, S_{\mu}(\boldsymbol{\lambda})$ have been defined, have rank $0 = r_0, \ldots, s_{\mu}$ over $\mathbb{R}(\boldsymbol{\lambda})$, and evaluate to $R_0(\mathbf{p}), \ldots, S_{\mu}(\mathbf{p})$ on substitution of $\boldsymbol{\lambda} = \mathbf{p}$, where $\mathbf{p}$ is the specific data point determined by (1.20). Let $\check{S}_{\mu}(\boldsymbol{\lambda})$ be the $s_{\mu} \times s_{\mu}$ submatrix of $S_{\mu}(\boldsymbol{\lambda})$ having the same row indices as $\check{S}_{\mu}(\mathbf{p})$. Our example guarantees that the determinant of $\check{S}_{\mu}(\boldsymbol{\lambda})$ does not vanish (identically in $\boldsymbol{\lambda}$). Hence, inversion of $\check{S}_{\mu}(\boldsymbol{\lambda})$ determines a suitable rational matrix $S_{\mu}^{\perp}(\boldsymbol{\lambda})$; and similar procedures as described in (i) above yield $R_{\mu+1}(\boldsymbol{\lambda})$, $S_{\mu+1}(\boldsymbol{\lambda})$. By induction it is now clear that the sequences $R_{\mu}(\boldsymbol{\lambda})$, $S_{\mu}(\boldsymbol{\lambda})$, $S_{\mu}^{\perp}(\boldsymbol{\lambda})$ can be defined in such a way that they evaluate to $R_{\mu}(\mathbf{p})$, $S_{\mu}(\mathbf{p})$, $S_{\mu}^{\perp}(\mathbf{p})$ at $\boldsymbol{\lambda} = \mathbf{p}$; hence, the same is true of $\check{S}_\mu(\boldsymbol{\lambda})$ and $T_\mu(\boldsymbol{\lambda})$. Since the latter have now been shown to have maximal rank compatible with their array size, it is clear that the evaluations $\check{S}_\mu(\mathbf{p}')$ and $T_\mu(\mathbf{p}')$ are defined and have maximal rank for all $\mathbf{p}'$ in some neighborhood of $\boldsymbol{\lambda} = \mathbf{p} \in \mathbb{R}^N$. It follows that the sequence $R_\mu(\mathbf{p}')$ is well defined and represents $\mathscr{R}_\mu(\mathbf{p}')$ for all such $\mathbf{p}'$. Finally, since the $R_\mu(\boldsymbol{\lambda})$ are rational, it follows that the last statement is true for all $\boldsymbol{\lambda} = \mathbf{p}'$ with $\mathbf{p}'$ in the complement of a proper algebraic variety in $\mathbb{R}^N$. That is, $d(\mathscr{R}_\mu(\mathbf{p})) = r_\mu(g)$, and the lemma is proved.

We now show that

$$r_{in} = n - \hat{q}_i, \qquad i \in \mathbf{k}, \tag{1.21}$$

and hence that

$$\mathscr{R}_i^n = \hat{\mathscr{K}}_i(g), \qquad i \in \mathbf{k}, \tag{1.22}$$

as claimed. Dropping the subscript i, we have from (1.15)

$$\begin{aligned} r_{\mu+1} &= n - n \wedge [n - n \wedge (r_\mu + m) + \hat{q}] \\ &= 0 \vee [n \wedge (r_\mu + m) - \hat{q}] \\ &\geq n \wedge (r_\mu + m) - \hat{q} \\ &\geq n \wedge (r_\mu + 1 + \hat{q}) - \hat{q} \qquad \text{(by (1.5))}. \end{aligned}$$

If $r_\mu < n - \hat{q}$ then $r_\mu + 1 + \hat{q} \leq n$, so

$$r_{\mu+1} \geq (r_\mu + 1 + \hat{q}) - \hat{q} = r_\mu + 1.$$

Since the $\mathscr{R}^\mu$ defined by (1.14) are nondecreasing, and since $d(\mathscr{R}^\mu) \leq d(\mathscr{K}) = n - \hat{q}$, it follows that $r_\mu \uparrow n - \hat{q}$ with convergence in at most $n - \hat{q} < n$ steps, and the claim (1.21) is established.

Finally, (1.4) implies that

$$\text{Rank} \begin{bmatrix} D_1 \\ \vdots \\ D_k \end{bmatrix} = \sum_{i=1}^{k} q_i(g) = \sum_{i=1}^{k} \text{Rank } D_i(g);$$

hence that

$$\left(\sum_{j \neq i} \text{Im } D_j'\right) \cap \text{Im } D_i' = 0(g), \qquad i \in \mathbf{k};$$

and so

$$\mathscr{K}_i + \hat{\mathscr{K}}_i = \mathscr{X}(g), \qquad i \in \mathbf{k}.$$

This combined with (1.22) shows that (1.2) is true generically. □

Remark 1. We recall from Section 0.16 the definition that a property Π is *well-posed at* $\mathbf{p} \in \mathbb{R}^N$ (or $\mathbf{p}$ is *well-posed relative* to Π) if $\Pi(\mathbf{p}') = 1$ at all $\mathbf{p}'$ in some neighborhood of $\mathbf{p}$. It is clear from the proof of Theorem 11.1 that if either one of conditions (1.4) or (1.5) fails, the solvability set for EDP, namely

$$\{\mathbf{p}: \Pi(\mathbf{p}) = 1\} \subset \mathbb{R}^N$$

is a subset of some (possibly trivial) proper variety in $\mathbb{R}^N$. It follows that no data point $\mathbf{p}$ can be well-posed in the sense of our definition. Of course, in a concrete application it may happen that not all perturbations of a given data point $\mathbf{p}$ are admissible, possibly because of constraints arising from definitional relations among certain state variables. It may then be true that $\mathbf{p}$ is well-posed in a restricted sense, namely in the topology determined by the class of admissible perturbations. Such cases require separate investigation.

Remark 2. From the proof of Theorem 11.1 we note that, if (1.4) and (1.5) are true, then

$$\mathscr{R}_i^* = \hat{\mathscr{K}}_i(g). \tag{1.23}$$

Thus equality holds at almost all $\mathbf{p}$, and the computational effort at such $\mathbf{p}$ is reduced accordingly.

11.2 State Space Extension Bounds

We shall now calculate the generic order of a dynamic compensator which achieves decoupling by state space extension. Our first result applies to the "naive" extension described in Section 9.6, with the corresponding bound n_a provided by Theorem 9.4:

$$n_a := \sum_{i=1}^{k} d(\mathscr{R}_i^*) - d\left(\sum_{i=1}^{k} \mathscr{R}_i^*\right). \tag{2.1}$$

Theorem 11.2. *Under the conditions of Theorem 11.1, and by use of the "naive" extension technique of Section 9.6, EDP is generically solvable by dynamic compensation of order no greater than*

$$n_a = (k-1)\left(n - \sum_{i=1}^{k} q_i\right). \tag{2.2}$$

Thus, if $n = 15$, $k = 2$, $q_1 = 3$, $q_2 = 5$ and $m \geq 6$, we have $n_a = 7(g)$. From now on, we write

$$q := \sum_{i=1}^{k} q_i.$$

PROOF. We first observe that, with $q \leq n$,

$$\sum_i \hat{\mathscr{K}}_i = \mathscr{X}(g).$$

Indeed,

$$\left(\sum_i \hat{\mathscr{K}}_i\right)^{\perp} = \left(\sum_i \bigcap_{j\neq i} \operatorname{Ker} D_j\right)^{\perp}$$

$$= \bigcap_i \sum_{j\neq i} \operatorname{Im} D_j'$$

$$= 0(g),$$

since the subspaces Im D_j' are generically independent. Furthermore,

$$\sum_i d(\hat{\mathscr{K}}_i) = \sum_i \left(n - n \wedge \sum_{j\neq i} q_j\right)(g) \qquad \text{(by (1.7))}$$

$$= \sum_i \left(n - \sum_{j\neq i} q_j\right) \qquad \text{(by (1.5))}$$

$$= kn - (k-1)q.$$

Therefore,

$$n_a = \sum_i d(\hat{\mathscr{K}}_i) - d\left(\sum_i \hat{\mathscr{K}}_i\right)(g) \qquad \text{(by (1.23) and (2.1))}$$

$$= kn - (k-1)q - n(g)$$

$$= (k-1)(n-q)(g),$$

as claimed. □

The bound (2.2) can be improved by exploiting the "efficient" extension technique described in Chapter 10. For this, we have the bound given by Theorem 10.1, namely

$$n_a^* := \sum_{i=1}^{k} d\left(\frac{\mathscr{R}_i^* + \mathscr{R}_0}{\mathscr{R}_0}\right) - d\left(\sum_{i=1}^{k} \frac{\mathscr{R}_i^* + \mathscr{R}_0}{\mathscr{R}_0}\right). \tag{2.3}$$

Here, in the notation of Chapter 10,

$$\mathscr{R}_0 := (\mathscr{R}_\bullet^* \cap \mathscr{V}_g)^{\vee}, \tag{2.4}$$

$$\mathscr{V}_g := \sup\{\mathscr{V} : \mathscr{V} \subset \check{\mathscr{R}}^* \ \&\ \exists F \in \mathbf{F}(\mathscr{V}),\ \sigma[(A+BF)|\mathscr{V}] \subset \mathbb{C}_g\} \tag{2.5}$$

and

$$\check{\mathscr{R}}^* := (\mathscr{R}_\bullet^*)^{\vee}. \tag{2.6}$$

To compute the generic value of n_a^* under the conditions of Theorem 11.1, we start by noting that

$$\check{\mathscr{R}}^* = (\hat{\mathscr{K}}_\bullet)^{\vee}(g) \qquad \text{(by (1.23))}$$

$$= \bigcap_{i=1}^{k} \sum_{j\neq i} \hat{\mathscr{K}}_j \qquad \text{(by (10.1.3))}$$

$$= \bigcap_i \sum_{j\neq i} \bigcap_{l\neq j} \mathscr{K}_l$$

$$= \bigcap_i \mathscr{K}_i \tag{2.7}$$

and thus

$$d(\check{\mathscr{R}}^*) = n - q(g).$$

We shall treat separately the cases $m < q$, $m = q$, and $m > q$.

Case 1. $m < q$. Let

$$\mathscr{V}^* := \sup \mathfrak{I}(\check{\mathscr{R}}^*).$$

Then $\mathscr{V}^* = \lim_\mu \mathscr{V}^\mu$, where

$$\mathscr{V}^0 = \check{\mathscr{R}}^*; \qquad \mathscr{V}^{\mu+1} = \check{\mathscr{R}}^* \cap A^{-1}(\mathscr{V}^\mu + \mathscr{B}), \qquad \mu \in \mathbf{n}.$$

Let $v_\mu := d(\mathscr{V}^\mu)$. By the same technique as in the proof of Theorem 11.1, we get that

$$v_{\mu+1} = 0 \vee [(n - q) + n \wedge (v_\mu + m) - n](g). \tag{2.8}$$

By iteration of (2.8) with $v_0 = n - q$, there follows

$$v_\mu = 0 \vee (n - m - 2^\mu(q - m))(g),$$

and therefore $\mathscr{V}^* = 0(g)$. From this and (2.5), $\mathscr{V}_g = 0(g)$, hence by (2.4), $\mathscr{R}_0 = 0(g)$. Then (2.3) yields

$$n_a^* = \sum_i d(\mathscr{R}_i^*) - d\left(\sum_i \mathscr{R}_i^*\right)(g) = (k - 1)(n - q)(g),$$

just as in the proof of Theorem 11.2.

Case 2. $m = q$. We have

$$d(\check{\mathscr{R}}^* + \mathscr{B}) = n \wedge (n - q + m)(g) = n,$$

and therefore $\check{\mathscr{R}}^* \in \mathfrak{I}(\mathscr{X})(g)$. Also,

$$d(\check{\mathscr{R}}^* \cap \mathscr{B}) = 0 \vee ((n - q) + m - n)(g) = 0,$$

and so $\check{\mathscr{R}}^*$ contains, generically, no c.s. other than zero. It follows that $\mathscr{V}_g$ is simply the "good" modal subspace of the map

$$(A + BF)|\check{\mathscr{R}}^*,$$

computed with any $F \in \mathbf{F}(\check{\mathscr{R}}^*)$.

Now $v_g := d(\mathscr{V}_g)$ does not possess a generic value. To reckon with its dependence on $\mathbf{p}$, we shall refine the definition of "generic" in a way which is *ad hoc* but suited to our purpose. Let

$$\mathbf{S} := \{\mathbf{p}: \mathbf{p} \in \mathbb{R}^N,\ d(\mathscr{B}(\mathbf{p})) = m\ \&\ \check{\mathscr{R}}^*(\mathbf{p}) \oplus \mathscr{B}(\mathbf{p}) = \mathscr{X}\}.$$

For $\mathbf{p} \in \mathbf{S}$, $v_g = v_g(\mathbf{p})$ is well defined and takes values $0 \le v_g \le n - q$. By the preceding discussion, there is some proper variety $\mathbf{V} \subset \mathbb{R}^N$ for which $\mathbf{V}^c \subset \mathbf{S}$. Now let

$$\mathbf{S}_v := \mathbf{S} \cap \{\mathbf{p}: v_g(\mathbf{p}) = v\}.$$

Thus,

$$\mathbf{S} = \bigcup_{\nu=0}^{n-q} \mathbf{S}_\nu$$

and

$$\mathbf{V}^c = \bigcup_{\nu=0}^{n-q} (\mathbf{V}^c \cap \mathbf{S}_\nu).$$

If now ψ is a function on the integers, we write

$$n_a^* = \psi(\nu_g)(g) \tag{2.9}$$

to mean the following: there exists a proper variety $\tilde{\mathbf{V}} \subset \mathbb{R}^N$, such that at each $\mathbf{p} \in \mathbf{V}^c \cap \tilde{\mathbf{V}}^c$, we have

$$n_a^*(\mathbf{p}) = \psi(\nu_g(\mathbf{p})) \qquad (= \psi(\nu) \text{ for some } \nu,\ 0 \le \nu \le n-q).$$

We shall not explore in detail the structure of the component subsets $\mathbf{V}^c \cap \mathbf{S}_\nu$. However, if we assume that, in the usual topology of the complex plane, $\mathbb{C}_g^0 \cap \mathbb{R} \neq \varnothing$ and $\mathbb{C}_b^0 \cap \mathbb{R} \neq \varnothing$ (where $(^0)$ denotes interior), then it can be shown (Exercise 11.3) that for each ν the interior of $\mathbf{V}^c \cap \tilde{\mathbf{V}}^c \cap \mathbf{S}_\nu$ is nonempty. At an interior point the value of n_a^* given by (2.11) below, is locally constant.

Proceeding on this basis, we have by (1.23) and (2.7)

$$\breve{\mathscr{R}}^* \subset \bigcap_i \mathscr{R}_i^*, \qquad i \in \mathbf{k}(g)$$

so

$$\mathscr{V}_g \cap \mathscr{R}_i^* = \mathscr{V}_g, \qquad i \in \mathbf{k}(g),$$

and then

$$\mathscr{R}_0 = \mathscr{V}_g(g). \tag{2.10}$$

Finally,

$$\begin{aligned} n_a^* &= \sum_i d\left(\frac{\mathscr{R}_i^*}{\mathscr{R}_0}\right) - d\left(\sum_i \frac{\mathscr{R}_i^*}{\mathscr{R}_0}\right) \\ &= \sum_i d(\mathscr{R}_i^*) - d\left(\sum_i \mathscr{R}_i^*\right) - (k-1)d(\mathscr{R}_0) \\ &= (k-1)(n-q) - (k-1)\nu_g(g) \qquad \text{(by (2.1), (2.2) and (2.10))} \\ &= (k-1)(n-q-\nu_g)(g). \end{aligned} \tag{2.11}$$

Case 3. $m > q$. We have $d(\breve{\mathscr{R}}^* \cap \mathscr{B}) = m - q > 0(g)$, and we shall show that $\breve{\mathscr{R}}^*$ is generically a c.s. As in Case 2, $\breve{\mathscr{R}}^* + \mathscr{B} = \mathscr{X}(g)$, so $\breve{\mathscr{R}}^* \in \mathfrak{I}(\mathscr{X})(g)$. Thus, $\breve{\mathscr{R}}^* \in \mathfrak{C}(\mathscr{X})$ if and only if $\breve{\mathscr{R}}^* = \lim_\mu \mathscr{R}^\mu$, where

$$\mathscr{R}^0 = 0; \qquad \mathscr{R}^{\mu+1} = \breve{\mathscr{R}}^* \cap (A\mathscr{R}^\mu + \mathscr{B}), \qquad \mu \in \mathbf{n}.$$

Let $d(\mathscr{R}^{\mu}) = \rho_{\mu}(g)$. Then, as in the proof of Theorem 11.1,

$$\rho_0 = 0, \qquad \rho_{\mu+1} = 0 \vee [n - q + n \wedge (\rho_{\mu} + m) - n],$$

from which it follows that $\rho_{\mu} \uparrow n - q$, and the assertion follows. But now $\mathscr{V}_g = \check{\mathscr{R}}^*(g)$, and we need only set $v_g = n - q$ in (2.11) to obtain

$$n_a^* = 0(g).$$

We summarize results as

Theorem 11.3. *Under the conditions of Theorem 11.1, and by use of the "efficient" decoupling technique of Section 10.3, EDP is generically solvable by dynamic compensation of order no greater than*

$$n_a^* = \begin{cases} (k-1)(n-q), & m < q \\ (k-1)(n-q-v_g), & m = q \\ 0, & m > q. \end{cases}$$

Here the result for $m = q$ is interpreted according to (2.9).

Efficient extension may be generically more economical than naive extension, but only if $m \geq q$. By the conditions for generic solvability, $m \geq q$ must hold when, in particular, the decoupled outputs z_i ($i \in \mathbf{k}$) are all scalars.

As an example, again let $n = 15$, $k = 2$, $q_1 = 3$, and $q_2 = 5$. We must have $m \geq 6$ and then, generically,

$$\begin{aligned} n_a^* &= 7 && \text{if } m = 6 \text{ or } 7 \\ &= 7 - v_g \quad (0 \leq v_g \leq 7) && \text{if } m = 8 \\ &= 0 && \text{if } m \geq 9. \end{aligned}$$

11.3 Significance of Generic Solvability

The results in this chapter are a guide in identifying practical situations where dynamic decoupling is likely, in principle, to be feasible. Of course, the notions of "generic solvability" and "well-posedness" are purely qualitative. They furnish no information about how well conditioned the computations may be which determine a solution (F_e, G_e in Figure 9.2) at a well-posed data point, or about the sensitivity of a solution in a neighborhood of such a point.

A solution of EDP typically depends critically on the parameters of A and B: with F_e, G_e fixed, decoupling will, in general, break down if these parameters undergo small variations from the values employed in design. The solution is thus "finely tuned" and must be maintained by a supervisory control with the capability of adaptive readjustment. The real significance of our results is that they point to exactly this possibility, at least when the conditions of generic solvability are met. Adaptive decoupling poses challenging problems of numerical conditioning and stability of which the study has only recently begun.

11.4 Exercises

11.1. Supply the omitted (computational) details in the proof of Lemma 11.1.

11.2. With reference to Remark 1 after Theorem 11.1, develop a plausible example where EDP is well-posed only in a "restricted" sense. Study the sensitivity of F_e, G_e with respect to the admissible variations of A and B.

11.3. With reference to the discussion of Case 2 of Theorem 11.3 show that under the assumptions stated and for suitable $\mathbf{V}$ and $\tilde{\mathbf{V}}$, the set $\mathbf{V}^c \cap \tilde{\mathbf{V}}^c \cap \mathbf{S}_\nu$ has nonempty interior. Hint: Exploit the fact that if $A: n \times n$, $B: n \times m$ and $R: (n - m) \times m$, then with $\mathscr{R} = \operatorname{Im} R$, there follows $\mathscr{R} \oplus \mathscr{B} = \mathscr{X}(g)$ in the parameter space of points $\mathbf{p} = (A, B, R)$. Given ν $(0 \le \nu \le n - m)$ construct A, together with B, R of maximal rank and $F \in \mathbf{F}(\mathscr{R})$, such that, if

$$\sigma_1 := \sigma[(A + BF)|\mathscr{R}], \qquad \sigma := \sigma(A + BF),$$

one has $|\sigma_1 \cap \mathbb{C}_g^0| = \nu$ and $|\sigma \cap \mathbb{C}_b^0| = n - \nu$. Finally, show that the last two relations hold locally at $\mathbf{p} = (A, B, R)$.

11.4. From the viewpoint of genericity discuss the partial decoupling problems of Section 9.9.

11.5 Notes and References

The material in this chapter is based largely on Fabian and Wonham [1]. For additional applications of the genericity concept, to the combined problem of decoupling and disturbance rejection, see Fabian [1]. Adaptive decoupling has been studied by Yuan [1], [2] and Yuan and Wonham [1].

We take this opportunity to point out that, in the version of Theorem 11.3 given in the preliminary edition of this book (Wonham [8], p. 289), the formula for n_a^* in the case $m = q$ was stated incorrectly.

12 Quadratic Optimization I: Existence and Uniqueness

In previous chapters our objectives in system synthesis have been almost entirely qualitative: we have indeed imposed requirements like stability on the system spectrum, but in the main have sought to realize very general properties of signal flow, as in tracking or noninteraction. By contrast, in this chapter and the next we take a somewhat more quantitative approach to realizing good dynamic response. We describe a systematic way of computing linear state feedback which ensures "optimal" recovery from an impulsive disturbance acting at the system input. It will later be clear how to incorporate the method into the framework of synthesis techniques already presented.

Optimality will be understood as the minimization of a positive quadratic functional of system output. It is the quadratic structure which guarantees that the optimal feedback control is linear, hence relatively simple to analyze and implement. In addition, the optimal control is fairly easily calculated. Finally, experience has shown that good dynamic response is usually achievable if the quadratic functional is suitably chosen. For these three reasons, rather than any specific interpretation of quadratic cost as such, the method of quadratic optimization has been widely adopted.

12.1 Quadratic Optimization

We begin with the standard system

$$\dot{x}(t) = Ax(t) + Bu(t), \qquad t \geq 0, \tag{1.1}$$

$$z(t) = Dx(t), \qquad t \geq 0, \tag{1.2}$$

and regard $\mathscr{U}$, $\mathscr{X}$ and $\mathscr{Z}$ as inner product spaces over $\mathbb{R}$. Suppose

$$x(0+) = x_0.$$

We may interpret this initial condition as arising from an external disturbance of form $x_0 \delta(t)$ appearing implicitly on the right side of (1.1). Stated loosely, our problem is to choose $u(t)$, $t \geq 0$, such that the system output $z(\cdot)$ is steered from its initial "disturbed" value $z(0+) = Dx_0$ to its "desired" regulated value $z = 0$, over a suitable recovery interval $[0, T)$, which may be infinite. To define the optimization problem we must further specify the class of admissible control functions $u(\cdot)$: $[0, T) \to \mathscr{U}$, and the cost attached to any particular $u(\cdot)$.

In our formulation we shall set $T = +\infty$; admit *a priori* controls which are essentially arbitrary; and attach to $u(\cdot)$ the cost

$$J(u) := \int_0^\infty [z(t)'z(t) + u(t)'Nu(t)]\, dt, \tag{1.3}$$

where $N \geq 0$ and $z(\cdot)$ is determined by (1.1) and (1.2). In fact, these are the only known conditions which, subject to mild technicalities, guarantee the following desirable result: the optimal (minimal cost) control can be implemented by linear time-invariant state feedback

$$u(t) = Fx(t), \qquad t \geq 0,$$

such that the closed-loop system map $A + BF$ is stable. Here, F is independent of x_0.

As our final goal is an optimal feedback control we shall rigorously define, in Section 12.2, an optimization problem in which the admissible controls are (possibly nonlinear) state feedback laws. Meanwhile there is something to be gained from a heuristic treatment which leads quickly to our main analytic tool, the functional equation of dynamic programming.

12.2 Dynamic Programming: Heuristics

Write

$$M := D'D, \tag{2.1}$$

$$L(x, u) := x'Mx + u'Nu.$$

Assuming an optimal control exists, introduce the *value function*

$$V(x) := \min_{u(\cdot)} \int_0^\infty L[x(t), u(t)]\, dt, \qquad x(0) = x. \tag{2.2}$$

Thus, $V(x)$ is the minimal cost expressed as a function of the initial state $x(0) = x$. Write

$$x(t) = \xi(t; x, u(\cdot))$$

for the solution of (1.1) with control $u(\cdot)$ and initial condition $x(0) = x$. Fix x, suppose $u^0(\cdot)$ is optimal on $[0, \infty)$, and let $\tau > 0$. We claim that the control function

$$u^0(t), \qquad \tau \le t < \infty, \tag{2.3}$$

is also optimal, relative to the state $\xi(\tau; x, u^0(\cdot))$ from which the system departs at time τ. Indeed, for any $u(\cdot)$,

$$J(u) = \left(\int_0^\tau + \int_\tau^\infty\right) L[\xi(t; x, u(\cdot)), u(t)]\, dt. \tag{2.4}$$

The first integral depends only on $u(t)$ for $0 \le t \le \tau$, and the second only on $x(\tau)$ together with $u(t)$ for $t > \tau$. Suppose we know $u^0(t)$ for $0 \le t \le \tau$: $x(\tau)$ is now determined. The second integral must then be a minimum when evaluated at the function (2.3), or we arrive at a contradiction. Now, because the dynamic equation (1.1) and the function $L(\cdot, \cdot)$ are invariant under shift of the origin of time, we have

$$\min_{\substack{u(t)\\ \tau \le t < \infty}} \int_\tau^\infty L[x(t), u(t)]\, dt = V[x(\tau)]. \tag{2.5}$$

Combining (2.2), (2.4) and (2.5), and expanding notation a little for clarity, we can write

$$V(x) = \min_{\substack{u(t)\\ 0 \le t \le \tau}} \left[\int_0^\tau L[\xi(t; x, u(s), 0 \le s \le t), u(t)]\, dt \right.$$
$$\left. + V[\xi(\tau; x, u(s), 0 \le s \le \tau)]\right]. \tag{2.6}$$

Equation (2.6) expresses the celebrated, "intuitively obvious" *principle of optimality*. We get a very convenient version of (2.6) in differential form by letting $\tau \downarrow 0$. For this, assume $u(\cdot)$ and $V(\cdot)$ are smooth and write $u := u(0)$. Then,

$$\xi(\tau; x, u(\cdot)) = x + \tau(Ax + Bu) + o(\tau),$$

$$\int_0^\tau L[\xi(t; x, u(\cdot)), u(t)]\, dt = \tau L(x, u) + o(\tau),$$

and

$$V[\xi(\tau; x, u(\cdot))] = V(x) + \tau(Ax + Bu)'V_x(x) + o(\tau),$$

where V_x is the first partial derivative of V. With these substitutions in (2.6) a formal passage to the limit yields

$$\min_u \,[(Ax + Bu)'V_x(x) + L(x, u)] = 0. \tag{2.7}$$

We refer to (2.7) as *Bellman's equation*. It says, in effect, to minimize the expression bracketed, regarded as a function of the variable u, with x and V_x as parameters. Suppose the minimizing u is

$$u = \omega(x, V_x). \tag{2.8}$$

Substituting (2.8) in (2.7), we obtain a first-order partial differential equation for V:

$$[Ax + B\omega(x, V_x)]'V_x + L[x, \omega(x, V_x)] = 0. \tag{2.9}$$

Now solve (2.9) for $V = V(x)$, compute $V_x(x)$, and finally obtain from (2.8) the *optimal feedback control law*

$$\varphi^0(x) = \omega[x, V_x(x)]. \tag{2.10}$$

The beauty of this approach lies in its intuitive directness, and the fact that it leads to a feedback control. In addition, it suggests a computational procedure, though as yet (2.9) is innocent of boundary conditions to render the solution (if any) unique. But rather than try to rigorize these matters directly, we shall redefine the problem precisely, show that (2.7) and (2.8) are *sufficient* conditions for optimality, and compute a reasonably explicit solution. This program will satisfy better the demands of logic.

12.3 Dynamic Programming: Rigor

From now on we confine attention to (possibly nonlinear) *state feedback controls*

$$u(t) = \varphi[x(t)].$$

Then, (1.1) becomes

$$\dot{x}(t) = Ax(t) + B\varphi[x(t)]. \tag{3.1}$$

Introduce the class Φ of *admissible controls* φ, characterized by the following properties:

i. The function $\varphi: \mathscr{X} \to \mathscr{U}$ is continuous.
ii. For every initial state $x(0) \in \mathscr{X}$ the differential equation (3.1) has a unique solution $x(\cdot)$ defined (and continuously differentiable) for $0 \le t < \infty$.
iii. For every initial state $x(0)$, the solution $x(\cdot)$ of (3.1) has the property

$$x(t) \to 0, \qquad t \to \infty.$$

Observe that if $F: \mathscr{X} \to \mathscr{U}$ is such that $A + BF$ is stable, the linear control $\varphi(x) = Fx$ belongs to Φ.

Existence and uniqueness of the solution of (3.1) are guaranteed if, for instance, $\varphi(x)$ grows no faster than $|x|$ as $|x| \to \infty$ and satisfies a uniform

Lipschitz condition in every ball $|x| \leq r$ $(r > 0)$; however, we shall not need such conditions explicitly. The stability condition (iii) is formally stronger than the condition of output regulation, namely

$$z(t) = Dx(t) \to 0, \qquad t \to \infty;$$

but it is technically convenient, and natural in the applications customarily made of the optimization technique.

Next, we introduce the *cost functional*

$$J: \mathscr{X} \times \Phi \to [0, \infty],$$

defined by

$$J(x, \varphi) := \int_0^\infty L[x(t), \varphi(x(t))]\, dt. \tag{3.2}$$

In the integrand, $x(\cdot)$ is the solution of (3.1) with $x(0) = x$. A control $\varphi^0 \in \Phi$ is *optimal* if $J(x, \varphi^0) < \infty$ for all $x \in \mathscr{X}$ and if

$$J(x, \varphi^0) \leq J(x, \varphi), \qquad x \in \mathscr{X},\ \varphi \in \Phi. \tag{3.3}$$

Our first technical assumption guarantees that Φ is nonempty and that $\varphi \in \Phi$ exists such that $J(x, \varphi) < \infty$ for all x.

A.1. *The pair* (A, B) *is stabilizable.* In fact, if $A + BF$ is stable and $\varphi(x) = Fx$,

$$J(x, \varphi) = \int_0^\infty x' e^{t(A+BF)'}(M + F'NF)e^{t(A+BF)}x\, dt$$

$$\leq c|x|^2, \qquad x \in \mathscr{X}, \tag{3.4}$$

for some constant c. Thus, we can now (rigorously) define the *value function*

$$V^0(x) := \inf\{J(x, \varphi): \varphi \in \Phi]. \tag{3.5}$$

Clearly,

$$0 \leq V^0(x) \leq c|x|^2.$$

Next, we shall demonstrate a sufficient condition for optimality of an admissible control.

Theorem 2.1 (Optimality Criterion). *Suppose there exist a control* $\varphi^0 \in \Phi$ *and a function* $V: \mathscr{X} \to \mathbb{R}$ *with the properties:*

i. $V(\cdot)$ *is continuously differentiable for* $x \in \mathscr{X}$. (3.6)

ii. *For some constant* c,

$$0 \leq V(x) \leq c|x|^2, \qquad x \in \mathscr{X}. \tag{3.7}$$

iii. $[Ax + B\varphi^0(x)]'V_x(x) + L[x, \varphi^0(x)] = 0, \qquad x \in \mathscr{X}.$ (3.8)

iv. $(Ax + Bu)'V_x(x) + L(x, u) \geq 0, \qquad x \in \mathscr{X},\ u \in \mathscr{U}.$ (3.9)

Then

$$V(x) = V^0(x) \tag{3.10}$$

and φ^0 *is optimal.*

PROOF. Write $\xi(t; x, \varphi)$ for the solution of (3.1) with $x(0) = x$. Then (3.6) and (3.8) imply

$$-\frac{d}{dt}\{V[\xi(t; x, \varphi^0)]\} = L[\xi(t; x, \varphi^0), \varphi^0(\xi(t; x, \varphi^0))], \qquad t \geq 0.$$

Since $\xi(\cdot\,; x, \varphi^0)$ satisfies (3.1), and $\varphi^0(\cdot)$ is continuous, we can integrate to obtain

$$V(x) = V[\xi(t; x, \varphi^0)] + \int_0^t L[\xi(s; x, \varphi^0), \varphi^0(\xi(s; x, \varphi^0))]\, ds.$$

Since $\xi(t; x, \varphi^0) \to 0$ $(t \to \infty)$, we have from (3.7) and the definition (3.2) of J,

$$V(x) = J(x, \varphi^0), \qquad x \in \mathscr{X}. \tag{3.11}$$

Let $\varphi \in \Phi$. Setting $u = \varphi(x)$ in (3.9) and integrating along the path $\xi(\cdot\,; x, \varphi)$ we get similarly

$$V(x) \leq V[\xi(t; x, \varphi)] + \int_0^t L[\xi(s; x, \varphi), \varphi(\xi(s; x, \varphi))]\, ds.$$

Letting $t \to \infty$,

$$V(x) \leq J(x, \varphi). \tag{3.12}$$

Inequality (3.3) follows by (3.11) and (3.12), so that φ^0 is optimal and (3.10) is true. □

Observe that conditions (3.8) and (3.9) are equivalent to Bellman's equation (2.7) or explicitly (2.8) and (2.9). We shall compute a solution of (2.7) with properties (3.6) and (3.7), and such that the corresponding control φ^0 given by (2.8) is admissible. Then Theorem 12.1 will establish that φ^0 is optimal. Let us assume

A.2. $N > 0$. (3.13)

Recalling (2.1), we find from (2.7) and (2.8)

$$\omega(x, V_x) = -\tfrac{1}{2}N^{-1}B'V_x \tag{3.14}$$

and then (2.9) becomes

$$(Ax)'V_x - \tfrac{1}{4}V_x'BN^{-1}B'V_x + x'Mx = 0. \tag{3.15}$$

We now make the inspired guess that optimal control is linear and note from (3.4) that V must necessarily be quadratic. Put

$$V(x) = x'Px, \qquad x \in \mathscr{X},$$

where P is symmetric. Substitution in (3.15) yields the matrix quadratic equation

$$A'P + PA - PBN^{-1}B'P + M = 0. \tag{3.16}$$

To solve (3.16) we need

A.3. *The pair* (D, A) *is detectable.* Then, we have

Theorem 12.2. *Under assumptions A.1, A.2 and A.3, the matrix quadratic equation* (3.16) *has a unique solution* P^0 *in the class of symmetric, positive semidefinite maps. Furthermore, the map*

$$A - BN^{-1}B'P^0$$

is stable.

The proof will be deferred to Section 12.4. Assuming this result, it is clear that

$$V(x) = x'P^0x \tag{3.17}$$

satisfies conditions (i) and (ii) of Theorem 12.1. On the basis of (2.10) and (3.14) define

$$\varphi^0(x) := -N^{-1}B'P^0x. \tag{3.18}$$

By Theorem 12.2 the control (3.18) yields a stable system map, and therefore φ^0 is admissible. By (3.14), φ^0 and V satisfy conditions (iii) and (iv) of Theorem 12.1, so $V = V^0$, the value function, and φ^0 is optimal.

We shall demonstrate that optimal control is even unique. By (3.17), (3.18) and a short computation,

$$\begin{aligned}(Ax + Bu)'V_x^0(x) + L(x, u) - [Ax + B\varphi^0(x)]'V_x^0(x) - L[x, \varphi^0(x)]\\ = (u + N^{-1}B'P^0x)'N(u + N^{-1}B'P^0x)\end{aligned} \tag{3.19}$$

for all $x \in \mathscr{X}$ and $u \in \mathscr{U}$. Denote the right side of (3.19) by $\Delta(u - \varphi^0(x))$. Clearly $\Delta \geq 0$, and $\Delta = 0$ if and only if $u = \varphi^0(x)$. Let $\varphi \in \Phi$. If $\varphi \neq \varphi^0$ there is a state $x_0 \in \mathscr{X}$ and by continuity, a neighborhood $\mathfrak{N}$ of x_0 such that $\varphi(x) \neq \varphi^0(x)$ for $x \in \mathfrak{N}$. By (3.8) and (3.19)

$$[Ax + B\varphi(x)]'V_x^0(x) + L[x, \varphi(x)] = \Delta[\varphi(x) - \varphi^0(x)].$$

Put $x = \xi(t; x_0, \varphi)$ in (3.20) and integrate on $[0, \infty)$. The result is

$$\begin{aligned}J(x_0, \varphi) &= \int_0^\infty L[\xi(t; x_0, \varphi), \varphi(\xi(t; x_0, \varphi))]\, dt\\ &= V^0(x_0) + \int_0^\infty \Delta[\varphi(\xi(t; x_0, \varphi)) - \varphi^0(\xi(t; x_0, \varphi))]\, dt.\end{aligned} \tag{3.21}$$

By continuity of $\xi(\cdot\,; x_0, \varphi)$ there exists $\delta > 0$, such that $\xi(t; x_0, \varphi) \in \mathfrak{N}$ for $0 \le t \le \delta$, hence the integrand in (3.21) is strictly positive on $[0, \delta]$. Therefore $J(x_0, \varphi) > V^0(x_0)$ and so φ cannot be optimal.

Summarizing results, we have

Theorem 12.3. *If (A, B) is stabilizable, (D, A) is detectable and $N > 0$, an optimal feedback control φ^0 exists and is unique in the class of admissible controls. In addition, $\varphi^0(x)$ is linear in x, and the corresponding closed loop system matrix is stable.*

We turn finally to a constructive proof of Theorem 12.2 which yields an algorithm for computing P^0 and thus φ^0.

12.4 Matrix Quadratic Equation

To prove Theorem 12.2 we recall Proposition 0.6 on the convergence of a bounded monotone sequence of symmetric maps, and the results of Sections 3.6 and 3.10 on detectability. In addition, we need three preliminary lemmas.

Lemma 12.1. *If $Q \ge 0$ and A is stable, the linear equation*

$$A'P + PA + Q = 0$$

has a unique solution P, and $P \ge 0$.

PROOF. If P is a solution

$$-\frac{d}{dt}(e^{tA'}Pe^{tA}) = -e^{tA'}(A'P + PA)e^{tA} = e^{tA'}Qe^{tA}$$

for all t. Integrating and using stability of A,

$$P = \int_0^\infty e^{tA'}Qe^{tA}\,dt \ge 0. \tag{4.1}$$

On the other hand, the integral in (4.1) is clearly a solution. □

Lemma 12.2 (Lyapunov Criterion). *Suppose $P \ge 0$, $Q \ge 0$, $(\sqrt{Q}, A)$ is detectable and*

$$A'P + PA + Q = 0. \tag{4.2}$$

Then A is stable. If $(\sqrt{Q}, A)$ is observable, then actually $P > 0$.

PROOF. From (4.2) there results the identity

$$P = e^{tA'}Pe^{tA} + \int_0^t e^{sA'}Qe^{sA}\,ds, \qquad t \ge 0.$$

Since $(\sqrt{Q}, A)$ is detectable, Proposition 3.2 asserts that the integral

$$Q(t) = \int_0^t e^{sA'} Q e^{sA}\, ds, \qquad t \geq 0$$

is bounded only if A is stable. Since

$$0 \leq Q(t) \leq P$$

the first conclusion follows. For the second, we recall from Lemma 3.1 that $Q(t) > 0$ if $t > 0$. □

Lemma 12.3. *Let P be symmetric, $N > 0$ and $F^0 = -N^{-1}B'P$. Write*

$$\psi(F) := (A + BF)'P + P(A + BF) + F'NF.$$

Then

$$\psi(F) - \psi(F^0) = (F - F^0)'N(F - F^0),$$

i.e. F^0 minimizes $\psi(F)$.

The proof is a simple computation.

Proof (of Theorem 12.2). We rewrite (3.16) in the form of two simultaneous equations for P and the state feedback F determined by (3.18):

$$(A + BF)'P + P(A + BF) + M + F'NF = 0 \tag{4.3}$$

$$F = -N^{-1}B'P. \tag{4.4}$$

The point of this maneuver is that (4.3) is *linear* in P for fixed F. This suggests that we construct a sequence $\{F_k, P_k;\ k = 1, 2, \ldots\}$ as follows:

1. Choose F_1, so that $A + BF_1$ is stable.
2. Having chosen $F_1, \ldots, F_k$ obtain P_k from

$$(A + BF_k)'P_k + P_k(A + BF_k) + M + F_k'NF_k = 0. \qquad (4.3)_k$$

3. Define

$$F_{k+1} := -N^{-1}B'P_k. \qquad (4.4)_k$$

It will be shown that the sequence P_k is well-defined, $P_k \geq 0$ and $P_k \downarrow$. By Lemma 12.1, P_1 is uniquely determined by $(4.3)_1$ and $P_1 \geq 0$. Suppose $P_1, \ldots, P_k$ are defined and nonnegative. Then, F_{k+1} is determined by $(4.4)_k$. By Lemma 12.3,

$$\begin{aligned}(A + BF_{k+1})'P_k + P_k(A + BF_{k+1}) + M + F_{k+1}'NF_{k+1}\\ &= (A + BF_k)'P_k + P_k(A + BF_k) + M + F_k'NF_k\\ &\quad - (F_k - F_{k+1})'N(F_k - F_{k+1})\\ &= -(F_k - F_{k+1})'N(F_k - F_{k+1}) \qquad (\text{by } (4.4)_k)\\ &= -Q_k, \text{ say,}\end{aligned}$$

where $Q_k \geq 0$. Thus,

$$(A + BF_{k+1})'P_k + P_k(A + BF_{k+1}) + M + Q_k + F'_{k+1}NF_{k+1} = 0. \quad (4.5)$$

By Theorem 3.6(ii), the pair

$$(\sqrt{M + Q_k + F'_{k+1}NF_{k+1}},\ A + BF_{k+1})$$

is detectable. Since $P_k \geq 0$ by assumption, Lemma 12.2 asserts that $A + BF_{k+1}$ is stable. Then Lemma 12.1 ensures that $P_{k+1} \geq 0$ is determined by $(4.3)_{k+1}$.

Subtracting $(4.3)_{k+1}$ from (4.5), we get

$$(A + BF_{k+1})'(P_k - P_{k+1}) + (P_k - P_{k+1})(A + BF_{k+1}) + Q_k = 0$$

and again by Lemma 12.1, $P_k - P_{k+1} \geq 0$. Thus $0 \leq P_k \downarrow$, and by Proposition 0.6,

$$P^0 := \lim P_k, \qquad k \uparrow \infty,$$

exists. Then

$$F^0 := \lim F_{k+1} = -N^{-1}B'P^0$$

exists as well. Taking the limit in $(4.3)_k$, we have that F^0, P^0 satisfy (4.3) and (4.4). By Theorem 3.6(ii), the pair

$$(\sqrt{M + F^{0\prime}NF^0},\ A + BF^0)$$

is detectable. Then, by Lemma 12.2, $A + BF^0$ is stable.

It remains to prove uniqueness. Suppose $\tilde{P}$, $\tilde{F}$ satisfy (4.3), (4.4) and $\tilde{P} \geq 0$. Write

$$\tilde{A} = A + B\tilde{F}, \qquad A^0 = A + BF^0.$$

By Lemma 12.3

$$A^{0\prime}\tilde{P} + \tilde{P}A^0 + M - Q + F^{0\prime}NF^0 = 0, \tag{4.6}$$

where

$$Q = (\tilde{F} - F^0)'N(\tilde{F} - F^0) \geq 0.$$

Also

$$A^{0\prime}P^0 + P^0A^0 + M + F^{0\prime}NF^0 = 0. \tag{4.7}$$

Subtracting (4.6) from (4.7) yields

$$A^{0\prime}(P^0 - \tilde{P}) + (P^0 - \tilde{P})A^0 + Q = 0. \tag{4.8}$$

Since A^0 is stable, Lemma 12.1 implies $P^0 \geq \tilde{P}$. Also, (4.3) for $\tilde{P}$, $\tilde{F}$ together with Theorem 3.6(ii) and Lemma 12.2, implies that the pair

$$(\sqrt{M + \tilde{F}'N\tilde{F}},\ \tilde{A})$$

is detectable and $\tilde{A}$ is stable. By an argument symmetric to the one leading to (4.8) we conclude that $\tilde{P} \geq P^0$, and thus finally $\tilde{P} = P^0$. □

It is interesting to see what happens if we drop the assumption that (D, A) is detectable. Suppose the optimization problem is given by

$$\dot{x} = u, \qquad \int_0^\infty u(t)^2\, dt = \min,$$

where x and u are scalars. Then (3.16) becomes $P^2 = 0$, so $P^0 = 0$, and $\varphi^0(x) = -P^0 x = 0$. Of course, the resulting system is not stable, and φ^0 is not admissible. On the other hand, the admissible controls

$$\varphi_\epsilon(x) = -\epsilon x, \qquad \epsilon > 0$$

yield

$$J(x, \varphi_\epsilon) = \frac{\epsilon x^2}{2},$$

so that

$$V^0(x) = \inf J(x, \varphi) = 0.$$

Thus an optimal control does not exist.

Again, consider the problem

$$\dot{x} = x + u, \qquad \int_0^\infty u(t)^2\, dt = \min.$$

From (3.16), $2P - P^2 = 0$, and we have two nonnegative solutions $P_1^0 = 0$, $P_2^0 - 2$, giving controls

$$\varphi_1^0(x) = 0, \qquad \varphi_2^0(x) = -2x.$$

Clearly, $J(x, \varphi_1^0) = 0$, but φ_1^0 is not admissible. On the other hand, φ_2^0 is admissible and so by Theorem 12.1 it is optimal.

We conclude that detectability is not necessary for the existence of an optimal control, but if detectability is absent existence may fail; and the same can be said about uniqueness of a nonnegative solution of (3.16). On the other hand, it can be shown (Exercise 12.3) that $(\sqrt{M}, A)$ is necessarily detectable if (3.16) has exactly one nonnegative solution P^0, where P^0 has the property that $A - BN^{-1}B'P^0$ is stable.

12.5 Exercises

12.1. For the scalar system

$$\dot{x} = ax + u$$

$$\int_0^\infty (mx^2 + u^2)\, dt = \min \qquad (m > 0),$$

solve the quadratic equation (3.16) explicitly, and also by successive approximation as in the proof of Theorem 12.2. Show that the latter technique is simply Newton's method. What can be said about the rate of convergence? Extend your discussion to the general case.

12.2. Show that if the hypothesis of detectability in Theorem 12.2 is strengthened to observability then P^0 is positive definite.

12.3. Verify the last assertion of Section 12.4 by showing that if $(\sqrt{M}, A)$ is not detectable, there exists a solution $P \geq 0$ of (3.16), such that

$$A - BN^{-1}B'P$$

is not stable. Hint: Write

$$\mathcal{N} := \bigcap_{i=1}^{n} \operatorname{Ker}(DA^{i-1})$$

for the unobservable subspace of (D, A), and $\mathcal{X}^+$ for the subspace of unstable modes of A. Show that if $(\sqrt{M}, A)$ is not detectable, then

$$\mathcal{N}^+ := \mathcal{N} \cap \mathcal{X}^+ \neq 0.$$

Check that $\mathcal{N}^+$ is A-invariant and that $\bar{D}$ exists such that the diagram commutes:

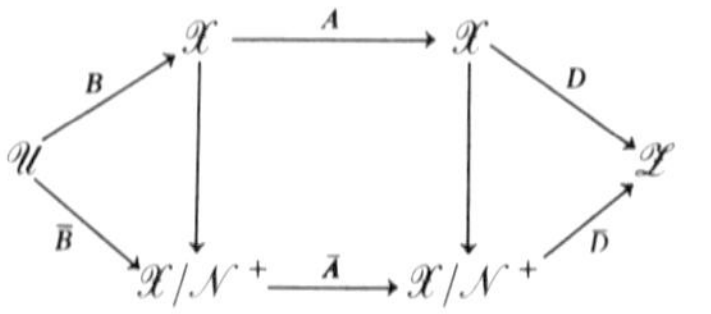

Verify that the triple $(\bar{D}, \bar{A}, \bar{B})$ is detectable and stabilizable if (A, B) is stabilizable. Next, write

$$\mathcal{X} = \mathcal{N}^+ \oplus \hat{\mathcal{X}}$$

and show that the corresponding matrices have the form

$$A = \begin{bmatrix} A_1 & A_3 \\ 0 & A_2 \end{bmatrix}, \qquad B = \begin{bmatrix} B_1 \\ B_2 \end{bmatrix}, \qquad D = [0 \quad D_2].$$

It is now easy to verify that (3.16) has a solution P with the stated properties.

12.4. Show that if (A, B) is stabilizable and (D, A) is detectable, the same is true with A replaced by $A - \alpha 1$, $\alpha > 0$; but the converse is false.

12.5. Consider the linear regulator problem with cost functional

$$\int_0^\infty e^{2\alpha t} L[x(t), u(t)]\, dt, \tag{5.1}$$

where α is a real constant and L is defined as in the text. Reasoning as in Section 12.1, show formally that the value function satisfies

$$(Ax)'V_x - \tfrac{1}{4}V_x' BN^{-1}B'V_x + 2\alpha V + x'Mx = 0.$$

Find the corresponding matrix quadratic equation and obtain the counterparts of Theorems 12.2 and 12.3.

12.6. Show that if the regulator problem of Exercise 12.5 is solvable for some α, then it is solvable for any $\alpha' < \alpha$, in the sense that linear state feedback exists, such that (5.1), with exponent α', is minimized. But is it necessarily true that the closed loop system matrix is stable?

12.7. *Discrete-time optimal regulation.* Consider the discrete-time system

$$x(t+1) = Ax(t) + Bu(t)$$
$$z(t) = Dx(t)$$

defined for $t = 0, 1, 2, \ldots$. Say A is *stable* if $|\lambda| < 1$ for all $\lambda \in \sigma(A)$. Investigate the optimal regulator problem in this setting, and obtain as the counterpart of (3.16) the equation

$$A'PA - P - A'PB(N + B'PB)^{-1}B'PA + M = 0. \tag{5.5}$$

In parallel with the text, prove:

Theorem 12.2′. *Let* $(\sqrt{M}, A)$ *be detectable and* (A, B) *be stabilizable. Then* (5.5) *has a unique solution* P^0 *such that* $P^0 \geq 0$. *Also,*

$$A - B(N + B'P^0B)^{-1}B'P^0A$$

is stable.

In the course of the proof you will need the following.

Lemma 12.1′. *If* $Q \geq 0$ *and* A *is stable, the linear equation*

$$A'PA - P + Q = 0$$

has a unique solution P, *and* $P \geq 0$. Hint: Use the *spectral radius formula:*

$$\max\{|\lambda| : \lambda \in \sigma(A)\} = \lim_{k \to \infty} |A^k|^{1/k}.$$

Lemma 12.2′. *Let* $P \geq 0$, $Q \geq 0$, $(\sqrt{Q}, A)$ *detectable, and* $A'PA - P + Q = 0$. *Then,* A *is stable.*

12.6 Notes and References

The "linear-quadratic" regulator has been a topic of longstanding interest in control theory (cf. Newton, Gould and Kaiser [1], Kalman [2]), both for the reasons mentioned at the beginning of this chapter and because the problem can often be solved in more general settings: as with partial differential equations (Lions [1]) or with account taken of random noise (e.g. Wonham [4]). In accordance with the scope of this book we have restricted attention to the infinite process time interval, as it is this which leads to a time-invariant control. For a more general treatment see especially Lee and Markus [1], Bellman [2], and Anderson and Moore [1]

The proof of Theorem 12.2 follows Wonham [2]; the method used there, sometimes called "quasilinearization" or "approximation in policy space," had been extensively discussed by Bellman [1], Kalaba [1], and Kleinman [1, 2]. An alternative approach to the matrix quadratic equation was developed by Potter [1], who expresses P in terms of the eigenvectors of the "Hamiltonian" matrix

$$\begin{bmatrix} A & -BN^{-1}B' \\ -D'D & -A' \end{bmatrix}$$

discussed in Chapter 13. A variety of results are known which relate stabilizability of (A, B), detectability or observability of (D, A), and properties of the Hamiltonian matrix: see Martensson [1] and Kučera [1]. In particular, the remark that Theorem 12.2 is true with (D, A) merely detectable (rather than observable) is due to Kučera [1], as is the result of Exercise 12.3.

13 Quadratic Optimization II: Dynamic Response

The approach to quadratic optimization described in the previous chapter has been widely advertised as a systematic technique to achieve good transient response with reasonable computational effort. While not generally disputed, this claim is based more on numerical experience than compelling theoretical arguments. Nevertheless, some precise information is available about the qualitative behavior of the closed loop system as a function of the weighting matrices M and N of the cost functional (12.1.3). In this chapter we present a selection of the simpler results, referring the reader to the literature for supplementary developments.

13.1 Dynamic Response: Generalities

We recall from Theorem 12.3 that the optimization problem was posed in such a way that the optimal system is necessarily stable, so that

$$z(t) = Dx(t) \to 0, \qquad t \to \infty. \tag{1.1}$$

On intuitive grounds we might expect that if the weighting matrix N is diminished (thus attaching a lower penalty to large values of $|u(t)|^2$) convergence in (1.1) would be speeded up. This is only broadly true: as N decreases, some eigenvalues of $A + BF$ may actually move to the right in the complex plane. Furthermore, attempts to improve convergence in (1.1) may lead to impracticably large values of the elements of F as well as large amplitude peaks in $|z(\cdot)|$. Very roughly, the problem is to arrange that $A + BF$ be such that convergence in (1.1) is rapid, and damped to avoid excessive overshoot and oscillation, while keeping the norm of F small. The last requirement is needed to prevent saturation of actuating elements by the control signals, in response to initial disturbances (values of $x(0+)$) which the system would normally encounter. A systematic procedure taking all these constraints into account is not yet available.

13.2 Example 1: First-Order System

Let $n = 1$ and A, B, M, N be scalars, which we write with lower case letters. Solving the quadratic equation (12.3.16), we get

$$\begin{aligned} p^0 &= \frac{\sqrt{a^2 + n^{-1}b^2m} + a}{n^{-1}b^2} \\ f^0 &= -\frac{\sqrt{a^2 + n^{-1}b^2m} + a}{b} \\ a + bf^0 &= -\sqrt{a^2 + n^{-1}b^2m}. \end{aligned}$$

As the control weighting n decreases to zero the value function p^0x^2 does the same, i.e. performance becomes perfect; the optimal gain f^0 increases in magnitude to $+\infty$, as does speed of response as measured by the eigenvalue $a + bf^0$. Conversely, heavy control weighting (large n) may result in sluggish response. Obviously the influence of m is reciprocal to that of n.

13.3 Example 2: Second-Order System

Let

$$\dot{x}_1 = x_2, \qquad \dot{x}_2 = u, \qquad z = (x_1, \sqrt{m_2}x_2),$$
$$L(x, u) = x_1^2 + m_2x_2^2 + nu^2.$$

Thus,

$$A = \begin{bmatrix} 0 & 1 \\ 0 & 0 \end{bmatrix}, \quad B = \begin{bmatrix} 0 \\ 1 \end{bmatrix}, \quad D = \begin{bmatrix} 1 & 0 \\ 0 & \sqrt{m_2} \end{bmatrix},$$
$$M = \begin{bmatrix} 1 & 0 \\ 0 & m_2 \end{bmatrix}, \qquad N = n.$$

Let

$$P^0 = \begin{bmatrix} p_1 & p_3 \\ p_3 & p_2 \end{bmatrix}.$$

Solving (12.3.16), we get

$$p_1 = \sqrt{m_2 + 2\sqrt{n}},$$
$$p_2 = \sqrt{n(m_2 + 2\sqrt{n}},$$
$$p_3 = \sqrt{n},$$

$$F^0 = -N^{-1}B'P^0 = -\left(\frac{1}{\sqrt{n}}, \sqrt{\frac{m_2 + 2\sqrt{n}}{n}}\right).$$

From this,

$$\sigma(A + BF) = -\sqrt{\frac{m_2 + 2\sqrt{n}}{4n}} \pm \sqrt{\frac{m_2 - 2\sqrt{n}}{4n}}.$$

If $m_2 > 0$, P^0 does not vanish as $n \downarrow 0$, and only one eigenvalue λ of $A + BF$ is unbounded:

$$\lambda \sim -1/\sqrt{m_2}, \; -\sqrt{m_2/n}.$$

The feedback gains become large:

$$F^0 \sim -(1/\sqrt{n}, \sqrt{m_2/n});$$

nevertheless, as $n \downarrow 0$ the state-variable "peaking index"

$$\sup_{t \geq 0} |\exp t(A + BF^0)| \tag{3.1}$$

remains bounded, showing that $|z(t)|$ is bounded as $n \downarrow 0$, uniformly for $t \geq 0$ and $|x(0)| = 1$. It is clear that the weighting factor m_2 attached to the derivative $x_2 = \dot{x}_1$ inhibits very fast response, as this would call for nearly impulsive velocities, and the square of a unit impulse has (formally) infinite integral.

If $m_2 = 0$, we have

$$\lambda = \sqrt{1/2\sqrt{n}} \;\; (-1 \pm i).$$

As $n \downarrow 0$ response becomes arbitrarily fast, necessarily at the expense of a high peaking index (3.1), which can be shown to behave unboundedly as $n^{-1/4}$. On the other hand, the "damping ratio," defined for a second-order system as

$$\zeta = -\frac{\lambda_1 + \lambda_2}{2|\lambda_1 \lambda_2|^{1/2}},$$

satisfies the inequality $\zeta \geq 1/\sqrt{2}$ for all $m_2 \geq 0$ and $n > 0$. It follows that $\sigma(A + BF^0)$ is confined to the left-plane sector enclosed by the rays

$$\arg \lambda = \frac{3\pi}{4}, \frac{5\pi}{4};$$

and this is a standard specification restricting "overshoot" associated with rapid oscillation.

13.4 Hamiltonian Matrix

Let $\lambda_0 \in \sigma(A + BF^0)$, with eigenvector ξ. Recalling (12.3.16) and (12.4.4), we have

$$\begin{aligned}\lambda_0 \xi &= (A + BF^0)\xi \\ &= A\xi - BN^{-1}B'\eta,\end{aligned} \tag{4.1}$$

where $\eta = P^0\xi$. Thus,

$$\begin{aligned}\lambda_0 \eta &= P^0(A + BF^0)\xi \\ &= (P^0A - P^0BN^{-1}B'P^0)\xi \\ &= (-M - A'P^0)\xi \\ &= -M\xi - A'\eta.\end{aligned} \tag{4.2}$$

From (4.1) and (4.2), there results

$$H\begin{bmatrix}\xi \\ \eta\end{bmatrix} = \lambda_0 \begin{bmatrix}\xi \\ \eta\end{bmatrix},$$

where H is the *Hamiltonian matrix*

$$H := \begin{bmatrix} A & -BN^{-1}B' \\ -M & -A' \end{bmatrix}. \tag{4.3}$$

Write $A^0 := A + BF^0$. From (12.3.16) and (12.4.4) it is quickly verified that H satisfies the identity

$$\begin{bmatrix} 1 & 0 \\ P^0 & -1 \end{bmatrix}(\lambda 1 - H)\begin{bmatrix} 1 & 0 \\ P^0 & 1 \end{bmatrix} = \begin{bmatrix} \lambda 1 - A^0 & BN^{-1}B' \\ 0 & -\lambda 1 - A^{0\prime} \end{bmatrix}. \tag{4.4}$$

Taking determinants on both sides of (4.4), we get

$$\det(\lambda 1 - H) = \det(\lambda 1 - A^0)\det(\lambda 1 + A^0). \tag{4.5}$$

From (4.5) there follows

$$\begin{aligned}\sigma(H) &= \sigma(A^0) \cup \sigma(-A^0) \\ &= \sigma(A^0) \cup (-\sigma(A^0)).\end{aligned}$$

Thus $\sigma(H)$ is symmetric about the imaginary, as well as the real, axis of $\mathbb{C}$, and the characteristic polynomial of H can be written as a polynomial in λ^2. Since A^0 is stable, $\sigma(A^0)$ is that part of $\sigma(H)$ lying in the open left-half complex plane.

To exploit this observation we compute $\det(\lambda 1 - H)$ in terms of the open-loop transfer matrix

$$G(\lambda) := D(\lambda 1 - A)^{-1}BN^{-1/2}, \tag{4.6}$$

where $D'D = M$. Set

$$\pi(\lambda) := \det(\lambda 1 - A), \qquad \pi^0(\lambda) := \det(\lambda 1 - A^0).$$

From (4.3), (4.5), and standard determinantal manipulations (Exercise 13.7) there results

$$\pi^0(\lambda)\pi^0(-\lambda) = (-1)^n \det(\lambda 1 - H)$$
$$= \pi(\lambda)\pi(-\lambda)\det[1 + G(\lambda)G(-\lambda)']. \tag{4.7}$$

Thus if $M \downarrow 0$, or if $N \geq \nu 1$ and $\nu \uparrow \infty$, the closed-loop poles tend, by (Rouché's) Theorem 0.3, to the roots of $\pi(\lambda)\pi(-\lambda)$ in $\Re e\ \lambda \leq 0$, that is, to the open-loop poles reflected, if necessary, in the imaginary axis. The behavior of the closed-loop poles as $N \downarrow 0$ is more complicated; a partial description will be given in the two sections to follow.

13.5 Asymptotic Root Locus: Single Input System

Suppose $B = b$ and set $N = \epsilon^2$. For simplicity we assume as well that (A, b) is controllable. By a similarity transformation, we can arrange that the matrices of A, b are in standard canonical form (1.15). It is now easy to compute the characteristic polynomial of H directly from (4.3). The result is

$$(-1)^n \det|\lambda 1 - H| = \epsilon^{-2}\theta(\lambda)'M\theta(-\lambda) + \pi(\lambda)\pi(-\lambda), \tag{5.1}$$

where $\pi(\lambda)$ is the ch.p. of A and $\theta(\lambda) := \text{col}(1, \lambda, \ldots, \lambda^{n-1})$. Write $\pi^0(\lambda)$ for the ch.p. of A^0. From (4.5) and (5.1) there follows

$$\pi^0(\lambda)\pi^0(-\lambda) = \epsilon^{-2}\theta(\lambda)'M\theta(-\lambda) + \pi(\lambda)\pi(-\lambda). \tag{5.2}$$

We are interested in the behavior of the roots of $\pi^0(\lambda)$ as $\epsilon \downarrow 0$, since this condition is equivalent to light weighting of the control and thus would be expected to encourage fast dynamic response. In this direction, we have the following asymptotic result.

Theorem 13.1. *Let*

$$\theta(\lambda)'M\theta(-\lambda) = \varphi(\lambda)\varphi(-\lambda), \tag{5.3}$$

where the roots s_i of $\varphi(\lambda)$ belong to the closed left-half complex s-plane. If $\deg \varphi = k$, then as $\epsilon \downarrow 0$, k of the roots of $\pi^0(\lambda)$ approach the fixed values s_i ($i \in \mathbf{k}$), while the remaining $n - k$ roots tend to infinity with asymptotic values

$$\epsilon^{-1/(n-k)} \exp\left[\frac{i\pi}{2(n-k)}(n - k + 1 + 2\nu)\right], \qquad \nu = 0, 1, \ldots, n - k - 1. \tag{5.4}$$

A proof is given at the end of this section.

The quantities (5.4) are simply the left-half plane roots of the equation

$$(-1)^{n-k}\lambda^{2(n-k)} + \epsilon^{-2} = 0.$$

The theorem says that for small ϵ the roots of $\pi^0(\lambda)$ (i.e. the closed loop poles) are nearly independent of the roots of $\pi(\lambda)$ (i.e. the eigenvalues of A), being determined essentially by the choice of the state weighting matrix M. The number k $(0 \le k \le n-1)$ of asymptotically finite closed loop poles is simply the highest order of derivative $x_1^{(k)} = x_{k+1}$ which is assigned positive weight in the cost functional. Next, (5.4) shows that the remaining $n-k$ closed loop poles are asymptotically uniformly distributed along a circular arc, terminating on rays at an angle $(n-k)^{-1}\pi/2$ with the imaginary axis. The corresponding factor in the squared real frequency response is

$$(\omega^{2(n-k)} + \epsilon^{-2})^{-1},$$

called in circuit theory a *Butterworth characteristic*. The time response of a Butterworth filter to an impulse is known to be well damped, and if ϵ is small, response is fast. However, there is no assurance that the asymptotically finite poles (roots of φ) are well damped: this is determined by the choice of M, i.e. of D. For example, if $n = 8$, $k = 6$ and

$$L(x, u) = \mu^2(x_1^2 + x_7^2) + u^2,$$

then

$$\theta(\lambda)'M\theta(-\lambda) = 1 + \lambda^{12},$$

which yields an unpleasantly oscillatory pole pair

$$\exp i\left(-\pi \pm \frac{5\pi}{12}\right).$$

To investigate output behavior in more detail, suppose

$$z(t) = \sum_{i=1}^{k+1} d_i x_i(t) = \sum_{i=1}^{k+1} d_i x_1^{(i-1)}(t),$$

so that $M = dd'$, where

$$d := \text{col}(d_1, \ldots, d_{k+1}, 0, \ldots, 0).$$

Write

$$\psi(\lambda) := \sum_{i=1}^{k+1} d_i \lambda^{i-1}. \tag{5.5}$$

From (5.3) and (5.5)

$$\psi(\lambda)\psi(-\lambda) = \varphi(\lambda)\varphi(-\lambda).$$

If $\psi(\lambda)$ happens to be stable then we have $\psi(\lambda) = \varphi(\lambda)$, the so-called minimum phase relation, and the transfer function from u to z is

$$\frac{\hat{z}}{\hat{u}} = \frac{\psi(\lambda)}{\pi^0(\lambda)} = \frac{\varphi(\lambda)}{\pi^0(\lambda)}.$$

According to Theorem 13.1, for small ϵ, $\varphi(\lambda)$ is approximately cancelled from $\pi^0(\lambda)$, and the transfer function is nearly Butterworth, as already noted. On the other hand, if $\psi(\lambda)$ has a factor

$$\lambda - \sigma, \sigma > 0; \qquad \text{or} \qquad (\lambda - \sigma)^2 + \omega^2, \sigma > 0,$$

then for small ϵ, $\hat{z}/\hat{u}$ will contain a factor close to

$$\frac{\lambda - \sigma}{\lambda + \sigma} \qquad \text{or} \qquad \frac{(\lambda - \sigma)^2 + \omega^2}{(\lambda + \sigma)^2 + \omega^2},$$

respectively. The component this factor adds to time response might well be slow or lightly damped.

The conclusion is that light weighting of the control may often yield but cannot guarantee good dynamic behavior of an arbitrarily chosen scalar output, in response to an initial perturbation of this output and its derivatives. Our analysis of the single-input single-output situation has revealed that the approach works best if the output to be quadratically minimized is in a minimum phase relation with the system state.

It is clear that a similar analysis applies in the dual situation where the regulated *output* is a scalar, i.e. rank $(M) = 1$ and rank (B) is arbitrary.

We conclude this section with a proof of Theorem 13.1. For this we need two preliminary results.

Lemma 13.1. *Let $\alpha(\lambda), \beta(\lambda) \in \mathbb{R}[\lambda]$. There exists $\varphi(\lambda) \in \mathbb{R}[\lambda]$ such that*

$$\alpha(\lambda)\alpha(-\lambda) + \beta(\lambda)\beta(-\lambda) = \varphi(\lambda)\varphi(-\lambda).$$

Furthermore, if $0 \leq M \in \mathbb{R}^{n \times n}$, with M symmetric, and if $\theta(\lambda) \in \mathbb{R}^n[\lambda]$, there exists $\varphi(\lambda) \in \mathbb{R}[\lambda]$ such that

$$\theta(\lambda)' M \theta(-\lambda) = \varphi(\lambda)\varphi(-\lambda).$$

PROOF. Write $\omega(\lambda) := \alpha(\lambda)\alpha(-\lambda) + \beta(\lambda)\beta(-\lambda)$. Since $\omega(-\lambda) = \omega(\lambda)$ and $\omega(\lambda) \geq 0$ for $\Re e\, \lambda = 0$, the prime factors of ω must be of form $\pi(\lambda)\pi(-\lambda)$ with $\pi(\lambda) \in \mathbb{R}[\lambda]$. The first statement is now clear, and the second follows by induction on the number of summands in the scalar product $(\sqrt{M}\,\theta(\lambda))'\sqrt{M}\,\theta(-\lambda)$. □

Remark. By swapping over prime factors if necessary it is clear that we can always arrange that the complex roots of $\varphi(\lambda)$ lie in $\Re e\, \lambda \leq 0$.

Lemma 13.2. *Let $\xi(\lambda), \eta(\lambda)$ be monic polynomials in $\mathbb{R}[\lambda]$, with*

$$n = \deg \xi > \deg \eta = m.$$

For $s \in \mathbb{C}$ and $\epsilon \geq 0$ let

$$\zeta(s, \epsilon) := \epsilon\xi(s) + \eta(s).$$

Then as $\epsilon \downarrow 0$, m of the roots of $\zeta(\cdot, \epsilon)$ tend to the roots of η, while the remaining $n - m$ roots tend asymptotically to

$$\epsilon^{-1/(n-m)} \times \{\text{roots of } t^{n-m} + 1 = 0\}.$$

Proof. Let s_0 be a root of η, and pick $\delta > 0$ such that no root of η other than s_0 lies in the disk $|s - s_0| \leq \delta$. Write $\mathfrak{C} := \{s\colon |s - s_0| = \delta\}$ and

$$q := \min_{s \in \mathfrak{C}} |\eta(s)|.$$

If $\epsilon_0 > 0$ is chosen such that

$$\epsilon_0 \max_{s \in \mathfrak{C}} |\xi(s)| < q,$$

then for all $|\epsilon| < \epsilon_0$, we have

$$|\epsilon\xi(s)| < q \leq |\eta(s)|, \qquad s \in \mathfrak{C}.$$

It follows by Rouché's Theorem that $\eta(s)$ and $\epsilon\xi(s) + \eta(s)$ have the same number of roots within $\mathfrak{C}$, i.e. if s_0 is of multiplicity σ, then for all ϵ sufficiently small, $\zeta(s, \epsilon)$ has exactly σ roots within $\mathfrak{C}$. Taking $\delta > 0$ arbitrarily small, we get that m roots of $\zeta(\cdot, \epsilon)$ approach the m roots of η as $\epsilon \to 0$. Next, let t^* be a fixed root of $t^{n-m} + 1 = 0$. It will be shown that $\zeta(\cdot, \epsilon)$ has a root $s(\epsilon)$ such that

$$\frac{s(\epsilon)}{\epsilon^{-1/(n-m)}t^*} \to 1 \text{ as } \epsilon \downarrow 0.$$

For this, let

$$\xi(s) = \mu(s)\eta(s) + v(s),$$

where $\deg \mu = n - m$, $\deg v \leq n - m - 1$. Then,

$$\zeta(s, \epsilon) = [\epsilon\mu(s) + 1]\eta(s) + \epsilon v(s),$$

so that

$$\hat{\zeta}(s, \epsilon) := \frac{\zeta(s, \epsilon)}{\eta(s)} = \epsilon\mu(s) + 1 + \frac{\epsilon v(s)}{\eta(s)}.$$

Put $t = \epsilon^{1/(n-m)}s$. Then simple computations verify that

$$\epsilon\mu(s) = t^{n-m} + O(\epsilon^{1/(n-m)}),$$

and

$$\frac{\epsilon v(s)}{\eta(s)} = O(\epsilon^{(m+1)/(n-m)}),$$

as $\epsilon \downarrow 0$, uniformly for $\frac{1}{2} \leq |t| \leq \frac{3}{2}$. It follows that

$$\begin{aligned}\tilde{\zeta}(t, \epsilon) &:= \hat{\zeta}(\epsilon^{-1/(n-m)}t, \epsilon) \\ &= t^{n-m} + 1 + O(\epsilon^{1/(n-m)})\end{aligned}$$

under the same conditions. Now, $|t^*| = 1$, and so for $\delta > 0$ small and fixed, there exists $\epsilon_0 > 0$, such that

$$|\tilde{\zeta}(t, \epsilon) - (t^{n-m} + 1)| < |t^{n-m} + 1|$$

for all t with $|t - t^*| = \delta$ and all $\epsilon, 0 \le \epsilon < \epsilon_0$. By Rouché's Theorem, $\tilde{\zeta}(t, \epsilon)$ has exactly one root, say $t^*(\epsilon)$, in $|t - t^*| < \delta$. Then

$$s^*(\epsilon) := \epsilon^{-1/(n-m)} t^*(\epsilon)$$

satisfies

$$\hat{\zeta}(s^*(\epsilon), \epsilon) = \tilde{\zeta}(t^*(\epsilon), \epsilon) = 0,$$

and also

$$|t^*(\epsilon) - t^*| = |\epsilon^{1/(n-m)} s^*(\epsilon) - t^*| < \delta,$$

so that

$$\left| \frac{s^*(\epsilon)}{\epsilon^{-1/(n-m)} t^*} - 1 \right| < \frac{\delta}{|t^*|} = \delta,$$

as we had to show. □

PROOF (of Theorem 13.1). Apply Lemma 13.1 to the polynomial $\theta(\lambda)' M \theta(-\lambda)$ and then Lemma 13.2 to the polynomial on the right side of (5.2).

13.6 Asymptotic Root Locus: Multivariable System

We retain the definition (4.6) of $G(\lambda)$, but effectively multiply N by ϵ^2 by replacing $G(\lambda)G(-\lambda)'$ by $\epsilon^{-2} G(\lambda)G(-\lambda)'$ in (4.7). Our objective is to describe the behavior of the roots of $\pi^0(\lambda)$ as $\epsilon \downarrow 0$. For this, let ρ be the rank of the rational matrix $G(\lambda)$ over the field $\mathbb{R}(\lambda)$ of rational functions of λ. For $\sigma \in \mathbf{p}$ define

$$\hat{\gamma}_\sigma(\lambda^2) := \sum \{\sigma \times \sigma \text{ principal minors of } G(\lambda)G(-\lambda)'\}. \tag{6.1}$$

It can be shown (Exercise 13.8) that none of the rational functions $\hat{\gamma}_\sigma(\lambda^2)$ is identically zero. Fix $\sigma \in \mathbf{p}$, and write $i := (i_1, \ldots, i_\sigma)$ etc. for the multi-index having $1 \le i_1 < i_2 < \cdots < i_\sigma \le n$, subject to dimensional compatibility with the matrices involved. Finally, write $G^i_j(\lambda)$ etc. for the minor of $G(\lambda)$ formed by selecting the entries having row index in the list i and column index in j.

Starting with the modified factor in (4.7), we have

$$\det[1 + \epsilon^{-2} G(\lambda)G(-\lambda)'] = 1 + \sum_{\sigma=1}^{\rho} \epsilon^{-2\sigma} \hat{\gamma}_\sigma(\lambda^2). \tag{6.2}$$

By (4.6), (6.1) and the Cauchy-Binet formula for minors,

$$\begin{aligned}\hat{\gamma}_\sigma(\lambda^2) &= \sum_i [G(\lambda)G(-\lambda)']_i^i \\ &= \sum_{i,j} G_j^i(\lambda)[G(-\lambda)']_i^j \\ &= \sum_{i,j} G_j^i(\lambda)G_j^i(-\lambda).\end{aligned}$$

Similarly, and by the rule for evaluating the minors of a matrix inverse,

$$\begin{aligned}G_j^i(\lambda) &= \sum_{k,l} D_k^i[(\lambda 1 - A)^{-1}]_l^k(BN^{-1/2})_j^l \\ &= \pi(\lambda)^{-1}\sum_{k,l}(-1)^{|k|+|l|}D_k^i(\lambda 1 - A)_{k'}^{l'}(BN^{-1/2})_j^l,\end{aligned} \tag{6.3}$$

where $|k| := k_1 + \cdots + k_\sigma$ and k' denotes the list of $n - \sigma$ indices complementary to k. By (6.3),

$$\gamma_j^i(\lambda) := \pi(\lambda)G_j^i(\lambda)$$

is a polynomial in $\mathbb{R}[\lambda]$. Thus,

$$\gamma_\sigma(\lambda^2) := \pi(\lambda)\pi(-\lambda)\hat{\gamma}_\sigma(\lambda^2) \in \mathbb{R}[\lambda^2]. \tag{6.4}$$

Now by (6.3), $G_j^i(\lambda)$ is of the form

$$G_j^i(\lambda) = \pi(\lambda)^{-1}[g_{ij}\lambda^{n-\sigma} + \cdots],$$

where the leading term $g_{ij}\lambda^{n-\sigma}$ is contributed by those terms in the double sum having $l' = k'$, i.e. $l = k$:

$$g_{ij} := \sum_k D_k^i(BN^{-1/2})_j^k = (DBN^{-1/2})_j^i. \tag{6.5}$$

Thus, formally

$$\gamma_\sigma(\lambda^2) = (-1)^{n-\sigma}\left(\sum_{i,j} g_{ij}^2\right)(\lambda^2)^{n-\sigma} + \cdots,$$

the remainder denoting terms of lower degree in λ^2. We shall write this as

$$\gamma_\sigma(\lambda^2) = g_\sigma(-\lambda^2)^{n-\sigma} + \cdots, \tag{6.6}$$

where

$$g_\sigma := \sum_{i,j} g_{ij}^2. \tag{6.7}$$

Collecting results, we have by (4.7), (6.2), (6.4) and (6.6) that the closed-loop poles are the roots in $\mathfrak{Re}\,\lambda < 0$ of the equation

$$\pi^0(\lambda)\pi^0(-\lambda) + \sum_{\sigma=1}^{\rho} \epsilon^{-2\sigma}[g_\sigma(-\lambda^2)^{n-\sigma} + \cdots] = 0. \tag{6.8}$$

To describe the asymptotic root locus as $\epsilon \downarrow 0$, we shall content ourselves with the case where the above formal analysis matches the actual situation, namely all the numbers g_σ are nonvanishing. By (6.5) and (6.7) this means simply that $\operatorname{rank}(DB) \geq \rho$, hence (Exercise 13.9) $\operatorname{rank}(DB) = \rho$. Our assumption is *a priori* plausible and is, in fact, valid for "generic" choices of the $q \times n$ matrix D and $n \times m$ matrix B, inasmuch as $\rho \leq \min(q, m)$. Of course, it may cease to hold for structures which in some sense are "special." In any event, we have

Theorem 13.2. *Let the rank of $G(\lambda)$ over $\mathbb{R}(\lambda)$ be ρ and assume that $\operatorname{rank}(DB) = \rho$. Then, as $\epsilon \downarrow 0$, $n - \rho$ of the closed-loop poles tend to the (symmetrized, $n - \rho$) roots in $\Re e\, \lambda \leq 0$ of the polynomial*

$$\gamma_\rho(\lambda^2) := \pi(\lambda)\pi(-\lambda) \sum_i [G(\lambda)G(-\lambda)']_{i\cdot}^{i}.$$

Here the sum is taken over all the $\rho \times \rho$ principal minors of $G(\lambda)G(-\lambda)'$. The remaining ρ closed-loop poles tend asymptotically to $-\epsilon^{-1}\nu_r$ $(r \in \mathbf{\rho})$, where $\nu_1, \ldots, \nu_\rho$ are the nonzero (hence, real and positive) eigenvalues of the map

$$N^{-1/2}B'D'DBN^{-1/2}. \tag{6.9}$$

PROOF. The proof follows the same lines as that of Theorem 13.1, and so need only be sketched. For the finite poles, multiply through (6.8) by $\epsilon^{2\rho}$ and let $\epsilon \downarrow 0$. The only term remaining is

$$g_\rho(-\lambda^2)^{n-\rho} + \cdots = \gamma_\rho(\lambda^2).$$

By Rouché's theorem, it follows that (6.8) has exactly one root in any fixed, small neighborhood of each root λ of γ_ρ, for all $\epsilon > 0$ sufficiently small; of course, if γ_ρ has any multiple roots this statement is interpreted in the obvious way. For the remaining ρ stable roots, multiply through (6.8) by $\epsilon^{2\rho}\lambda^{-2(n-\rho)}$ and set $\mu := \epsilon\lambda$. For the roots of interest one now has the equation

$$g(\mu^2) + h(\mu^2, \epsilon) = 0, \tag{6.10}$$

where

$$g(\mu^2) := \mu^{2\rho} + \sum_{\sigma=1}^{\rho} (-1)^\sigma g_\sigma \mu^{2(\rho-\sigma)}$$

and

$$h(\mu^2, \epsilon) := (-1)^n \mu^{-2(n-\rho)} \sum_{j=1}^{2n} \epsilon^j k_j(\mu^2),$$

the $k_j(\cdot)$ being polynomials. Recalling that $DBN^{-1/2}$ has rank ρ, it is easy to check (Exercise 13.11) that the characteristic polynomial of the map (6.9) is $\nu^{m-\rho}g(\nu)$. Thus, the roots in $\Re e\, \mu \leq 0$ of $g(\mu^2)$ are just the numbers $-\nu_r$ $(r \in \mathbf{\rho})$ as described. In particular, $g(0) \neq 0$. By use of this fact, and the observation that

$$h(\mu^2, \epsilon) \to 0, \qquad \epsilon \downarrow 0$$

uniformly for μ in any closed, bounded subset of $\mathbb{C} - \{0\}$, we conclude on the basis of Rouché's theorem that the zeros of $g(\mu^2)$ and of $g(\mu^2) + h(\mu^2, \epsilon)$ coincide in the limit $\epsilon \downarrow 0$. If a typical root of $g(\mu^2)$ is v^*, and the corresponding zero of (6.10) is $v(\epsilon)$, then we have $v(\epsilon) - v^* \to 0$ $(\epsilon \downarrow 0)$. Setting $\lambda(\epsilon) := \epsilon^{-1} v(\epsilon)$, we obtain the asymptotic relation

$$\frac{\lambda(\epsilon)}{v^*/\epsilon} \to 1, \qquad \epsilon \downarrow 0$$

that was to be proved. □

As an illustration of this result, let

$$A = \begin{bmatrix} 0 & 1 & 0 \\ 0 & 0 & 1 \\ 0 & 0 & 0 \end{bmatrix}, \qquad B = \begin{bmatrix} 0 & 0 \\ 1 & 0 \\ 0 & 2 \end{bmatrix}, \qquad D = \begin{bmatrix} 1 & 0 & 0 \\ 0 & 1 & 0 \end{bmatrix},$$

and $N = 1$. Then,

$$G(\lambda) = \lambda^{-3} \begin{bmatrix} \lambda & 2 \\ \lambda^2 & 2\lambda \end{bmatrix}$$

with

$$\rho = \text{Rank } G(\lambda) = \text{Rank}(DB) = 1.$$

From this,

$$\begin{aligned} \gamma_1(\lambda^2) &= \lambda^3(-\lambda)^3 \sum \{1 \times 1 \text{ principal minors of } G(\lambda)G(-\lambda)'\} \\ &= (\lambda^2 - 1)(\lambda^2 - 4), \end{aligned}$$

so that two poles of the optimal closed-loop system have loci that terminate as $\epsilon \downarrow 0$ on the real axis at $\lambda = -1, -2$. The remaining pole tends asymptotically to $-\epsilon^{-1}1$, the number 1 being the positive eigenvalue of $(DB)'DB$.

As another example let (D, A, B) be complete and $G(\lambda)$ be square and of full rank:

$$\rho = \text{Rank } G(\lambda) = \text{Rank}(DB) = m.$$

Then

$$\gamma_m(\lambda^2) = \pi(\lambda)\pi(-\lambda) \det[G(\lambda)G(-\lambda)'].$$

On transforming $G(\lambda)$ to its Smith-McMillan form, and referring to the discussion of transmission zeros in Section 5.5, we find (Exercise 13.12) that the roots of $\gamma_m(\lambda^2)$ in $\Re e\ \lambda \leq 0$ are just the transmission zeros of $G(\lambda)$ reflected, if necessary, in the imaginary axis. Thus, $n - m$ root loci terminate at these zeros, while the remaining m loci terminate at $-\infty$ on the real axis. This result reduces to that of Theorem 13.1 if, in (5.3), $\deg \varphi = n - 1$.

13.7 Upper and Lower Bounds on P^0

By use of Lemmas 12.1 and 12.3 it is easy to see that an upper bound for P^0 can be calculated by choosing any F such that $A + BF$ is stable and computing the corresponding matrix P from (12.4.3). As shown in Section 12.4, such upper bounds can be successively improved to yield P^0 in the limit.

It is interesting that a lower bound on P^0 can be computed provided we strengthen stabilizability of (A, B) to controllability, and detectability of (D, A) to observability. Under the latter condition it is easily verified (Exercise 12.2) that $P^0 > 0$, so that P^{0-1} exists. The following lemma shows that it suffices to compute an upper bound for P^{0-1}.

Lemma 13.3. *Let* $0 < Q^0 \leq Q$. *Then,* $Q^{0-1} \geq Q^{-1}$.

PROOF. Choose T orthogonal, such that

$$T'Q^0T = \text{diag } \Lambda;$$

then

$$\Lambda^{-1/2}T'Q^0T\Lambda^{-1/2} = 1.$$

Write $R := \Lambda^{-1/2}T'QT\Lambda^{-1/2}$ and choose S orthogonal, such that $S'RS = \text{diag } M$. Then $M \geq S'1S = 1$; clearly $M^{-1} \leq 1$; and the result follows by a simple computation. □

Write $Q := P^{-1}$ and multiply both sides of (12.3.16) by Q to obtain

$$(-A')'Q + Q(-A') - QMQ + BN^{-1}B' = 0. \tag{7.1}$$

By Theorem 12.2 and Exercise 12.2, (7.1) has a unique solution $Q^0 > 0$ in the class of positive semidefinite matrices, provided $(-A', \sqrt{M})$ is stabilizable and $(\sqrt{BN^{-1}B'}, -A')$ is observable. As these properties follow by the assumptions stated above, we conclude that $Q^0 = P^{0-1}$. Choosing K so that $-A' + \sqrt{M}\,K$ is stable, we solve for Q the linear equation obtained from (7.1), namely

$$(-A' + \sqrt{M}\,K)'Q + Q(-A' + \sqrt{M}\,K) + BN^{-1}B' + K'NK = 0.$$

Then, $0 < Q^0 \leq Q$ and by Lemma 13.3,

$$P^0 \geq Q^{-1}.$$

Although these bounding procedures offer little insight *per se* into the behavior of the solution as a function of the parameters, they are useful in computation, and as will be shown next, help to provide information on the stability margin of $A + BF^0$.

13.8 Stability Margin. Gain Margin

If $A: \mathscr{X} \to \mathscr{X}$ is stable, we define the *stability margin* α of A as the distance of $\sigma(A)$ from the imaginary axis:

$$\alpha := -\max\{\Re e\ \lambda: \lambda \in \sigma(A)\}.$$

A simple estimate of α is provided by the following.

Proposition 13.1. *If A is stable, $Q > 0$, P is symmetric, and*

$$A'P + PA + Q = 0, \tag{8.1}$$

then

$$\alpha \geq \tfrac{1}{2}|P|^{-1}|Q^{-1}|^{-1}$$

PROOF. By Lemma 12.2, $P > 0$. From (8.1)

$$(A + \beta 1)'P + P(A + \beta 1) + Q - 2\beta P = 0.$$

Again by Lemma 12.2, $A + \beta 1$ is stable if $Q - 2\beta P > 0$, that is, if (in obvious notation)

$$\begin{aligned} 2\beta &< \frac{\min \sigma(Q)}{\max \sigma(P)} \\ &= [\max \sigma(Q^{-1}) \max \sigma(P)]^{-1} \\ &= (|Q^{-1}|\,|P|)^{-1}. \end{aligned}$$ □

Applying this result to (12.4.3) and (12.4.4) we get for the stability margin α^0 of $A + BF^0$:

$$\alpha^0 \geq \tfrac{1}{2}|P^0|^{-1}|(M + P^0BN^{-1}B'P^0)^{-1}|^{-1}.$$

The result may be useful if an upper bound for P^0 is known: thus $P^0 \leq p1$ implies

$$\alpha^0 \geq \tfrac{1}{2}p^{-1} \min \sigma(M). \tag{8.2}$$

In classical control theory the term "gain margin" refers to the amount by which a loop gain parameter can be increased from its nominal value, without causing system instability. One example of such an estimate in the context of quadratic optimality is the following. Replace B by BK, where $K = \operatorname{diag}[k_1, \ldots, k_m]$, and for simplicity assume that $M > 0$ and N is diagonal. It will be shown that the optimal system designed on the assumption that $K = 1$ (the "nominal" value) remains stable over the parameter range

$$K \geq \tfrac{1}{2}1. \tag{8.3}$$

This result can be paraphrased by saying that the optimal system has infinite gain margin and 50% gain reduction tolerance at each control input channel. For the proof, write (12.4.3) as

$$(A + BKF^0)'P^0 + P^0(A + BKF^0) + M$$
$$+ F^{0\prime}NF^0 + [B(1 - K)F^0]'P^0 + P^0B(1 - K^0)F^0 = 0.$$

By Lemma 12.2, $A + BKF^0$ will certainly be stable provided

$$F^{0\prime}NF^0 + [B(1 - K)F^0]'P^0 + P^0B(1 - K)F^0 \geq 0.$$

Substitution of $F^0 = -N^{-1}B'P^0$ yields

$$P^0B[(K - \tfrac{1}{2}1)N^{-1} + N^{-1}(K - \tfrac{1}{2}1)]B'P^0 \geq 0$$

and (with N diagonal) this will hold if (8.3) is true.

13.9 Return Difference Relations

In this section we obtain an identity involving the frequency response of an optimal system. This will be interpreted, albeit artificially, as an indication of the insensitivity of system response to a small perturbation of the open loop system matrix A.

Consider (12.3.16), written as

$$-A'P - PA + PBN^{-1}B'P - M = 0. \tag{9.1}$$

For simplicity of notation replace B by $B\sqrt{N}$ (or set $N = 1$). Recall

$$M = D'D, \qquad F = -B'P,$$

and write $R(\lambda) := (\lambda 1 - A)^{-1}$. Then from (9.1) we obtain by successive manipulations:

$$(-\lambda 1 - A)'P + P(\lambda 1 - A) + PBB'P = M;$$

$$B'PR(\lambda)B + B'R(-\lambda)'PB + B'R(-\lambda)'PBB'PR(\lambda)B = B'R(-\lambda)'MR(\lambda)B;$$

$$[1 - FR(-\lambda)B]'[1 - FR(\lambda)B] = 1 + [DR(-\lambda)B]'[DR(\lambda)B]. \tag{9.2}$$

Define the *return ratio*

$$T(\lambda) := -FR(\lambda)B$$

and the *return difference*

$$\Phi(\lambda) := 1 - FR(\lambda)B. \tag{9.3}$$

Then, from (9.2)

$$\Phi(-\lambda)'\Phi(\lambda) = 1 + H(-\lambda)'H(\lambda), \tag{9.4}$$

where

$$H(\lambda) := DR(\lambda)B$$

is the open loop transfer matrix from u to z. Set $\lambda = i\omega$ in (9.4) and note that a matrix of the form $\Phi^*\Phi\colon \mathbb{C}^m \to \mathbb{C}^m$ is positive semidefinite relative to the complex inner product. There follows

$$\Phi^*(i\omega)\Phi(i\omega) \geq 1, \qquad \omega \in \mathbb{R}. \tag{9.5}$$

Equation (9.4) is the *return difference identity* and (9.5) is the *return difference inequality*. The term "return difference" originates in circuit theory; its use here is prompted by the signal flow graph, Fig. 13.1, where we put $u = Fx + v$.

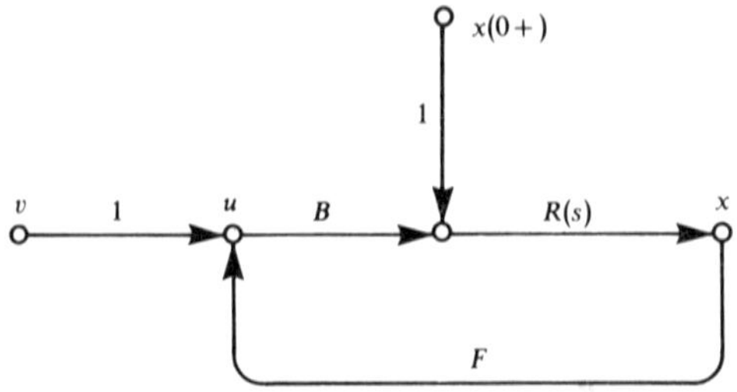

Figure 13.1 Signal Flow Graph: Closed-Loop System.

From the flow graph, we see that formally

$$\hat{x} = (1 - R(\lambda)BF)^{-1}R(\lambda)B\hat{v}.$$

Introduce the *sensitivity matrix*

$$S(\lambda) := (1 - R(\lambda)BF)^{-1}. \tag{9.6}$$

It is easily checked that the rational matrix inversion here is legal. By (9.3) and (9.6), the (trivial) identity

$$[1 - F'B'R(-\lambda)']F'F[1 - R(\lambda)BF] = F'[1 - B'R(-\lambda)'F'][1 - FR(\lambda)B]F$$

can be written

$$[S(-\lambda)^{-1}]'F'FS(\lambda)^{-1} = F'\Phi(-\lambda)'\Phi(\lambda)F,$$

and then by (9.5) there follows

$$\begin{aligned} F'F &= S^*(i\omega)F'\Phi^*(i\omega)\Phi(i\omega)FS(i\omega) \\ &\geq S^*(i\omega)F'FS(i\omega). \end{aligned} \tag{9.7}$$

It will be shown that (9.7) implies a qualitative distinction between the "closed-loop" graph of Fig. 13.1 and the "open-loop" graph of Fig. 13.2. In Fig. 13.2 the open loop control u_0 is defined by

$$\hat{u}_0(\lambda) = FS(\lambda)R(\lambda)x(0+), \tag{9.8}$$

so that for both graphs

$$\hat{x}(\lambda) = S(\lambda)R(\lambda)x(0+).$$

Suppose now that the system matrix A depends on a real parameter θ, with $A = A(\theta)$ continuously differentiable in a neighborhood of some nominal parameter value, say $\theta = 0$. Then, $R = R(\lambda, \theta)$, $S = S(\lambda, \theta)$, and

$$\frac{\partial S(\lambda, 0)}{\partial \theta} = S(\lambda, 0) \frac{\partial R(\lambda, 0)}{\partial \theta} BFS(\lambda, 0).$$

For the closed loop graph, $\hat{x}(\lambda) = \hat{x}_c(\lambda, \theta)$ say, and

$$\frac{\partial \hat{x}_c(\lambda, 0)}{\partial \theta} = \frac{\partial[S(\lambda, 0)R(\lambda, 0)]}{\partial \theta} x(0+)$$

$$= S(\lambda, 0) \frac{\partial R(\lambda, 0)}{\partial \theta} [BFS(\lambda, 0)R(\lambda, 0) + 1]x(0+). \tag{9.9}$$

For the open loop graph, let $\hat{x}(\lambda) = \hat{x}_0(\lambda, \theta)$. The derivative will be computed on the assumption that the open loop control (9.8) does not depend on θ:

$$\frac{\partial \hat{x}_0(\lambda, 0)}{\partial \theta} = \frac{\partial R(\lambda, 0)}{\partial \theta} [x(0+) + B\hat{u}_0(\lambda)]. \tag{9.10}$$

Comparison of (9.8)–(9.10) yields

$$\frac{\partial \hat{x}_c(\lambda, 0)}{\partial \theta} = S(\lambda, 0) \frac{\partial \hat{x}_0(\lambda, 0)}{\partial \theta}. \tag{9.11}$$

To exploit (9.11) we proceed formally. For θ small,

$$\delta x_c(t, \theta) = x_c(t, \theta) - x_c(t, 0) \doteq \theta \frac{\partial x_c(t, 0)}{\partial \theta},$$

$$\delta x_0(t, \theta) = x_0(t, \theta) - x_0(t, 0) \doteq \theta \frac{\partial x_0(t, 0)}{\partial \theta}.$$

Applying Parseval's theorem and using (9.7) and (9.11), we get for the first-order variations

$$\begin{aligned} 2\pi \int_0^\infty |F\, \delta x_0(t, \theta)|^2\, dt &= \int_{-\infty}^\infty |F\, \delta \hat{x}_0(i\omega, \theta)|^2\, d\omega \\ &\doteq \theta^2 \int_{-\infty}^\infty \left| F \frac{\partial \hat{x}_0(i\omega, 0)}{\partial \theta} \right|^2 d\omega \\ &\geq \theta^2 \int_{-\infty}^\infty \left| FS(i\omega) \frac{\partial \hat{x}_0(i\omega, 0)}{\partial \theta} \right|^2 d\omega \\ &= \theta^2 \int_{-\infty}^\infty \left| F \frac{\partial \hat{x}_c(i\omega, 0)}{\partial \theta} \right|^2 d\omega \\ &\doteq 2\pi \int_0^\infty |F\, \delta x_c(t, \theta)|^2\, dt. \end{aligned} \tag{9.12}$$

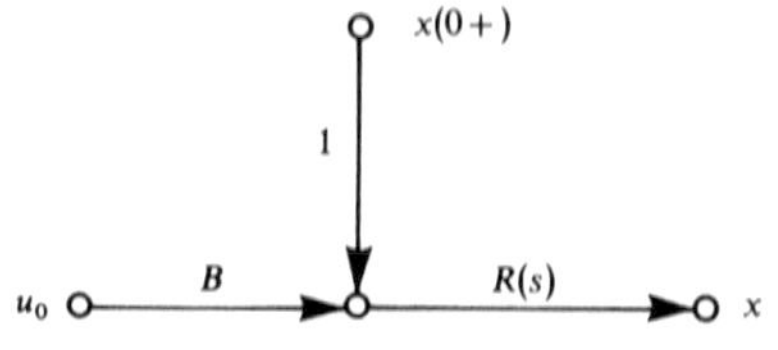

Figure 13.2 Signal Flow Graph: Open-Loop System.

The inequality (9.12) reveals that the closed loop graph (Fig. 13.1) is less "sensitive" (or no more so) than the open loop graph (Fig. 13.2), with respect to a small parameter change in A, and when sensitivities are measured by the indicated quadratic integral. Inasmuch as the latter measure was carefully selected to make the indicated inequality come out, one should not place great weight on the practical significance of the result. Perhaps the main contribution of the discussion is the link between state-variable methods and frequency-domain ideas derived from classical feedback theory.

13.10 Applicability of Quadratic Optimization

It should be clear from this chapter and the preceding that quadratic optimization is simply one technique for computing a feedback map F such that $A + BF$ is stable, given a stabilizable pair (A, B). As such, it does not by itself solve any of the basic structural problems of system synthesis. Indeed, we have seen in our study of noninteraction, and of regulation, that considerable algebraic preparation of a synthesis problem may be necessary before the issue of pole assignment in general, or stabilization in particular, can be properly dealt with.

After such preparation has been completed, we are typically in a position to stabilize several pairs (A, B), which arise as the system matrices for suitable independent sub-problems of the main problem we started with. It is at this stage that the quadratic optimization algorithm may prove useful, but only in competition with alternative stabilization techniques. To the extent that they are available, indices of performance based on more direct descriptions of transient response, explored with efficient algorithms for parameter optimization, should be considered as alternatives to quadratic optimization in numerical design.

13.11 Exercises

13.1. Regard the Hamiltonian matrix (4.3) as a map $H: \mathscr{X} \oplus \mathscr{X}' \to \mathscr{X} \oplus \mathscr{X}'$. Under the conditions of Theorem 12.2, the polynomials $\pi^0(\lambda)$, $\pi^0(-\lambda)$ are coprime, so that

$$\mathscr{X} \oplus \mathscr{X}' = \operatorname{Ker} \pi^0(H) \oplus \operatorname{Ker} \pi^0(-H).$$

Show that if

$$\begin{bmatrix} \xi_i \\ \eta_i \end{bmatrix}, \qquad i \in \mathbf{n},$$

is a basis for Ker $\pi^0(H)$, then P^0 can be represented as the $n \times n$ matrix

$$P^0 = [\eta_1 \cdots \eta_n][\xi_1 \cdots \xi_n]^{-1}.$$

Hint: Begin by showing from (4.4) that if $\left[\begin{smallmatrix} \xi \\ \eta \end{smallmatrix}\right]$ is an eigenvector of H in Ker $\pi^0(H)$, then $P^0 \xi = \eta$.

13.2. Referring to the Hamilton-Jacobi theory of first-order partial differential equations, show that the characteristic strips of (12.3.15) satisfy the ordinary differential system

$$\frac{d}{dt}\begin{bmatrix} \xi \\ \eta \end{bmatrix} = H \begin{bmatrix} \xi \\ \eta \end{bmatrix}. \tag{11.1}$$

13.3. Referring to the theory of the Bolza problem in the calculus of variations, consider the variational problem

$$\int_0^T [x(t)'Mx(t) + u(t)'Nu(t)]\,dt = \min,$$

with T free, and side condition

$$\dot{x}(t) = Ax(t) + Bu(t).$$

Show that the Euler equations for this problem are equivalent to (11.1).

13.4. Find a good estimate of the "peaking index"

$$\sup_{t \geq 0} |\exp[t(A + BF^0)]|$$

and relate it effectively to M and N.

13.5. A relation not well understood is the quantitative dependence of the sensitivity of a system on the topology of its signal flow graph. A major reason for synthesizing feedback configurations is that one may thereby achieve superior sensitivity performance, as compared to open-loop configurations with the same nominal transmission (cf. Chapter 8, and Section 13.9). Investigate quantitatively.

13.6. Using the fact that $A(\theta) + BF$ is stable for θ small, introduce appropriate extra hypotheses to rigorize the application of Parseval's theorem in deriving (9.12). Does the comparison of open and closed loop graphs make sense if $A(0)$ is unstable?

13.7. Verify the representation (4.7) using the general determinantal relations:

$$\det\begin{bmatrix} A & B \\ C & D \end{bmatrix} = \det A \cdot \det(D - CA^{-1}B), \qquad \det A \neq 0, \tag{11.2}$$

and

$$\det(1 + EF) = \det(1 + FE), \tag{11.3}$$

for arbitrary matrices of compatible dimension. Hint: For (11.2) multiply $\left[\begin{smallmatrix} A & B \\ C & D \end{smallmatrix}\right]$ on the left by

$$\begin{bmatrix} 1 & 0 \\ -CA^{-1} & 1 \end{bmatrix}.$$

For (11.3) note that

$$\begin{bmatrix} 1 & E \\ -F & 1 \end{bmatrix} \simeq \begin{bmatrix} 1 & -F \\ E & 1 \end{bmatrix} \tag{11.4}$$

under the orthogonal transformation $\left[\begin{smallmatrix} 0 & 1 \\ 1 & 0 \end{smallmatrix}\right]$, and apply (11.2) to both sides of (11.4).

13.8. It is known that the rank of a Hermitian matrix (over $\mathbb{C}$) is the size of its largest nonvanishing principal minor. Prove that the same is true for the rank of $G(\lambda)G(-\lambda)$ over $\mathbb{R}(\lambda)$. Hint: Note that $G(i\omega)G(-i\omega)'$ is Hermitian for all real ω at which $G(i\omega)$ is defined, and exploit analyticity.

13.9. If ρ is the rank of $D(\lambda 1 - A)^{-1}B$ over $\mathbb{R}(\lambda)$, show that $\rho \geq \text{rank}(DB)$.

13.10. Let $A: n \times n$. Show that

$$\det(\lambda 1 - A) = \lambda^n + \sum_{k=1}^{n} (-1)^k \alpha_k \lambda^{n-k},$$

where α_k is the sum of the principal minors of order k of A. In particular, $\alpha_1 = \text{tr}(A)$, $\alpha_n = \det(A)$.

13.11. Let C: $p \times q$. Show that the ch.p. of $C'C$ is

$$\det(\lambda 1 - C'C) = \lambda^q + \sum_{r=1}^{q} (-1)^r \gamma_r \lambda^{n-r},$$

where

$$\gamma_r = \sum_{j,k} [C_j^k(r)]^2$$

and $C_j^k(r)$ is the (j, k)th $r \times r$ minor of C. Hint: Use the standard formula (Exercise 13.10) for the coefficients of a characteristic polynomial, together with Cauchy-Binet.

13.12. Prove the statement concerning transmission zeros near the end of Section 13.6.

13.12 Notes and References

Algebraic properties of the Hamiltonian matrix are discussed by Potter [1], Martensson [1] and Kučera [1]; the result in Exercise 13.1 was first proved (in somewhat less generality) by Potter. Theorem 13.1 is due essentially to Chang [1]; see also Kalman [3] and, in the direction of a multivariable generalization, Tyler and Tuteur [1], and Kwakernaak and Sivan [1]. Theorem 13.2 first appeared in Wonham [8]; further results on asymptotic root loci of the optimal system are given by Kwakernaak [1]. A related issue is the behavior of the performance index, or equivalently $P = P(\epsilon)$, as

$\epsilon \downarrow 0$. This problem of "cheap control" has been discussed by Kwakernaak and Sivan [1], Godbole [1], Francis [2], and Francis and Glover [1]. Roughly speaking $P(\epsilon) \downarrow 0$, or the control action becomes "perfect," if the plant transfer matrix is right-invertible and is minimum phase in the appropriate sense. The connection of these ideas with singularly perturbed systems is explored by Young, Kokotović and Utkin [1].

The determinantal relations used in Sections 13.4 and 13.6 can be found in Gantmacher [1]: for the Cauchy-Binet theorem, p. 9; the rule for minors of a matrix inverse, p. 21; the identity (11.2), p. 45; and the formula for the coefficients of the ch.p., p. 70. The trick in Section 13.7 for obtaining a lower bound for P^0 is due to Bellman [3]; an alternative and neat proof of Lemma 13.3 is given by Bechenbach and Bellman [1]. The result on gain margin in Section 13.8 is due to Safonov and Athans [1]; more information on the tolerance of quadratically optimal systems to large parameter perturbations can be found in Wong and Athans [1], and Wong, Stein and Athans [1]. The return difference relations (9.4) and (9.5) are simple extensions of results of Kalman [3]. A general discussion of return difference and sensitivity in multivariable systems can be found in Cruz and Perkins [1], Perkins and Cruz [1], and Cruz [1]; see also Pagurek [1]. For insight into Exercise 13.5 consult Wierzbicki [1].

References

B. D. O. Anderson

[1] *Output-nulling invariant and controllability subspaces.* Preprints, Sixth Triennial World Congress, International Federation of Automatic Control (IFAC), Boston/Cambridge, Mass., 1975; Part 1B, paper no. 43.6.
[2] *A note on transmission zeros of a transfer matrix.* IEEE Trans. Aut. Control AC-21 (4), 1976, pp. 589–591.

B. D. O. Anderson, J. B. Moore

[1] *Linear Optimal Control.* Prentice-Hall, Englewood Cliffs, N.J., 1971.

B. D. O. Anderson, R. W. Scott

[1] *Output feedback stabilization—solution by algebraic geometry methods.* Proc. IEEE 65 (6), 1977, pp. 849–861.

A. Andronov, L. Pontryagin

[1] *Systèmes grossiers.* Dokl. Akad. Nauk SSSR 14, 1937, pp. 247–251.

G. Basile, G. Marro

[1] *Luoghi caratteristici dello spazio degli stati relativi al controllo dei sistemi lineari.* L'Elettrotecnica 55 (12), 1968, pp. 1–7.
[2] *Controlled and conditioned invariant subspaces in linear system theory.* J. Optimization Th. & Appl. 3 (5), 1969, pp. 306–315.
[3] *On the observability of linear time-invariant systems with unknown inputs.* J. Optimization Th. & Appl. 3 (6), 1969, pp. 410–415.

E. Beckenbach, R. Bellman

[1] *Inequalities.* Springer-Verlag, Berlin, 1961.

R. Bellman

[1] *Dynamic Programming*. Princeton University Press, Princeton, N.J., 1957.
[2] *Introduction to the Mathematical Theory of Control Processes*, Vol. 1: Linear Equations and Quadratic Criteria. Academic Press, New York, 1967.
[3] *Upper and lower bounds for the solutions of the matrix Riccati equation*. J. Math. Anal. & Appl. 17 (2), 1967, pp. 373–379.

S. P. Bhattacharyya

[1] *Output regulation with bounded energy*. IEEE Trans. Aut. Control AC-18 (4), 1973, pp. 381–383.

S. P. Bhattacharyya, J. B. Pearson, W. M. Wonham

[1] *On zeroing the output of a linear system*. Information and Control 20 (2), 1972, pp. 135–142.

G. E. P. Box, S. L. Andersen

[1] *Permutation theory in the derivation of robust criteria and the study of departures from assumption*. J. Roy. Stat. Soc. B-17 (1), 1955, pp. 1–26.

F. M. Brasch, Jr., J. B. Pearson

[1] *Pole placement using dynamic compensators*. IEEE Trans. Aut. Control AC-15 (1), 1970, pp. 34–43.

R. Brockett

[1] *The geometry of the set of controllable linear systems*. Res. Rpts. Aut. Control Lab., Fac. Engrg., Nagoya Univ., 24, 1977, pp. 1–7.

P. Brunovsky

[1] *A classification of linear controllable systems*. Kybernetika 6 (3), 1970, pp. 173–188.

M. F. Chang, I. B. Rhodes

[1] *Disturbance localization in linear systems with simultaneous decoupling, pole assignment, or stabilization*. IEEE Trans. Aut. Control AC-20 (4), 1975, pp. 518–523.

S. S. L. Chang

[1] *Synthesis of Optimum Control Systems*. McGraw-Hill, New York, 1961.

S. S. L. Chang, L. A. Zadeh

[1] *On fuzzy mapping and control*. IEEE Trans. Sys. Man and Cyber. SMC-2 (1), 1972, pp. 30–34.

E. M. Cliff, F. H. Lutze

[1] *Application of geometric decoupling theory to synthesis of aircraft lateral control systems*. J. of Aircraft 9 (11), 1972, pp. 770–776.

[2] *Decoupling longitudinal motions of an aircraft.* Proc. Fourteenth Joint Automatic Control Conf., 1973, pp. 86–91.

M. Cremer

[1] *A precompensator of minimal order for decoupling a linear multivariable system.* Int. J. Control 14 (6), 1971, pp. 1089–1103.

J. B. Cruz, Jr., (Ed.)

[1] *Feedback Systems.* McGraw-Hill, New York, 1972.

J. B. Cruz, Jr., W. R. Perkins

[1] *A new approach to the sensitivity problem in multivariable feedback system design.* IEEE Trans. Aut. Control AC-9 (3), 1964, pp. 216–223.

E. J. Davison

[1] *The robust control of a servomechanism problem for linear time-invariant multivariable systems.* IEEE Trans. Aut. Control AC-21 (1), 1976, pp. 25–34.

E. J. Davison, S. G. Chow

[1] *An algorithm for the assignment of closed-loop poles using output feedback in large linear multivariable systems.* IEEE Trans. Aut. Control AC-18 (1), 1973, pp. 74–75.

M. J. Denham

[1] *Stabilization of linear multivariable systems by output feedback.* IEEE Trans. Aut. Control AC-18 (1), 1973, pp. 62–63.

[2] *A necessary and sufficient condition for decoupling by output feedback.* IEEE Trans. Aut. Control AC-18 (5), 1973, pp. 535–537.

C. A. Desoer

[1] *Notes for a Second Course on Linear Systems.* Van Nostrand, New York, 1970.

B. W. Dickinson

[1] *On the fundamental theorem of linear state variable feedback.* IEEE Trans. Aut. Control AC-19 (5), 1974, pp. 577–579.

E. Fabian

[1] *Decoupling, Disturbance Rejection and Sensitivity.* Ph.D. Thesis, Dept. of Electrical Engineering, Univ. of Toronto, December, 1974.

E. Fabian, W. M. Wonham

[1] *Generic solvability of the decoupling problem.* SIAM J. Control 12 (4), 1974, pp. 688–694.

[2] *Decoupling and data sensitivity.* IEEE Trans. Aut. Control AC-20 (3), 1975, pp. 338–344.

[3] *Decoupling and disturbance rejection.* IEEE Trans. Aut. Control AC-20 (3), 1975, pp. 399–401.

P. L. Falb, W. A. Wolovich

[1] *Decoupling in the design and synthesis of multivariable control systems.* IEEE Trans. Aut. Control AC-12 (6), 1967, pp. 651–659.

T. E. Fortmann, D. Williamson

[1] *Design of low-order observers for linear feedback control laws.* IEEE Trans. Aut. Control AC-17 (3), 1972, pp. 301–308.

B. A. Francis

[1] *The linear multivariable regulator problem.* SIAM J. Control and Optimization 15 (3), 1977, pp. 486–505.
[2] *Perfect regulation and feedforward control of linear multivariable systems.* Proc. 1977 IEEE Conf. on Decision and Control, pp. 760–762.

B. A. Francis, K. Glover

[1] *Bounded peaking in the optimal linear regulator with cheap control.* IEEE Trans. Aut. Control AC-23 (4), 1978, pp. 608–617.

B. A. Francis, O. A. Sebakhy, W. M. Wonham

[1] *Synthesis of multivariable regulators: the internal model principle.* Appl. Maths. and Optimization 1 (1), 1974, pp. 64–86.

B. A. Francis, W. M. Wonham

[1] *The role of transmission zeros in linear multivariable regulators.* Int. J. Control 22 (5), 1975, pp. 657–681.
[2] *The internal model principle for linear multivariable regulators.* Appl. Maths. & Optimization 2 (2), 1975, pp. 170–194.
[3] *The internal model principle of control theory.* Automatica 12 (5), 1976, pp. 457–465.

F. R. Gantmacher

[1] *The Theory of Matrices*, Vol. 1. Chelsea, New York, 1959.

E. G. Gilbert

[1] *Controllability and observability in multivariable control systems.* SIAM J. Control 1 (2), 1963, pp. 128–151.
[2] *The decoupling of multivariable systems by state feedback.* SIAM J. Control 7 (1), 1969, pp. 50–63.

E. G. Gilbert, J. R. Pivnichny

[1] *A computer program for the synthesis of decoupled multivariable feedback systems.* IEEE Trans. Aut. Control AC-14 (6), 1969, pp. 652–659.

S. Godbole

[1] *Comments on 'The maximally achievable accuracy of linear optimal regulators and linear optimal filters'.* IEEE Trans. Aut. Control AC-17 (4), 1972, pp. 577.

W. H. Greub

[1] *Linear Algebra*. Third Edn., Springer-Verlag, New York, 1967.
[2] *Multilinear Algebra*. Springer-Verlag, New York, 1967.

R. Guidorzi

[1] *External reduction of linear multivariable systems*. Ricerche di Automatica 3 (2), 1972, pp. 113–120.

J. Hadamard

[1] *Lectures on Cauchy's Problem in Linear Partial Differential Equations*. Dover Publications, New York, 1952.

J. K. Hale

[1] *Ordinary Differential Equations*. Wiley-Interscience, New York, 1969.

M. L. J. Hautus

[1] *Controllability and observability conditions of linear autonomous systems*. Proc. Kon. Ned. Akad. Wetensch. Ser. A, 72, 1969, pp. 443–448.

M. Heymann

[1] *Pole assignment in multi-input linear systems*. IEEE Trans. Aut. Control AC-13 (6), 1968, pp. 748–749.
[2] *On the input and output reducibility of multivariable linear systems*. IEEE Trans. Aut. Control AC-15 (5), 1970, pp. 563–569.
[3] *Controllability subspaces and feedback simulation*. SIAM J. Control and Optimization 14 (4), 1976, pp. 769–789.

M. W. Hirsch, S. Smale

[1] *Differential Equations, Dynamical Systems and Linear Algebra*. Academic Press, New York, 1974.

I. M. Horowitz

[1] *Synthesis of Feedback Systems*. Academic Press, New York, 1963.

S. Hosoe

[1] *On a time-domain characterization of the numerator polynomials of the Smith McMillan form*. IEEE Trans. Aut. Control AC-20 (6), 1975, pp. 799–800.

N. Jacobson

[1] *Lectures in Abstract Algebra, Vol. 2: Linear Algebra*. Van Nostrand, Princeton, N.J., 1953.

C. D. Johnson

[1] *Accommodation of external disturbances in linear regulator and servomechanism problems*. IEEE Trans. Aut. Control AC-16 (6), 1971, pp. 635–644.

R. Kalaba

[1] *On nonlinear differential equations, the maximum operation, and monotone convergence.* J. Math. Mech. 8 (4), 1959, pp. 519–574.

R. E. Kalman

[1] *On the general theory of control systems.* In Automatic and Remote Control, [Proc. First Internat. Congress, International Federation of Automatic Control (IFAC), Moscow, 1960], Butterworth, London, 1961; Vol. 1, pp. 481–492.
[2] *Contributions to the theory of optimal control.* Bol. Soc. Matem. Mexicana (Ser. 2) 5, 1960, pp. 102–119.
[3] *When is a linear control system optimal?* J. Basic Engrg. 86 (1), 1964, pp. 51–60.
[4] *Kronecker invariants and feedback.* Proc. Conf. on Ordinary Differential Equations, NRL Mathematics Research Center, June 14–23, 1971; in C. Weiss (Ed.), Ordinary Differential Equations, Academic Press, New York, 1972, pp. 459–471.

R. E. Kalman, P. L. Falb, M. A. Arbib

[1] *Topics in Mathematical System Theory.* McGraw-Hill, New York, 1969.

R. E. Kalman, Y. C. Ho, K. S. Narendra

[1] *Controllability of linear dynamical systems.* Contrib. to Differential Equations 1 (2), 1962, pp. 189–213.

R. J. Kavanagh

[1] *The multivariable problem.* Progress in Control Engineering 3, 1966, pp. 94–129.

C. R. Kelley

[1] *Manual and Automatic Control.* Wiley, New York, 1968.

H. Kimura

[1] *Pole assignment by gain output feedback.* IEEE Trans. Aut. Control AC-20 (4), 1975, pp. 509–516.
[2] *A further result on the problem of pole assignment by output feedback.* IEEE Trans. Aut. Control AC-22 (3), 1977, pp. 458–463.
[3] *Geometric structure of observers for linear feedback control laws.* IEEE Trans. Aut. Control AC-22 (5), 1977, pp. 846–855.

G. Klein, B. C. Moore

[1] *Eigenvalue-generalized eigenvector assignment with state feedback.* IEEE Trans. Aut. Control AC-22 (1), 1977, pp. 140–141.

D. L. Kleinman

[1] *On the linear regulator problem and the matrix Riccati equation.* Rpt. ESL-R-271, Mass. Inst. Tech., Electronic Systems Lab., June, 1966.
[2] *On an iterative technique for Riccati equation computations.* IEEE Trans. Aut. Control AC-13 (1), 1968, pp. 114–115.

N. N. Krasovskii

[1] *On the stabilization of unstable motions by additional forces when the feedback loop is incomplete.* Appl. Maths. & Mech. 27 (4), 1963, pp. 971–1004.

[2] *On the stabilization of dynamical systems by supplementary forces.* Differential Equations 1 (1), 1965, pp. 1–9.

V. Kučera

[1] *A contribution to matrix quadratic equations.* IEEE Trans. Aut. Control AC-17 (3), 1972, pp. 344–347.

H. Kwakernaak

[1] *Asymptotic root loci of multivariable linear optimal regulators.* IEEE Trans. Aut. Control AC-21 (3), 1976, pp. 378–382.

H. Kwakernaak, R. Sivan

[1] *The maximally achievable accuracy of linear optimal regulators and linear optimal filters.* IEEE Trans. Aut. Control AC-17 (1), 1972, pp. 79–86.

C. E. Langenhop

[1] *On the stabilization of linear systems.* Proc. Amer. Math. Soc. 15 (5), 1964, pp. 735–742.

A. J. Laub, B. C. Moore

[1] *Calculation of transmission zeros using QZ techniques.* Automatica 14 (6), 1978, pp. 557–566.

E. B. Lee, L. Markus

[1] *Foundations of Optimal Control Theory.* Wiley, New York, 1967.

G. Lee, D. Jordan

[1] *Pole placement with feedback gain constraints.* Proc. 1975 IEEE Conf. on Dec. and Control, pp. 188–190.

S. Lefschetz

[1] *Differential Equations: Geometric Theory.* Second Edn., Interscience, New York, 1959.

J. L. Lions

[1] *Optimal Control of Systems Governed by Partial Differential Equations.* Springer-Verlag, New York, 1971.

D. G. Luenberger

[1] *Observing the state of a linear system.* IEEE Trans. Military Electronics MIL-8, 1964, pp. 74–80.

[2] *Observers for multivariable systems.* IEEE Trans. Aut. Control AC-11 (2), 1966, pp. 190–197.

A. G. J. Macfarlane, N. Karcanias

[1] *Poles and zeros of linear multivariable systems: a survey of the algebraic, geometric and complex-variable theory.* Int. J. Control 24 (1), 1976, pp. 33–74.

S. MacLane, G. Birkhoff

[1] *Algebra.* Macmillan, New York, 1967.

M. Marcus

[1] *Finite Dimensional Multilinear Algebra, Part 1.* Dekker, New York, 1973.

K. Mårtensson

[1] *On the matrix Riccati equation.* Information Sciences 3 (1), 1971, pp. 17–50.

S. K. Mitter, R. Foulkes

[1] *Controllability and pole assignment for discrete-time linear systems defined over arbitrary fields.* SIAM J. Control 9 (1), 1971, pp. 1–7.

B. C. Moore

[1] *On the flexibility offered by state feedback in multivariable systems beyond closed loop eigenvalue assignment.* IEEE Trans. Aut. Control AC-21 (5), 1976, pp. 689–692.

B. C. Moore, A. J. Laub

[1] *Computation of supremal* (A, B)*-invariant and* (A, B)*-controllability subspaces.* IEEE Trans. Aut. Control AC-23 (5), 1978, pp. 783–792.

B. S. Morgan, Jr.

[1] *The synthesis of linear multivariable systems by state variable feedback.* IEEE Trans. Aut. Control AC-9 (4), 1964, pp. 405–411.

A. S. Morse

[1] *Output controllability and system synthesis.* SIAM J. Control 9 (2), 1971, pp. 143–148.

[2] *Structural invariants of linear multivariable systems.* SIAM J. Control 11 (3), 1973, pp. 446–465.

A. S. Morse, W. M. Wonham

[1] *Decoupling and pole assignment by dynamic compensation.* SIAM J. Control 8 (3), 1970, pp. 317–337.

[2] *Triangular decoupling of linear multivariable systems.* IEEE Trans. Aut. Control AC-15 (4), 1970, pp. 447–449.

[3] *Status of noninteracting control.* IEEE Trans. Aut. Control AC-16 (6), 1971, pp. 568–580.

I. H. Mufti

[1] *On the observability of decoupled systems.* IEEE Trans. Aut. Control AC-14 (1), 1969, pp. 75–77.

[2] *A note on the decoupling of multivariable systems.* IEEE Trans. Aut. Control AC-14 (4), 1969, pp. 415–416.

P. Murdoch

[1] *Observer design for a linear functional of the state vector.* IEEE Trans. Aut. Control AC-18 (3), 1973, pp. 308–310.

C. V. Negoita, M. Kelemen

[1] *On the internal model principle.* Proc. 1977 IEEE Conf. on Decision and Control, pp. 1343–1344.

M. M. Newmann

[1] *Design algorithms for minimal-order Luenberger observers.* Electronics Letters 5 (17), 1969, pp. 390–392.

G. C. Newton, L. A. Gould, J. F. Kaiser

[1] *Analytical Design of Linear Feedback Controls.* Wiley, New York, 1957.

B. Noble

[1] *Applied Linear Algebra.* Prentice-Hall, Englewood Cliffs, N.J., 1969.

B. Pagurek

[1] *Sensitivity of the performance of optimal control systems to plant parameter variations.* IEEE Trans. Aut. Control AC-10 (2), 1965, pp. 178–180.

W. R. Perkins, J. B. Cruz, Jr.

[1] *The parameter variation problem in state feedback control systems.* J. Basic Engrg. 87 (1), 1965, pp. 120–124.

V. M. Popov

[1] *The solution of a new stability problem for controlled systems.* Automatic and Remote Control 24 (1), 1963, pp. 1–23.

[2] *Hyperstability and optimality of automatic systems with several control functions.* Rev. Roum. Sci. Electrotech. et Energ. 9, 1964, pp. 629–690.

B. Porter

[1] *Synthesis of Dynamical Systems.* Nelson, London, 1969.

B. Porter, R. Crossley

[1] *Modal Control: Theory and Applications.* Taylor & Francis, London, 1972.

J. E. Potter

[1] *Matrix quadratic solutions.* SIAM J. Appl. Maths. 14 (3), 1966, pp. 496–501.

Z. V. Rekasius

[1] *Decoupling of multivariable systems by means of state variable feedback.* Proc. Third Allerton Conf. on Circuit and System Theory, 1965, pp. 439–447.

H. H. Rosenbrock

[1] *State-Space and Multivariable Theory.* Wiley, New York, 1970.

M. G. Safonov, M. Athans

[1] *Gain and phase margin for multiloop LQG regulators.* IEEE Trans. Aut. Control AC-22 (2), 1977, pp. 173–179.

O. A. Sebakhy, W. M. Wonham

[1] *A design procedure for multivariable regulators.* Automatica 12 (5), 1976, pp. 467–478.

L. Shaw

[1] *Pole placement: stability and sensitivity of dynamic compensators.* IEEE Trans. Aut. Control AC-16 (2), 1971, pp. 210.

L. M. Silverman

[1] *Decoupling with state feedback and precompensation.* IEEE Trans. Aut. Control AC-15 (4), 1970, pp. 487–489.

L. M. Silverman, H. J. Payne

[1] *Input-output structure of linear systems with application to the decoupling problem.* SIAM J. Control 9 (2), 1971, pp. 199–233.

H. W. Smith, E. J. Davison

[1] *Design of industrial regulators: integral feedback and feedforward control.* Proc. IEE 119 (8), 1972, pp. 1210–1216.

O. J. M. Smith

[1] *Feedback Control Systems.* McGraw-Hill, New York, 1958.

P. W. Staats, Jr., J. B. Pearson

[1] *Robust solution of the linear servomechanism problem.* Automatica 13 (2), 1977, pp. 125–138.

G. Strang

[1] *Linear Algebra and Its Applications.* Academic Press, New York, 1976.

T. Takamatsu, I. Hashimoto, Y. Nakai

[1] *A geometric approach to multivariable control system design of a distillation column.* Preprints, Seventh Triennial World Congress, International Federation of Automatic Control (IFAC), Pergamon, Oxford (U.K.), 1978, pp. 309–317.

R. Thom

[1] *Structural Stability and Morphogenesis.* Benjamin, Reading (Mass.), 1975.

E. C. Titchmarsh

[1] *The Theory of Functions.* Oxford Univ. Press, London, 1932.

H. S. Tsien

[1] *Engineering Cybernetics.* McGraw-Hill, New York, 1954.

J. S. Tyler, Jr., F. B. Tuteur

[1] *The use of a quadratic performance index to design multivariable control systems.* IEEE Trans. Aut. Control AC-11 (1), 1966, pp. 84–92.

I. N. Voznesenskii

[1] *On controlling machines with a large number of controlled parameters.* Avtomatika i Telemekhanika, 1938, nos. 4–5, pp. 65–78.

B. L. Van der Waerden

[1] *Algebra.* Ungar, New York, 1970.

J. von Neumann

[1] *Probabilistic logics and the synthesis of reliable organisms from unreliable components.* In C. E. Shannon & J. McCarthy (Eds.), Automata Studies, Annals of Math. Stud. No. 34, Princeton Univ. Press, Princeton, N.J., 1956; pp. 43–98.

M. E. Warren, A. E. Eckberg

[1] *On the dimensions of controllability subspaces: A characterization via polynomial matrices and Kronecker invariants.* SIAM J. Control and Optimization 13 (2), 1975, pp. 434–445.

A. Wierzbicki

[1] *Differences in structures and sensitivity of optimal control systems.* Archiwum Automatyki i Telemechaniki 15 (2), 1970, pp. 111–135.

P-K. Wong, M. Athans

[1] *Closed-loop structural stability for linear-quadratic optimal systems.* IEEE Trans. Aut. Control AC-22 (1), 1977, pp. 94–99.

P-K. Wong, G. Stein, M. Athans

[1] *Structural reliability and robustness properties of optimal linear-quadratic multivariable regulators.* Preprints, Seventh Triennial World Congress, International Federation of Automatic Control (IFAC), Pergamon, Oxford (U.K.), 1978, pp. 1797–1805.

W. M. Wonham

[1] *On pole assignment in multi-input controllable linear systems.* IEEE Trans. Aut. Control AC-12 (6), 1967, pp. 660–665.

[2] *On a matrix Riccati equation of stochastic control.* SIAM J. Control 6 (4), 1968, pp. 681–697.

[3] *Algebraic methods in linear multivariable control.* Proc. 1970 Canadian Conf. on Aut. Control, 1970.
[4] *Random Differential Equations in Control Theory.* In A. T. Bharucha-Reid, Ed., Probabilistic Methods in Applied Mathematics, Vol. 2, pp. 132–212; Academic Press, New York, 1970.
[5] *Dynamic observers: geometric theory.* IEEE Trans. Aut. Control AC-15 (2), 1970, pp. 258–259.
[6] *Algebraic methods in linear multivariable control.* In A. S. Morse (Ed.), System Structure, IEEE Cat. no. 71C61-CSS, New York, 1971, pp. 89–96.
[7] *Tracking and regulation in linear multivariable systems.* SIAM J. Control 11 (3), 1973, pp. 424–437.
[8] *Linear Multivariable Control: A Geometric Approach.* Lecture Notes in Economics and Mathematical Systems, Vol. 101. Springer-Verlag, New York, 1974.
[9] *Towards an abstract internal model principle.* IEEE Trans. Sys. Man and Cyber. SMC 6 (11), 1976, pp. 735–740.
[10] *On structurally stable nonlinear regulation with step inputs.* Systems Control Rpt. No. 7710, Systems Control Group, Dept. of Electl. Engrg., Univ. of Toronto, July, 1977.

W. M. Wonham, A. S. Morse

[1] *Decoupling and pole assignment in linear multivariable systems: a geometric approach.* SIAM J. Control 8 (1), 1970, pp. 1–18.
[2] *Feedback invariants of linear multivariable systems.* Automatica 8 (1), 1972, pp. 93–100.

W. M. Wonham, J. B. Pearson

[1] *Regulation and internal stabilization in linear multivariable systems.* SIAM J. Control 12 (1), 1974, pp. 5–18.

E. E. Yore

[1] *Optimal decoupling control.* Proc. 1968 Joint Automatic Control Conference, pp. 327–336.

K. K-D. Young, P. V. Kokotović, V. I. Utkin

[1] *A singular perturbation analysis of high-gain feedback systems.* IEEE Trans. Aut. Control AC-22 (6), 1977, pp. 931–938.

P. C. Young, J. C. Willems

[1] *An approach to the linear multivariable servomechanism problem.* Int. J. Control 15 (5), 1972, pp. 961–979.

J. S-C. Yuan

[1] *Adaptive Decoupling Control of Linear Multivariable Systems.* Ph.D. Thesis, Dept. of Electrical Engineering, Univ. of Toronto, June, 1976.
[2] *Structural instability of a class of decoupling solutions.* IEEE Trans. Aut. Control AC-22 (5), 1977, pp. 843–846.

J. S-C. Yuan, W. M. Wonham

[1] *An approach to on-line adaptive decoupling.* Proc. 1975 IEEE Conf. on Decision and Control, pp. 853–857.

Index: Relational and Operational Symbols

Index: Letter Symbols

Index: Synthesis Problems

Subject Index

Applications of Mathematics

Vol. 1
Deterministic and Stochastic Optimal Control
By **W.H. Fleming** and **R.W. Rishel**
1975. ix, 222p. 4 illus. cloth

Vol. 2
Methods of Numerical Mathematics
By **G.I. Marchuk**
1975. xii, 316p. 10 illus. cloth

Vol. 3
Applied Functional Analysis
By **A.V. Balakrishnan**
1976. x, 309p. cloth

Vol. 4
Stochastic Processes in Queueing Theory
By **A.A. Borovkov**
1976. xi, 280p. 14 illus. cloth

Vol. 5
Statistics of Random Processes I
General Theory
By **R.S. Liptser** and **A.N. Shiryayev**
1977. x, 394p. cloth

Vol. 6
Statistics of Random Processes II
Applications
By **R.S. Liptser** and **A.N. Shiryayev**
1978. x, 339p. cloth

Vol. 7
Game Theory
Lectures for Economists and Systems Scientists
By **N.N. Vorob'ev**
1977. xi, 178p. 60 illus. cloth

Vol. 8
Optimal Stopping Rules
By **A.N. Shiryayev**
1978. x, 217p. 7 illus. cloth

Vol. 9
Gaussian Random Processes
By **I.A. Ibragimov** and **Y.A. Rosanov**
1978. approx. 290p. cloth